Land Tax in Australia

Land Tax in Australia defines how land tax operates and is administered across state and local government in Australia. International expert Vince Mangioni reviews the current status and emerging trends in these taxes in Australia and compares them with the UK, USA, Canada, Denmark and New Zealand. Using substantial original research, the author sets out what Australia must do through practice and policy to reform and bring this tax into the twenty-first century.

The need for fiscal reform and strengthening the finances of Australia's sub-national government is long overdue and is key in reforming Australia's federation. These reforms aim to minimise the taxpayer revolts encountered in previous attempts at land tax reform, while improving tax effort in line with other advanced OECD countries.

This book provides an essential resource for all property professionals working in development, valuation, law, investment, as well as accountants, tax economists and government administrators. It is highly recommended for students studying property, taxation, legal and social science courses.

Vince Mangioni is Associate Professor at the University of Technology Sydney, Australia. He is an internationally recognised expert on recurrent land and property taxation and frequently speaks at international conferences and industry workshops on this subject.

"Recurrent property taxes are a very important source of revenue used to support the provision of local services in many countries around the world. However, they are often unpopular and can be controversial. Part of the reason for their unpopularity is that they often lack transparency and are frequently misunderstood both by those who are called upon to pay the tax and by the politicians who are responsible for the property tax system. At the present time, land tax in Australia is very much in the headlines and the subject of debate both locally and nationally; there is considerable pressure for reform of land tax, but there is also significant resistance. I am very pleased therefore that Vince Mangioni has written this book to assist in clarifying the key issues and providing international case studies that will assist policy-makers and the public in understanding and modernising the current land tax system in Australia."

– Paul Sanderson JP LLB (Hons) FRICS FIRRV, President,
International Property Tax Institute

Land Tax in Australia

Fiscal reform of sub-national government

Vince Mangioni

Supported by

First published 2016
by Routledge
2 Park Square, Milton Park, Abingdon, Oxon OX14 4RN

and by Routledge
711 Third Avenue, New York, NY 10017

Routledge is an imprint of the Taylor & Francis Group, an informa business

British Library Cataloguing-in-Publication Data
A catalogue record for this book is available from the British Library

Library of Congress Cataloging-in-Publication Data
Mangioni, Vincent, author.
Land tax in Australia : fiscal reform of sub-national government / Vincent Mangioni.
pages cm
Includes bibliographical references and index.
1. Land value taxation—Law and legislation—Australia. 2. Real property tax—Law and legislation—Australia. 3. Real property and taxation—Australia. I. Title.
KU2830.L36M36 2016
336.220994—dc23
2015023187

ISBN: 978-1-138-83125-4 (hbk)
ISBN: 978-1-315-73666-2 (ebk)

Typeset in Baskerville
by Apex CoVantage, LLC

Contents

Tables

Figures

Acknowledgements

A special acknowledgement and note of thanks is extended to the Australian Property Institute (API) for their support in undertaking this research. The New South Wales Division was instrumental in the valuation research comprising the simulations and focus groups in Parts 2 and 4 of this book. A note of thanks is extended to the National Office for their encouragement and support of the recognition that land and property tax reform will make as part of the broader tax reforms of Australia.

Acknowledgement is extended to the following people for their assistance and support:

Associate Professor Garrick Small, Central Queensland University
Professor Kauko Viitanen, Aalto University
Professor Frances Plimmer, United Kingdom
Professor Neil Warren, Professor of Taxation UNSW
Dr Margaret McKerchar, (Professor of Taxation UNSW – Retired)
Connie Fair, British Columbia Valuations Canada
Larry Hummel, MPac Toronto Canada
Paul Sanderson, President International Property Tax Institute
Chris Bennett, International Association of Assessing Officers
Department of Finance New York
Los Angeles County Assessor's Office
Professor Heather MacDonald, School of Built Environment UTS
Professor Desley Luscombe, Faculty of Design Architecture and Building
Associate Professor Roberta Ryan, Centre for Local Government UTS
Robert Mangioni (lawyer to the stars)

Ministry of Taxation Denmark Skat
Ministry of Taxation Finland Skat
Royal Institution of Chartered Surveyors
Valuers-General of Australia and New Zealand
(In particular: Neill Sullivan, VG of NZ and Phil Western, former VG of NSW)

A special thank you to my family: Antonella, Zachary and Philippa

About the author

Dr Vince Mangioni is an Associate Professor in Property Economics and Development in the School of Built Environment, Faculty of Design Architecture and Building at the University of Technology Sydney (UTS). His research focuses on land tax and the role it plays in financing state and local (sub-national) government in Australia. His PhD is in taxation and focuses on the rating and taxing of land, and his subsequent research focuses on the fiscal reform of sub-national government. Vince was an advisor and presenter to Australia's Future Tax System (2010), also known as the Henry Review, on state land tax and local government rating and was involved in the review of the rating and taxing of land in Queensland. He is a statutory valuer and undertook his training at the NSW Valuer-General's Office during the 1990s in the rating and taxing of land.

Vince has published widely, both nationally and internationally, on land and property taxation and has been a visiting fellow at the Australian School of Taxation (Atax), UNSW and a visiting researcher at the School of Real Estate and Surveying, Aalto University Helsinki. He has reviewed rating and taxing systems internationally, including those in Denmark, Finland, New Zealand, Canada and the United States. He has consulted with leading tax economists and local government experts at OECD World Headquarters Paris, in reviewing the re-emerging importance of recurrent land taxation for sub-national government in Europe following the Global Financial Crisis. Vince has consulted to local government in New South Wales and Victoria in reviewing and advising on rating systems and assessing their contribution to the financial sustainability of local government. He is an associate researcher at the Centre for Local Government and Centre for Contemporary Design Practices at the University of Technology Sydney.

Overview

The emphasis of this book is on the fast-emerging need for tax reform in Australia and the contribution land tax should make as part of the tax reform agenda. This book positions land tax within the broader tax system, brings forward the need to consider the framework in which this tax operates and opens the debate for de-siloing the tax system. While vertical fiscal imbalance is put forward as one of the key planks of reform, the overriding emphasis is that, as a low tax country, Australia must improve its overall tax effort, of which recurrent land taxation is one of the key areas which must contribute to this reform.

Debate now builds over which individual taxes should contribute more the total tax take in Australia. While much focus is directed towards the Goods and Services Tax, a consumption based tax, there has been little focus on the taxation of capital and in particular land, which must equally contribute to the tax reform agenda. As a sub-national (state and local) government tax, Australia has significant room to improve revenue from this source in bringing this tax in line with other advanced OECD countries. Rather than all the heavy lifting of tax reform incumbent on the Commonwealth, sub-national government has capacity to contribute to tax reform from the bottom up through modernising its recurrent tax on land.

The reform of land tax as proposed in this book will encourage reorganisation and reform the administration of this tax across state and local government in Australia. While debate has recently centred on constitutional recognition of local government, the greater imperative is managing the limitations the states impose on local government revenue raising capacity from land tax. At present, the states are unable to evolve and reform their own recurrent land tax due to its salience, while revenue from this source at the local government level varies significantly across Australia and, in particular, its capital cities.

International case studies are used to demonstrate how land tax has been used through earmarking to education in the United States and Canada, to which a similar rationale in the earmarking of this tax to infrastructure is an option for Australia. These examples show how recurrent land tax (referred to as the property tax) internationally has been divested to local government in both unitary and federated structures in advanced economies of the world. It is demonstrated that at the state level in Australia, state land tax applies to a narrow number of taxpayers and property, that the governing legislation is over-focused on how to manage the vast exemptions, concessions and allowances, which have distorted the efficiency and robustness of this revenue source.

The need to modernise Australia's tax system has never been greater and is needed to bring Australia's tax system into the 21st century. Focus must centre on alignment of government, tax effort and reforming a number of key taxes of which recurrent land tax is central. There is no one single tax reform fix for Australia, but a raft of reforms are needed, of which this book addresses one of the important tax policies. Some readers will find the recommendations made to be a brutal assault on the great Australian icon, the principal place of residence and on residential property in general. This icon, however, escapes capital gains tax, state land tax, conveyance stamp duty on inter-generational transfer, and in some states is subject to rate capping and pegging, resulting in a manifestly low tax contribution from this source.

In summary, the book ultimately makes its contribution through a critique of land tax, how it is currently administered, the manifestly poor contribution it makes to Australia's tax effort and the methods used to assess this tax. In conclusion, it provides a framework for change and makes a number of key recommendations for the reforms needed to bring this tax into the 21st century.

Book structure

Part 1: Examines the evolution of property tax policy in Australia post the 1970s and the emerging direction over the next three decades. Emphasis is placed on the need to raise more revenue from efficient taxes such as recurrent land tax, while reducing mobility taxes which include stamp duty. These taxes are considered barriers to entering the property market and impact decisions to trade up and down, particularly in the housing market. It further addresses how infrastructure was the initial rationale for the imposition of this tax and how again this is the focus for earmarking revenue from this source. This section provides important historic trends and the fast-emerging trajectory of this tax and revenues within Australia and, to a limited extent, abroad. It lays the foundation underpinning the rationale for recurrent land tax reform.

Part 2: Reviews the operation of land taxation (state and local government) and the bases on which this tax is assessed across the states and territories of Australia. It critiques cases and responses in the review of land taxes and addresses the emerging challenges in the valuation of land in each state used to assess the tax. It addresses the operation and administration, concessions, allowances and exemptions from land tax across each jurisdiction. It demonstrates the distortionary impact of the current two-tier land tax system in Australia, which is supported by vast statutory and administrative frameworks which govern these. It further shows how many of the well-intended concessions are now outdated and require prompt reform. The emerging direction of local government rating is analysed, with emphasis on more densely populated agglomerations. It demonstrates the emerging rifts between ratepayers of single dwellings versus those in multi-unit development in some states. This part of the book is of specific importance to tax practitioners, including lawyers, accountants and valuers who are engaged by both property owners as well as those commissioned by government to administer the tax and valuation systems.

Part 3: Examines the application of land taxes abroad through an international comparison of the operation of land and property taxes and the valuation processes in several countries which still tax land among other bases on which this tax is assessed. This provides an important basis for

Australian policy makers, tax economists and government administrators in understanding how this tax and valuation systems operate abroad. It sets out factors which have precipitated tax revolts and subsequent reforms resulting in a more efficient design of property tax policy internationally. While different countries' tax systems are unique to their circumstances, having evolved over many centuries, this comparison assists in analysing the technical and operational aspects of land and property taxation. It provides a basis for understanding what works and what needs reform, and how countries have and are currently responding to the need for land tax reform.

Part 4: The book concludes by defining the important emerging direction of land taxation and the looming policy decisions confronting government. The design required to maximise revenue from recurrent land tax, while achieving greater harmonisation and uniformity across jurisdictions of Australia, is a complex work in progress. This part links key elements of the three previous parts in designing principles, policy and practices required in modernising recurrent land tax in Australia in the 21st century. While Parts 2 and 3 highlight the disparities of land tax across Australia and the administration and application of this tax abroad, Part 4 sets out two options for reform. The first option is the minimalist reform, which addresses factors which are not negotiable. The second and more fundamental option commences the debate for reform by de-siloing tax from the tiers of government. This allows a more purposeful account of revenue subsidiarity between the Commonwealth and sub-national government, while the reform of sub-national government revenue is introduced and transitioned over the next decade.

Terms and acronyms

Acronyms

ABS	Australian Bureau of Statistics
AFTS	Australia's Future Tax System
API	Australian Property Institute
COAG	Council of Australian Government
FHOG	First Home Owners Grant
GDP	Gross Domestic Product
GST	Goods and Services Tax
LGA	Local Government Area
OECD	Organisation for Economic Cooperation and Development
VFI	Vertical Fiscal Imbalance

Terms

Fiscal Federalism	The collection and control of tax revenue by central government
Central / National government	Commonwealth government
Highest and best use	Use that is physically possible, financially feasible, maximally productive and legally permissible
Sub-national government	State and local government
Recurrent land tax	Local government council rates and states land tax
Valuer	Appraiser or chartered surveyor who determines the value of land for land tax

Part 1

Status quo and the evolving challenge

1 Advance Australia fair

Introduction

Governments use tax policy to fund the delivery of essential services to the public, the provision of infrastructure either directly or through private public partnerships, and for the broader economic imperative of shaping and influencing drivers of the economy. In contrast to the use of monetary policy, which may have broader implications for investment and economic growth, fiscal alternatives through well-designed tax policy are used by intuitive government to achieve sustainable public policy and benefits for the electorate.

Over the past two decades, taxation has become more complex and has evolved to take on a number of new forms in Australia. These include taxes on consumption in the form of a goods and services tax (GST) and, more recently, the inclusion of carbon price mechanisms, prior to which capital gains and a raft of transaction taxes, in the form of stamp duties, evolved. While the latter tax has operated for several years, its economic impact has evolved into a mobility tax, creating a lock-in effect on housing and acting as an inhibitor for those attempting to enter the housing market, as well as those wishing to trade either on the way up or on the way down in the homeownership cycle.

As interest in fiscal policy gained momentum during the beginning of the 21st century and prior to the Global Financial Crisis, the Commonwealth commissioned an international comparison of Australia's taxes, benchmarking our taxes against other countries (Hendy & Warberton 2006). The study provides a comprehensive snapshot of where Australia ranked on the OECD world stage at that point. In summary, the study showed Australia to be a low-taxing country among the OECD economies on the measure of tax effort as a percentage of GDP; however, on a tax per capita basis it sits around the OECD average. The report highlights that, while the overall mix between direct and indirect tax is broadly in line with other OECD countries, the composition of Australia's fine grain tax mix differs. The status of Australia's ranking among OECD countries is shown in Table 1.1.

Table 1.1 Ranking as a percentage of tax to GDP 2010

Country	*Total Tax: GDP*	*World Ranking*
Denmark	47.6	1
Sweden	44.2	2
France	44.07	3
Belgium	44.06	4
Australia	26.51	30
Korea	25.91	31
United States	24.1	32
Chile	21.21	33
Mexico	19.72	34
OECD Average	**34.12**	

Source: OECD Tax Statistics 2010

It is stated that Australia's indirect tax differs with lower value-added and sales taxes and relies more heavily on property transaction taxes, of which the latter is imposed and collected by the states. It is further added that Australia does not levy any wealth, estate inheritance or gift taxes. It is demonstrated later, in Chapter 2, that a distinction exists in what is referred to as 'property and transaction taxes' and that recurrent land taxes are low in Australia among the advanced OECD economies, while in contrast, it is transaction taxes on property that are high.

It is demonstrated that recurrent land tax in Australia is low compared with advanced OECD economies and that opportunity exists for Australia to improve its tax effort from this source while reducing less efficient transaction taxation on property. It is further shown that Australia's dual land tax system, which operates at the local and state government levels, has the capacity to be recalibrated and exceed the tax revenue foregone from potential reductions in conveyance stamp duty. The dual land tax system which currently operates in Australia is unique compared with those abroad, and, with the encouragement and initial involvement of central government, Australia is in a position to reform its outdated mix of land and property taxes.

In summary, the transformation of Australia's tax system is not the sole responsibility of central government to reform; however, central government plays an important financial mentoring role in assisting sub-national government reform itself. It further has the role in aiding the restructure of the Australian Federation, just as the states played their role in 1901 when they elected to federate.

Structure of government and evolution of Australia's tax system

Australia is a federated structure of government comprising six states, two territories and more than 500 local governments. The three-tier system of

Australian government is an outcome of Federation in 1901, when the six British colonies – New South Wales, Western Australia, Queensland, Victoria, South Australia and Tasmania – united to form the Commonwealth of Australia. Up until the 1850s, each colony was run by a non-elected governor appointed by the British Parliament.

By 1860, all the colonies, apart from Western Australia, had been granted partial self-government by Britain (Western Australia became self-governing in 1890). Each had its own written constitution, parliament and laws, although the British Parliament retained the power to make laws for the colonies and could overrule laws passed by the colonial parliaments. By the end of the 19th century, many colonists felt a national government was needed to deal with issues such as defence, immigration and trade (Commonwealth Australia 2014).

For Federation to happen, it was necessary to find a way to unite the colonies as a nation with a central or national government, while allowing the colonial parliaments to maintain their authority. The Australian Constitution, which sets out the legal framework by which Australia is governed, resolved this issue by giving Australia a federal system of government. This means power is shared between federal and state governments.

State parliaments are subject to the national constitution as well as their own state constitution. A federal law overrides any state law inconsistent with it. In practice, the two levels of government cooperate in many areas where states and territories are formally responsible, such as education, transport, health and law enforcement. Income tax is levied federally, and debate between the levels of governments about access to revenue and duplication of expenditure functions is a perennial feature of Australian politics.

In 1992 the federal government established the Council of Australian Governments (COAG), which includes the prime minister, state premiers, chief ministers and the president of the Australian Local Government Association. COAG meets twice a year to discuss intergovernmental matters of which tax reform is a work in progress. Its objective is to develop and implement national policy reforms requiring cooperative action between the three levels of government: national, state or territory and local. In furthering the development of local government, in 2008 the federal government set up the Australian Council for Local Government so that it could work directly with local government (Commonwealth Australia 2014).

Although a fast-evolving tier of government, local governments are created by legislation at the state and territory level and are not a constitutional level of government. They do not have tax-raising powers of their own; all tax-raising powers and planning decisions are controlled and are at the discretion of state government across Australia. Referenda were held in 1974 and 1988 to provide constitutional recognition to local government, which both failed (Pearson 1994). In contrast to its traditional role of roads, rates and rubbish, local government is progressively moving into housing, early education and health services.

The challenge confronting state and local government relations on taxation resembles those confronting state and commonwealth government 60 years earlier as the Commonwealth asserted its dominance on tax policy during World War II. The states are acutely aware and remember their battle with the Commonwealth 40 years after Federation on income tax revenue as an own tax revenue source. The states do not want to repeat this exercise with local government by granting constitutional recognition and wrestling over the limited own source taxes available to sub-national government in Australia.

Before World War II, both the Commonwealth and state government collected income tax. In 1942, in order to fund the war effort, the federal government became the sole collector of income tax. It did this by passing laws which raised the federal tax rate and gave some of the proceeds back to the states on the condition that they drop their income tax. States receive this money in the form of funding grants. Technically a state could still collect its own income tax, but this would mean its people would be taxed twice and the state would forfeit its funding grants. Warren (2004) highlights the consequences of state government maintaining income taxes at that point, which resulted in a dollar for dollar reduction in grant revenue to the states from the commonwealth for every dollar of income tax they continued to raise.

Four states – Western Australia, Victoria, South Australia and Queensland – challenged the legislation allowing the Commonwealth to collect income taxes. The High Court ruled it was valid on the grounds that section 51 (ii) of the Constitution gives the Commonwealth Parliament power to make laws relating to taxation, even though in practice the legislation removed a state power. It also ruled that, under section 96 of the Constitution, the federal government could attach conditions to funding grants, and therefore it was legal to only give compensation to states that stopped collecting income tax. Intended as a wartime measure, the arrangement has remained in place ever since. As a result, the states are now more dependent on federal government funding from grants.

Australia is a young and modern nation with a small, strong and evolving economy and an egalitarian society which prides itself on equality and the well-being of the community. Taxation is important to Australians and, in addition to simply financing the country, it is an important mechanism for ensuring a contribution is made for earning and prosperity. The unification of Australia under a federated structure centralised financial control and the overarching management of the nation with the Commonwealth by achieving a consistent and harmonious approach to taxation.

Governance and oversight of the tax system by the newly formed Commonwealth was an important cornerstone of Federation, which allowed tax revenue equalisation and distribution across Australia. This objective served Australia well through the early and mid-20th century and in particular during both World Wars, where migration to the cities and financing the war effort were well coordinated. During the initial urbanisation phases of Australia's capital cities, population growth followed each of the World Wars and fiscal federalism achieved its objectives in addressing the needs of a fast-evolving and geographically centralising nation.

The evolution of fiscal arrangements across the tiers of government is set out in Table 1.2, which highlights the relatively small percentage of taxation raised by state and local government known as *fiscal imbalance* (Warren 2004:39). The fiscal imbalance from the collection of own source taxes results in the states providing more than half of the services and infrastructure in Australia, while raising less than 17 per cent of its revenue from own source taxes. Local government raises 3 per cent of revenue from own source taxes and is responsible for 4.5 per cent of all expenditure. The shortfall between own source taxes and expenditure by state and local government comprises grants from the Commonwealth. This arrangement has generally been successful for several decades; however, in the 21st century, with more demand for services and infrastructure required from state government, a more complex rationale requires tax burden and responsibility to be spread across all tiers of government.

The collection of tax revenue by federated central governments internationally is set out in Table 1.3, which shows that Australia's tax system is well out of line with other OECD countries, with central government collecting twice as much revenue as central government in the United States and Canada. This is a matter to be addressed in the review of the tax system, as it is clear that either decentralisation of taxation is warranted or, conversely, that the tax revenues collected by state and local government require reform. It will be demonstrated that a combination of these two factors is the most plausible outcome in bringing the tax system into the 21st century. While the Commonwealth has control of revenue in Australia, it is not beyond sub-national government contributing to the recalibration of own source revenue progressively over the next decade.

Taxes in Australia may be dissected by name or by reference to economic grouping in which government manages fiscal and economic policy. In addressing the economics of taxation, all taxes are grouped and reviewed for the purposes of distinguishing their objectives and efficiencies, as shown in Figure 1.1 as Labour, Consumption and Capital. The grouping of these

Table 1.2 Percentage share of taxation revenue by sphere of government over past two decades

	Commonwealth	*State*	*Local*
1990–91	79.0%	17.4%	3.6%
2000–01	81.9%	15.1%	3.0%
2010–11	80.5%	16.0%	3.5%

Source: ABS Cat. No. 5506.0 Taxation Revenue Australia

Table 1.3 Tax collected by central government

United States	40.6%
Canada	41.5%
OECD Ave 2012	54.5%
Belgium	56.4%
Austria	66.3%
Australia	81.7%
Mexico	81.9%

Source: OECD Tax Statistics

taxes is important to government, particularly central government, in maintaining taxation equilibrium, employment incentives and access to housing across Australia. In addition to the traditional and historic economic rationale for taxing land due to its limited supply, neutrality and visibility, a further important rationale has emerged for its resurgence as a tax on capital.

It is noted in Figure 1.2 that recurrent land taxes which comprise state land tax and local rates, which are combined in this figure, are the fourth highest tax source of the top 10 taxes by revenue in Australia. It is highlighted by Australia's Future Tax System (AFTS) (2008) that, while the 10 taxes in Figure 1.2 account for approximately 90 per cent of total tax revenue in Australia, the remaining 10 per cent of tax revenue is derived from 115 different taxes. In Figure 1.2, at the sub-national level of government, recurrent land taxes are identified as being a tax with potential for increases in revenue, while conveyance stamp duty is a tax to be reduced. At the Commonwealth level, personal income and company taxes are identified as a tax to be further moderated, with increases in revenue from the GST, which sits under the category of consumption, which along with capital are the target categories for increases in tax revenue.

Among the factors which tax revenue collected under the category of Labour in Figures 1.1 results from Australia's aging population, as is the case in many OECD countries. The progressive decline in the ratio of Australians working to those aged over 65 (retirement age) between 1970, where there were 7.5 workers for every Australian over 65, is projected to decline to 2.5 working Australians for every Australian over the age of 65 by 2056. As at 2010, the ratio was 5 working Australians for every Australian over 65 (ABS cat. no. 3222.0). This has resulted in maintaining tax revenue from income steady and, where possible, reducing taxes on labour to retain Australians in the workforce longer and to attract targeted labour and professions from abroad. This dynamic will impact the age at which access to the aged pension is gained, which has steadily increased across all economies of the world. This may further impact superannuants going forward in accessing preserved superannuation entitlements.

As highlighted in Figure 1.2, income tax represents the largest source of tax revenue in Australia, and is above the OECD average, while company

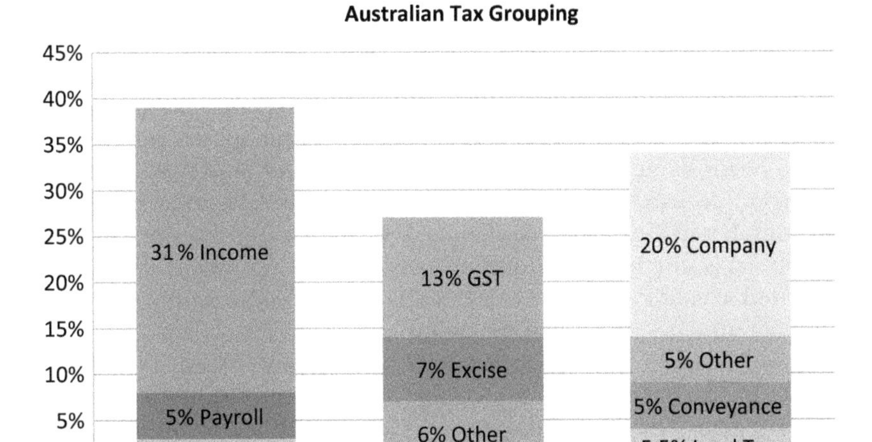

Figure 1.1 Tax mix by categories of labour, consumption and capital

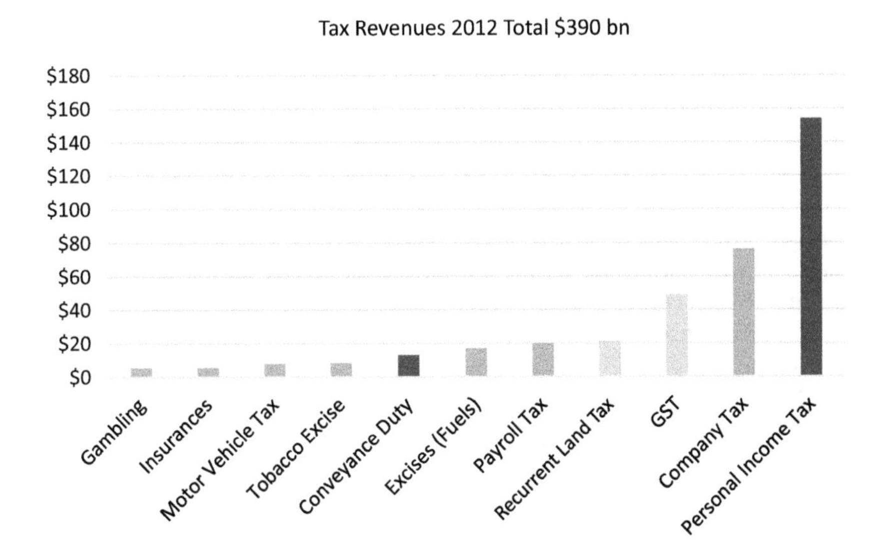

Figure 1.2 Top 10 taxes by revenue

Source: ABS Cat. No. 5506.0 Taxation Revenue Australia

taxes by contrast are low. Company tax rates are maintained internationally competitive as companies make an important contribution in maintaining high employment. The importance of maintaining competitive labour taxes moves the tax pendulum towards increasing taxes on consumption and capital and away from labour. This raises the question as to whether increases in tax revenue are to be spread wide through higher consumption taxes or higher capital taxes of which land is the primary asset.

While the priority of government is a strong economy which generates taxes from labour, tax revenues from income will be impacted by an aging population, which is compounded by increased expenditure on maintaining the needs of an aging population. This is further impacted by decreasing workforce participation rates and tax revenue collected from labour. The complex question that confronts government over the next 30 years in Australia is whether the additional source of tax revenue needed will be generated from consumption or capital taxes, or a combination of the two, in maintaining tax revenue in Australia.

Australia's aging population has been on the government's radar for a number of years and has been the subject of earlier inquiries into Australia's tax system. The Asprey Inquiry (1975) focused heavily on income tax and the necessity to maintain incentives to work and to ensure tax was not a disincentive as more Australians crept into higher tax brackets. The Meade Inquiry (1977), a comprehensive review of direct taxation in the United Kingdom, addressed similar matters to Asprey as to which income taxes were high on this inquiry's agenda.

During the mid-1970s, personal income tax indexation was introduced; this automatically adjusted the tax thresholds to account for wages growth which was far more robust before the introduction of the Wage Accords of the 1980s, which tied increases in wages to productivity. This indexation was abolished in 1981 when the government at that time realised that they had lost control over tax revenue adjustment and the ability to use adjustments of tax thresholds to their advantage. This control gives national government the opportunity to sell tax and other reforms at opportune times, being an important trade-off mechanism with the electorate.

Smith (2009) points out the progress of reforms in income tax since the 1980s, particularly in reducing the top personal tax rate, which was 60 per cent in the early 1980s and is now 46.5 per cent including the Medicare levy. The income level at which the top tax rate commences was 2.4 times average earnings in 2014, which is close to twice the level of the early 1980s. While adjustment to income tax has occurred since the 1980s, it continues as a work in progress for central government as bracket creep through wages growth continually impacts more taxpayers who move into higher tax brackets.

Inquiries into tax reform and government expenditure

The process of reviewing Australia's tax system and sources of tax revenue, in conjunction with planning the restructure of government and the relative efficient allocation of its functions across its tiers, is a work in progress. Several inquiries have resulted in constructive tax reform options and opportunities to bring the national and sub-national structure of government into the 21st century. The decoupling of the repetitive functions and operational processes across the tiers of government will need to be streamlined while retaining a level of policy oversight. This will require a balancing act, a structured recalibration as the Federation evolves and moves into the provision of fluent operational services and infrastructure policy reform.

Two recently conducted reviews of Australia's tax system and the processes of government have provided an important macro analysis and laid conceptual platforms for reforms. A summary of these reviews sets out the issues to be addressed and provides a level of design for the way forward. The more prescriptive and micro application of potential reforms are discussed in Part 4; however, a summary of the key points raised by AFTS (2010) and National Commission of Audit (2014) provide a context to bridge where Australia has come from and where it is going. This is then followed by a discussion of some of the recent tax policy reforms of the past two decades.

Australia's future tax system review

The most recent review of Australia's tax system by AFTS (2010) is a comprehensive root and branch review which examined the major taxes imposed in Australia (excluding GST) and the movement of tax revenue between the tiers of government under Australia's current federated structure. The review defines the most relevant factors impacting Australia over the next 30 years. These factors provide important context in driving the direction of change and addressing the specific remedies needed to reform our tax system in meeting a number of fast emerging challenges.

Among the key challenges to be addressed in re-shaping the tax system is sustainability which has environmental and institutional dimensions. The first deals with the imperative mitigation of CO_2 emissions and the contribution that fiscal reform would make in the encouragement of industries which provide clean energy solutions and contribute to the economy. The second element of institutional sustainability encompasses tax reforms needed to evolve alternate tax bases resulting from increasing urbanisation of Australia's cities. While not specified in the review, these may also extend through carefully crafted tax reforms on foreign-owned primary production land and the protection of Australia's rural food bowl.

Increasing globalisation with freer flows in investment and labour aid the reform of tax policy in maintaining Australia's competitiveness while ensuring robust tax revenues. Within Australia, both demographic and geographic change are among the greatest factors to impact the Australian economy and hence the fiscal direction and financial management of the economy. As one of the world's fastest urbanising countries, with over two thirds of Australia's population living within its capital cities, the financing of public infrastructure is looming as a critical priority to be addressed. While part of this challenge can be addressed through private-public partnerships and user pay models, fiscal finance remains the primary source of funding for traditional public infrastructure over the next 30 years.

In focusing on tax mix, one of the key consultation questions of the Henry Review asks whether Australia's tax transfer system appropriately deals with property and wealth and whether new approaches should be introduced. This supplements the question as to whether state and local government taxes are efficient in their current form and whether these taxes could be improved, which in turn would contribute to improving Vertical Fiscal Imbalance (VFI) that exists between national and sub-national government. With the Commonwealth raising over 80 per cent of tax revenue in Australia, more could be done to encourage the states to increase tax revenues from existing undertaxed sources while reforming less efficient taxes.

Among the key challenges to be addressed in this review results from the store of wealth and savings held in one's home.

While a complex and multi-dimensional part of the Australian identity, housing continues to fracture into two main groups, those who own their home and those who aspire to own their home. While much of the challenge revolves around housing supply in the capital cities of Australia, the existence of transfer taxes, better known as conveyance stamp duty, is the second largest single factor impacting mobility and homeownership in Australia, of which supply is the primary factor.

With two thirds of the cost of trading up or down to an alternate property attributed to conveyance stamp duty, the holding period of homes in the capital cities of Australia has increased over the past decade. This has resulted from an under- and over-utilisation of housing stocks which impact first homebuyers and those waiting to trade up to larger housing held by empty nesters. The underpinning issue impacting housing mobility, which contributes to housing affordability, is bracket creep in transaction taxes as property prices continue to increase.

This revenue source is not one easily surrendered by the states, being a one-off payment and far less visible to alternate sources of tax revenue derived from recurrent annual taxes on land. While housing supply remains the primary factor impacting housing affordability, of which the states are final arbiters on approval of supply, they are also in a position to dismiss any argument that removing or replacing transfer mobility taxes would greatly influence housing affordability. A mid-point does exist, however, which supports the case that increased housing mobility would encourage freer trading of property, a point to be addressed in Part 4.

The final matter resulting from the AFTS (2010) tax review is on tax harmonisation of sub-national taxes imposed within and across states and local government boarders. This applies to local government rating as well as land taxes imposed by the states. Tax base harmonisation is important in maintaining neutrality and transparency of taxes, which underpins the integrity of the state's tax system. This is of particular importance in the expansion of recurrent land taxes as high profile taxes due to their visibility, a point further expanded on in Chapter 2 and Part 4.

National commission of audit

The review and refinement of government and its functions, services and administration is an important and indeed mandatory self-assessment mechanism for maintaining a sustainable and relevant framework under the principles of 'Good Government'. The National Commission of Audit (2014) embarked on such a review examining the major programmes of government, monitoring government finances and reviewing the fiscal outlook of Australia. Its findings resulted in 64 recommendations covered within six broad categories, of which the key categories addressed are approaches to government, new fiscal rules and reforming the Federation.

The latter of these categories have been examined from various perspectives in which Warren (2006), in undertaking a benchmarking analysis of Australia's intergovernmental fiscal arrangements, emphasises three key areas for reform: i) assigning functions in a federation, ii) benefits and costs of a federation and iii) tax assignment in a federal system. While much focus has centred on taxation reform, Warren (2006) underpins reform by asserting the importance and analysing the benefits of firstly assigning functions in a federation. Among the key points identified for reform are decentralisation and the assignment of primary expenditure responsibility to one level of government and tax assignment which follows responsibility to ensure sub-national tax instruments are adequate to meet commitments.

The National Commission of Audit (2014) has identified significant overlap between the activities of the Commonwealth and states and the recommendation that these be reviewed. The Commission refers to the importance of the principle of 'subsidiarity' for policy and services, as far as practicable, to be delivered by the level of government closest to the people receiving the benefit of those services. It further highlights the need to reduce duplication and, where that cannot be avoided, that cooperation between the relevant spheres of government applies. In undertaking this recommendation, it is important that local government also participates as it continues to evolve in the provision of services, as it is the closest sphere of government to the taxpaying public.

The above point ties in with the proposed funding arrangements of sub-national government and reduces the current vertical fiscal imbalance which is particularly relevant to Australia. In increasing revenue-raising capacities to be more in line with expenditures, the National Commission of Audit (2014) recommends the sharing of the Commonwealth's income tax base. This would be achieved by the Commonwealth correspondingly reducing its income tax percentage rate by the percentage applied by the states; this way, revenue raised by the states would be better hypothecated by the states. However, as local government is further perceived to be closer to the taxpaying public, it plays an important role in participating in the decentralisation of the income tax base and is even better positioned to justify revenue hypothecation on behalf of itself and indeed the states to some degree.

While the Commission raises the above important and highly relevant initiatives, the assignment of specific tasks between the Commonwealth and states, and between the states and their operational arm of local government, is long overdue. This has become of particular relevance in the relationship between state and local government, of which these two tiers of sub-national government collectively develop, manage and oversee more than 80 per cent of all infrastructure in Australia, while collectively raising less than 20 per cent of the tax revenue. This factor is of importance for the tax paying public to understand; while the case for tax reform is important, the preceding priority of defining the structure of government and allocating both service and capital responsibilities is the starting point.

In the case of the GST, the National Commission of Audit (2014) further suggests the need to allocate revenue from this tax source on an equal per capita basis. In addition to addressing consumption taxes, taxes on capital and more specifically recurrent property tax, the distribution of this revenue also needs to be allocated on a capita basis, both by the states and local government. The specifics of how this would apply are addressed in Part 4. Both the National Commission of Audit (2014) and AFTS (2010) make structural and fundamental recommendations for the reform of government. The Commission highlights the importance of defining the roles and responsibilities across the tiers of government, as well as making recommendations for the sharing of income tax revenue between the states and sub-national government. AFTS (2010) makes a significant contribution through recommendations in the recalibration of Australia's tax mix among other factors which improve housing mobility and hence affordability.

Tax reforms of the late 20th and early 21st centuries

While income tax remains the dominant source of revenue in Australia, maintaining a competitive workforce requires competitive income tax policy. During the late 1980s, and again in the late 1990s, two important tax policies were formulated in progressing Australia's savings and fiscal development. The introduction of the Compulsory Superannuation Guarantee Levy (CSGL), which is a hypothecated tax, currently sits at 9.5 per cent of income and the introduction of the GST currently sitting at 10 per cent. These are progressive tax reforms of the later part of the 20th century. Both of these reforms were revolutionary and contributed in positioning Australia in addressing the challenges of an aging and progressively retiring population.

At the commencement of the GFC 2007/08, tax policy and revenues were on the table again with further additions to the tax base through the proposed Mineral Resources Rent Tax (MRRT) and carbon pricing. Strong opposition to the MRRT has been mounted in which there have been mixed responses to its success and potential longevity. The economics behind this tax is strong and compelling, while the politics reside as the pressure point driving its sustainability. Mining has been the subject of the demise of two post–WWII prime ministers amid lobbying against reforms to first nationalise mining during the 1970s and, more recently, to impose a tax on mineral resources.

The Whitlam Government fell, resulting from attempts to nationalise mining, which was to be achieved through a series of offshore loans, which later became known as the loans affair of 1974–75. The loans affair embarrassed the Whitlam Government and exposed it to claims of impropriety. The Malcolm Fraser-led opposition used its numbers in the Senate to block the government's budget legislation in an attempt to force an early general election, citing the loans affair as an example of 'extraordinary and reprehensible' circumstances (National Archives of Australia 2009).

The Rudd Government, which presided over the inception of the MRRT, was the second prime ministership to fall; however, unlike the Whitlam Labour Government which was removed from office, Rudd was replaced with Gillard who narrowly retained power at the following 2010 election.

Lobbying against the MRRT resulted in its provisions being pared back with considerably less revenue raised than was initially anticipated. Exemptions and concessions created significant revenue carve-outs which would have impacted miners who had not budgeted for this tax at the varying stages of their exploration, development and extraction phases. The idea of the MRRT, like all tax reform, was well intended; however, longer and progressive phase-in is needed for this tax to be built into mining companies' estimates and business models. The MRRT is an important tax source for Australia and is sustainable in Australia's geographic location with its major trading partners and stable political economy.

The second and more subliminal issue with this tax is the tussle between the tiers of government imposing mining taxes in the various forms in which it is imposed. At the time of Federation, the states demarked the ownership of land into fee simple ownership, representing the right to use land subject to planning laws with rights of succession, whilst it retained the rights to minerals and resources below the ground. The states privatised the exploration and extraction process in exchange for exploration licenses and royalties paid on resources extracted from the land. To this end, a tax already exists on mining in Australia imposed by the states.

Similar to the Commonwealth and state tussle over income tax revenue of the 1940s, the mining tax emerges as a similar tussle between these tiers of government in Australia this century. This tax, while an important source of revenue, raises questions over the independence of government of all persuasions and its ability to reform fiscal policy and, more importantly, the decentralisation of Australia's tax system. While the Grants Commissions play an important role in redistributing revenues across the tiers of government, Australia's highly centralised tax system is yet to be resolved while the states remain in an adolescent fiscal phase of development. This specific matter is addressed in further detail in Part 4.

The most recent and important tax reform of the 21st century is combating environmental degradation. The science is clear that the industrial revolution has impacted the atmosphere and this must be combated through the imposition of carbon price mechanisms. Carbon price mechanisms operate in two primary forms; the first is as a tax and the second as an emissions trading scheme. The tax option is the least favourable, as its label as a tax is not saleable to the electorate, while the latter, as an emissions trading scheme, is far more saleable and palatable to the electorate. In summary, the first is a pure tax for the right to pollute, while the latter involves the purchase of credits issued by government for the right to pollute and the ability to trade those credits among polluters or to trade those credits on an exchange. In either form, these mechanisms sit under the category as a tax on consumption.

The primary divide in the debate on consumption taxation runs along a scale of how consumption tax revenue is to be derived. At one end of the scale is the carbon tax; at the opposite end is no carbon tax and an increase in the GST rate, or including all forms of food currently exempt within the GST, or a combination of the two. In a simple analogy, the carbon tax or emissions trading scheme (ETS) is imposed on the main polluters and the impact of such impost filters through to the consumer through any increase in the cost of goods or services provided. The impact of the tax was offset for low-income earners through changes to the income tax brackets between the years of 2011/12 and 2012/13 onwards as shown in Table 1.4. Noted in the adjustment of the thresholds is the increase in the tax-free threshold and in the second band in which a tax reduction of $1,078 is expended, thus compensating taxpayers across the lower tax brackets.

This did not augur well for some taxpayers, who did not feel sufficiently compensated while the rates at the upper end of the tax scale remained static. In contrast to the immediate past two compensatory adjustments made for the introduction of the GST in 2000 and the economic stimulus package of 2008, in which a combination of payments were made directly to some taxpayers and adjusting tax thresholds at the lower end did little to benefit those taxpayers at the upper end of the tax rates. This epitomises the perception of the benefits of a cheque in the hand versus incremental benefits received through adjustments to tax indexation. The visibility of the cheque in the hand for the impost of the GST to pensioners was broadly greeted with the perception that, '***I have received my cheque, I have been compensated***'.

Regardless of whether the cheque represented parity of compensation, its visibility largely gave the perception of parity. In summary, the rationale for the approach adopted in the compensation for the carbon tax through

Table 1.4 Income tax rates and threshold comparison

Tax rates 2010–11 & 2011–12		*Tax rates 2012–14*	
Taxable income	*Tax on this income*	*Taxable income*	*Tax on this income*
$0–$6,000	Nil	$0–$18,200	Nil
$6,001–$37,000	15c for each $1 over $6,000	$18,201–$37,000	19c for each $1 over $18,200
$37,001–$80,000	$4,650 plus 30c for each $1 over $37,000	$37,001–$80,000	$3,572 plus 32.5c for each $1 over $37,000
$80,001–$180,000	$17,550 plus 37c for each dollar $80,000	$80,001–$180,000	$17,547 plus 37c for each $1 over $80,000
$180,001 and over	$54,550 plus 45c for each $1 over $180,000	$180,001 and over	$54,547 plus 45c for every $1 over $180,000

Source: Author

adjustments in the tax rates and thresholds at the lower end of the tax scale, as shown in Table 1.4, did not gain the same reception. While most reform to the tax rates has been at the upper end of the scale, the much needed restructuring of the lower end of the tax rates, and in particular thresholds, is significant in compensating many low income earners. The trebling of the nil threshold from $6,000 to $18,000 in Table 1.4 impacted many taxpayers at the lower end of the scale, with no adjustment for those on incomes above $80,001.

Taxation design and economic reform

The focus now turns to economic growth as the primary stimulus for the generation of tax revenue within the existing tax categories. This provides an important context for determining which taxes are least likely to impact or distort the economy and provides a platform to further analyse the fundamentals that underpin recurrent land taxes and how these are applied. Much of the rationale used to rank taxes has largely relied on qualitative ranking against the principles of *good tax design*, which is not to be discounted. However, Arnold (2008) provides empirical analysis of the tax mix that best underpins economic stimulation using OECD economies as the backdrop of a study which ranks the status of taxes in achieving this objective.

It is highlighted that specificity of circumstances impacting individual countries, such as differences in economic cycles and micro economic policy, are not always the same; however, the modelling used does provide a robust outcome for determining which taxes are best and which are least suited in stimulating aggregate economic growth. The study examines the relationship between tax structures and economic growth by entering indicators of the tax structure into a set of panel growth regressions for 21 OECD countries, for which both the accumulation of physical and human capital are accounted.

The results of the analysis suggest that income taxes are generally associated with lower economic growth compared with taxes on consumption and property. More precisely, the findings allow the establishment of a ranking of tax instruments with respect to their relationship to economic growth. Property taxes, and particularly recurrent taxes on immovable property, seem to be the most growth-friendly, followed by consumption taxes, with personal income taxes ranked last.

A comparison of the impact on GDP per capita by levying income, consumption and recurrent property taxes shows improvement in long-run GDP per capita, resulting from working and earning higher income with no increasing income tax burden. Conversely, where increases in recurrent property taxes or consumption taxes apply, no negative impact is noted on long-run GDP. Further, it is shown that increases in income tax does impact the long run GDP and that significant negative impact results from increasing income tax (Arnold 2008).

The outcomes of Arnold's (2008) study are confirmed by the Business Council of Australia (2013), which asserts that the architecture of a reformed tax system should recognise that a system whereby the most inefficient state taxes are abolished has the potential to deliver significant increases in productivity. They further recognise that tax foregone must be replaced through alternate tax sources to ensure that the states and territories have access to adequate revenue. Consumption and land-based taxes are identified as being more efficient taxes.

While the findings of this study highlight the economic benefits of taxing property and consumption in contrast to income, in practice property and consumption are at opposite ends of the taxpayer psyche. In contrast to economic factors, social factors prevail in determining the tax mix and composition of a country. The social factors are aptly addressed in the classic definition of the 'Art of Taxation' as defined by Jean Baptiste Colbert (1619–1683) – French economist and Minister of Finance under King Louis XIV of France: 'The art of taxation consists in so plucking the goose as to obtain the largest possible amount of feathers with the smallest possible amount of hissing'.

The visibility of taxation is the most salient feature which impacts its acceptability. Consumption tax, and specifically the goods and services tax, is well subsumed within the cost of goods and services. For most (PAYEE), taxpayers' consumption taxes are near invisible, with retailers and service providers the primary tax collection agency. Similarly for income tax revenue, employers are the tax collection agency on behalf of government, in which the tax is deducted for most employees prior to wages and salaries being paid.

In contrast, land tax and, to a slightly lesser degree, local council rates are the most visible, least accepted and sometimes confronting of all taxes in Australia and indeed internationally. To this end, sub-national government is left collecting a minority of the tax revenue in Australia and among what is left to collect are the most salient and unpopular taxes with taxpayers. The two primary reasons for the salience of land tax are firstly its visibility, in which most taxpayers are not accustom to directly paying tax; this underpins the second reason, which raises the question the purpose of the tax. High on the taxpayer radar is the overriding question as to whether these taxes are consolidated revenue taxes or quid pro taxes for services rendered. Most consumers cannot fathom the payment of taxes into consolidated revenue as an immediate annexure to the tax paid, and knowing the return is an important connection made.

It is the pondering of this question which compels many taxpayers to further explore the rationale for the existence of land tax, questioning the operation of the tax and understanding their tax assessments. As a multifaceted tax, land taxes, and in particular council rating, are neither simple nor always transparent. The manufacture of value on which these taxes are assessed are conceptually simple, however complex in practice. Value as it presently stands, and assessed on a property-by-property basis, remains the contestable element for challenging land taxes. While unpopular, land taxes have remained acceptable to taxpayers, as frequently reviewed values are the arbiter in distinguishing the amount of tax paid between taxpayers.

Summary

This chapter has provided a review of Australia's tax system and the categories of taxes used by government in managing the economy. As one of the most centralised tax systems internationally, Australia has scope to decentralise its tax revenue to lower tiers of government, in which the states play an important role in contributing to this reform through the realignment of their existing tax revenues and with support from central government. A number of compelling reasons have been identified for maintaining competitive income taxation, and this has progressively been achieved since the 1980s with ongoing reforms to income tax thresholds and rates. With the top tax rates being progressively reduced from 60 to 46.5 cents in the dollar over the past 30 years and recent reforms at the lower end of the tax scale, maintaining a strong economy and incentive to work is government's priority.

It is highlighted in principle that recurrent land tax is an efficient alternative source of tax revenue to conveyance stamp duty which progressively impacts housing mobility tax, a factor contributing to the efficient use of existing housing stock. It has been identified by the OECD, as well as Australia's peak business body, that land taxes, and to a lesser degree consumption taxes, are the taxes to target in increasing Australia's tax effort, which is one of the lower taxed countries as a percentage of GDP. While land tax meets the economic profile as a tax of the future, its salient visibility and complexity in the assessment of the tax base impacts its acceptability by the taxpayer. Recalibrating revenue from conveyance stamp duty to recurrent land tax and progressively increasing the latter is a priority for sub-national government in Australia; however, it is viewed as a zero sum gain by the states.

It has been shown that land tax in Australia operates at the sub-national level and is imposed by both state and local government. While it has been suggested that further tax revenue be derived from this source, several questions now emerge as to which of these tiers of sub-national government is the best place to increase revenue from this tax. To what extent and how much further this tax base may be increased and, most importantly, to what extent a tax on land is still relevant in highly urbanised agglomerations is dealt with in Parts 2 and 4. In summary, 'Are the principles which underpin a tax on land still relevant?' and 'What are the emerging challenges in the evolution of this tax in the 21st century?' are questions to be answered in making this transition.

Through the restructure of the fiscal arrangements from land tax between the tiers of government, opportunity exists in Australia to fund critical infrastructure. These opportunities extend to include the recalibration of tax mix and the reduction of inefficient taxes which impact housing mobility and affordability. It further highlights the benefits of reform to be achieved from the decentralisation of Australia's tax system and addressing Australia's vertical fiscal imbalance in order to bring Australia's tax system into the 21st century.

2 Evolution economics and status quo of taxing land

Introduction

The taxation of land in its various forms as a source of government revenue can be traced back to 5000 B.C. in Egypt, where taxes were determined on the produce of land. The Athenian Empire of ancient Greece also achieved success during the earlier part of its reign between 530–468 B.C., where the tax was determined on the value of land based on its productivity. In his conquests through Persia, India and Egypt between 356–323 B.C., Alexander the Great implemented property taxes to assist in the financial restoration of these economies and the promotion of employment and labour in rebuilding services and infrastructure (Carlson 2004).

Between 200 B.C.–300 A.D., the Roman Empire developed and introduced the first value-based system of taxing land. The primary feature of this system was to tax land not on what it did produce, but on what it could produce. This approach represented the initial adoption of the highest and best use principle and was used to combat the emerging food shortage problem caused by farmers who were not utilising land to its maximum potential. The Medieval Period (500–1425 A.D.) was a period of notoriety in England for the administration and organisation of taxes on land. During the reign of King William I (1066–1087), the first national and orderly recording of wealth and assets was undertaken in England. The 'Doomsday Book', which was completed in 1086, was a detailed and comprehensive audit of the assets owned in England at that time and had the clear intention of providing an accurate and concise record of the assets on which a tax could be levied (Daw 2002).

During the medieval period, the survey of land was significant in that it formulated value by reference to land attributes and provided a static and reliable basis for the assessment of taxes on immovable property. The hearth tax (also known as the chimney tax) was introduced in England and Wales in 1662. The negative impact of this tax, which was based on the number of fireplaces in a property, required internal inspections of property by tax assessors and was viewed as a tax on warmth and food. As chimneys were essential for cooking and maintaining a habitable environment in winter months, this approach to taxing property fell out of favour and was abolished by King William III in 1689.

Following the removal of the hearth tax, the window tax was introduced and operated in England for over 150 years until it was repealed in 1851 and replaced by a house duty. Properties with more windows were charged more tax; the tax itself was seen as easily assessable from the exterior of the property. Opposition to the tax mounted as it was seen as a tax on light and air and, in mitigating the impact of the tax, taxpayers would board up windows to minimise its impact, which distorted the neutrality of the base on which the tax was assessed.

From the beginning of European settlement in the United States (US) in 1607, and in Plymouth, Massachusetts in 1620, property taxes on land, buildings and personal property were imposed to pay for churches and religious education. As property taxes grew, the Plymouth Council was directed at the request of its community to publish lists of taxpayers, their assets and tax payable. This pressure grew from suspicions of inequitable assessments, abatements and residency fraud and the movement of assets from one residence to another (Carlson 2004).

Between 1834 and 1896, uniformity in taxation became a major constitutional issue in the US and 31 states adopted constitutional uniformity clauses. These clauses were intended to improve and ensure consistency in values on which the tax was determined. An ideological divide between the North and South saw property taxes fall out of favour in the South where larger estates were held by the wealthy. As the necessity for property taxes grew, a residential frontage tax was introduced in New Orleans. This method was met with the development of the shotgun house, a long narrow house about 12 feet wide with a hallway to one side and rooms located in tandem off the hallway. As the property tax increased for two-story houses, the *camel-back* (with the second story at the rear) was developed. In many parts of New Orleans the camel-back house did not qualify as a second story and avoided the higher second story property tax (McAlester & McAlester 1997).

Difficulties were encountered in the assessment of improvements and in determining the improved value of land over the centuries across jurisdictions; land as the basis for determining a recurrent tax was the subject of much debate from the 1700s. It consumed the interest of many economists, both in support and, to a lesser degree, in opposition to its use. David Ricardo (1772–1823), John Stuart Mill (1806–1873) and Henry George (1839–1897) each contributed to the case for land to be used as the basis on which a recurrent tax should be assessed.

The concept of land tax as an economic rent was founded by David Ricardo, who defined rent with clarity. While much of Ricardo's work was exemplified within a rural context, his understanding of the potential difficulties in separating economic rent from rent paid for land which included soil fertility and other agricultural improvements was evident. Further, Ricardo viewed land tax as one that should fall on the landlord and not impact the cost of production. To this end, he attributed the imposition of land tax based on the land production, in that it should not be applicable to unproductive land (McLean 2004).

The inseparability factor highlighted by Ricardo was also acknowledged by Henry George and was criticised as a fatal concession by the Henry George paradigm. Despite George's position, one of the greater contributions to be assigned to Henry George was his recognition that location is as much a source of 'intramarginal surpluses as variations were in the fertility of different plots of land'. He further recognised the importance of taxing land based on its capacity to produce and distinguished the point that land should be taxed if it were idle but yet had the capacity to produce (George 1879).

The work of Henry George related to a different time and circumstances to those observed by Ricardo. While most of Ricardo's work related to soil fertility and produce from land in the 17th century, George witnessed during the 18th century the shortage of housing and land speculation achieved through land banking and the withholding of land ripe for use in housing and industry during the beginning of the industrial revolution. It was from this time forward that land value taxation and economic rent was viewed as much a function of location as the earlier position assigned by Ricardo as being the 'indestructible powers of the soil'.

Henry George brought land value taxation to the fore in an urban context and perhaps made the greatest contribution to its place in modern urbanised society. The work of John Mill is not to be disregarded, as his thesis acknowledged the impact of population growth and its impact on the transition of land uses from rural to urban use and alternatives for this uplift in value to be captured through taxation. In advancing the work of Ricardo, Mill suggested the use of capturing the uplift in value through incremental betterment taxation, which was in fact one form of land value taxation.

Two important points emerge from a review of land tax in its various forms. The first is that a tax levied on built attributes of land may be distorted through the alteration of improvements by the taxpayer, which impacts the neutrality and economic efficiency of the tax. The second point emerged in light of the assessment of the tax, not on what the land actually produced, but on its potential to produce, which may not be reflected in either the use of the land or the intensity of its use at the time of assessment of the tax. This point is one which separates a tax on land into two ideological spheres. The first is as a consumption tax based on the services the property is perceived to utilise. The second is applied to land as capital and assessed as an economic rent which further captures any increase in value, a point to be addressed in further detail later in this chapter in the review of land tax in Australia today.

Economics of taxing land and the urban challenge

Ricardo's work exemplified highest and best use on economic rent from land determined on its productivity, or potential productivity, over and above its socially beneficial value. This approach distinguishes land value over and above the marginal site value for all land across the whole economy. While ideal in theory, Ricardo recognised that the determination of economic rent encompassed a difficulty in the inability to separate the return from land versus that component of return derived from improvements which brings land into production, which defined the practical limitations of his theory. In a rural context, improvement was achieved through soil fertility and, in essence, represented the first step towards the necessity of improving land in order for it to produce and achieve an economic rent.

In adding a further dimension to Ricardo's theory, Johann Von Thunen (1783–1850), a spatial economist, took a broader view on highest and best use by examining land use patterns within a location. Land uses are grouped within the context of an overall urban system, 'offering paradigms used to identify land use changes over time and potential directions of a cities' growth'. This is important in the assessment of land value taxation, as incremental increases in value are additionally captured by this tax from one valuation cycle to the next. In contrast, the Ricardian premise is more closely aligned with appraisal practice and is more micro-oriented, allowing a comparative approach between alternate sites or property, based on specific site attributes (Dotzour et al 1990).

While Von Thunen, in principle, embraced returns from land, this return was more broadly focused on the distance of land in production from markets in which the produce would sell. To this end, Von Thunen observed that in an evolving city, the most valuable land was a market-determined phenomenon in which aggregated uses of land were formed based on the distance of land to locations in which the produce would sell. From this perspective, Von Thunen determined that the orderly formation of land was established based on the demand for the produce, plus the cost and time required to transport that produce to markets (Dotzour et al 1990).

While both the Von Thunen and Ricardo paradigms have strengths, each also has weaknesses. Ricardo's theory suffered from the lack of being able to distinguish the return from unimproved land with that of land improvements which brought it into production. The Ricardo paradigm, however, reflected the value of land's location within in its yield or return. Conversely, Von Thunen's paradigm suffered from an absence in differences between the attributes of the land itself and variability in topography, soil fertility and changes in demand or price for commodities from one property to the next. In summary, Von Thunen's model assumed parity of topography, fertility and the removal of any physical barriers, such as rivers or terrain that impacted the land (Alonso 1964).

In optimising the strengths of each paradigm, a combination of the Von Thunen and Ricardian paradigms has a place in the assessment of land tax in an urban environment. The integration of these two paradigms creates a framework which accounts for the strengths of each, while concurrently reducing the remaining weaknesses of the other (Wilson 1995 and Dotzour et al 1990). They each embrace the principle of highest and best use of land; however, in a modern articulate society principle must be translated into practice and further made transparent to the taxpayer as to how this concept applies to their tax obligation.

In this regard, land tax and, more specifically, the valuation of land is to be evidence based in addition to its zoning and permissible use. This means that the highest and best use of land must be clearly defined and a demand test applied, of which demand reflects the dominant evolving use of land in a specific location. This is in conjunction with the current zoning and permissible use of the land, which has been addressed in part by the courts in Australia in transitional use locations and will be examined in more detail in Chapter 3 under the valuation of land. In principle, however, in an urban context the dominant use of land is reflected by reference to the improvements on land in highly urbanised locations, in which the valuer must determine that use by reference to land uses and the improvements that define that use.

The starting point for the determination of land value based on highest and best use in highly urbanised locations unavoidably requires reference to improvements on land which defines that use. This is the case in many urban locations, resulting from the absence of land sales, which include sales where improvements are demolished after sale. This basis of value has potential to suffer from the misconception by some taxpayers that they are taxed on the actual improvements on their land, rather than the improvements which should be on the land where that use is not highest or best, or in cases where their improvements are not maximally productive.

The traditional economics supporting land as the most efficient base on which to assess a tax is the rationale that land is fixed in supply as set out in Figure 2.1. However, traditional economics has formulated this position based on the limited supply of land within a given location. In a modern urban context, this economic principle must be further expanded to account for the progressive vertical subdivision and use of airspace in the fast-emerging, vertical urban agglomerations. While land has traditionally been considered fixed in supply, supply is vastly impacted by increasing floor space ratios which permit the vertical development of land. This is shown in Figure 2.1, in which S represents the supply of land and has traditionally been viewed as fixed in supply.

In contrast to tradition, the limited supply of land has been expanded to include the vertical use of land which further includes land use both above and below the land's surface. To this end, the tradition of viewing land purely from a horizontal and rural context as a medium of production has moved to

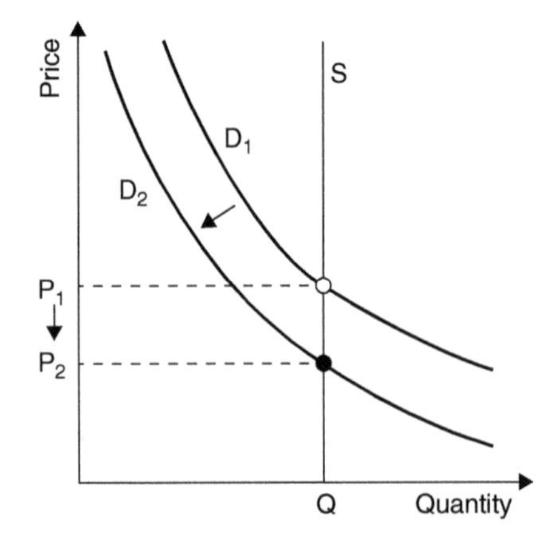

Figure 2.1 Land demand and supply curve

facilitate more sophisticated land use, with units of production in the form of goods and services. This factor does not discount the theory of taxing economic rent; however, it does provide additional challenges in measuring the value of land and defining at which point the economic rent that exists exceeds the unit of production and how this is determined in practice.

The rationale for determining economic rent in the modern metropolis is compounded by the cost of bringing land into production. This goes well beyond the traditions of soil fertility and requires vast structures, to which tax write-offs in the form of depreciation are needed to sufficiently justify the replacement of these developments. In essence, the traditional supply or quantity of land is fixed horizontally; however, as innovative methods of building evolve, along with vertical subdivision of airspace and more sophisticated methods of holding interests in land, the foundation of land as the base of this tax must also evolve in modernising this tax.

The discussion on the economics of taxing land does not diminish the rationale or acceptance for its use in highly urbanised locations. It defines the challenges in ensuring that land as the base of the tax and, more specifically, the process used to determine the value of land sufficiently accounts for the complexities encountered in its measurement in modern society. This is of particular importance in ensuring that this tax aligns with the principles of good tax design.

Land tax: from settlement to the present in Australia

This section addresses the evolution of land tax in Australia and the diverging rationale between its impost as a capacity-to-pay versus a benefits-received tax. It further explains the fiscal relationship between state and local government and analyses the impact on revenues from land tax across sub-national government. It defines the challenges confronting two tiers of government that collect revenue from the same tax base and demonstrates that, under the current dual imposition of this tax, it is more accepted by taxpayers when imposed by local government. This is in contrast to its imposition by state government, a rationale that has progressively evolved since Federation.

Land tax commenced in Australia in 1884 (Smith 2005) and continues to predominantly operate as a tax on land imposed by the states and is imposed on a number of different bases by local government. Australia is one of the few countries that impose a recurrent tax on land value by two tiers of government. Table 2.1 sets out the evolution of government, the evolving uses and taxation of land which facilitates its development. The last column of this table sets out the perceived rationale, being the least defined but often most controversial aspect of the tax. The top half of Table 2.1 shows that between 1788 and the late 1880s, land tax was administered by the states (formerly referred to as colonies), which was the initial single tier of government.

Table 2.1 Evolution and structure of government and property tax

Gov't	*Period*	*Purpose*	*Mechanism / Base*	*Rationale*
State/Colonies	(1788 – 1850) Initial use and development	Promote initial development and productivity of land	Planning laws permitting development Taxation mechanism (Land Value Tax) Reflects potential highest and best use	Encourage the development, use and productivity of land Finance the initial provision of services and infrastructure
	(1850 – late 1800s) Stable settlement	Subdivision and break-up of large estates	Benefits tax	
	1884 local gov't formed under Municipalities Act 1884 (1901 – present) Federation Redevelopment/ reurbanisation and expanding city	Finance provisions for existing and new services Redevelop and changes in land use patterns	Planning laws permitting changes in use and redevelopment Taxation mechanism (Land Value Taxation) Highest and best use	Extend the provisions of services and infrastructure Transition Facilitate land use change
Commonwealth State Local	1950s – today	Finance provisions for existing and new services	State Land Tax	Earmarked to services (perceived)

Source: Author

This was a simple structure in which land tax was established as a means of providing revenue for services and the settlement and expansion of Australia's colonies (Daly 1982). Land tax was introduced to fund the establishment of towns and associated infrastructure, including roads and community facilities, and at this point its understanding as a tax for services was established (Brennan 1971). During the pre-Federation colonisation of Australia, there was little debate on the rationale for the payment of land taxes, as the importance of bringing land into production was the initial purpose for alienating land and was well accepted by settlers. In the mid-1800s, legislative provisions were enacted in each colony for local government to be formed and in the late 1800s local government was given powers to levy land tax in conjunction with the states. Local government rationalised the imposition of rates for the maintenance of roads, street lighting and rubbish collection, services that were informally managed by the colonies.

After Federation, land tax was levied by each of the three tiers of government in Australia (Smith 2005). In 1910, the states ceased taxing land, leaving this source of revenue to the Commonwealth and local government to collect, while the states collected income tax. By 1907, Warren (2004) highlights that each state had introduced income taxes. During the initial years of Federation, land tax became a burden on homeowners and the first inquiry into the tax was undertaken. Poor coordination of the tax, which was then assessed on the unimproved value of land, resulted in the states and Commonwealth using different values for the same property in the same year of assessment.

The states progressively reintroduced land tax after the Commonwealth took over the collection of income taxes in 1942 (Simpson & Figgis 1998). A subtle divergence emerged in the rationalisation and acceptance of land tax by taxpayers from the period between the pre- and post-Federation period of Australia. This is traced back to the initial alienation of land which was achieved through grant, sale or leasehold interests in land. With each of these options, the state retained the right to collect a rent, impose tax, regulate the use of land and retain the first option to purchase the land back from landholders. In the initial period of alienation, it was apparent that increases in the value of land were attributed directly to services provided which increased the value of land, in conjunction with the produce from land.

As land traded among settlers, a market developed and the price for land was progressively determined. As land traded several times, the nexus between the increased value resulting from the provision of infrastructure became more disparate and the rationale for the payment of land tax moved to a tax for the use and maintenance of services. When the value of infrastructure progressively became capitalised into land values, the nexus between the added value it contributed became less obvious to taxpayers, who in turn questioned the purpose of its imposition.

Table 2.2 Concessions for pensioners and first homebuyers

State	*Legislation*	*Provision*	*Housing type*
Victoria	Duties Act 2000 – Division 5	Pensioner and first homeowner exemptions and concessions	New
New South Wales	Duties Act 1997 – Pt 8 Div 1	First new home	New
Queensland	Duties Act 2001 – Div 3	Concessions for homes and first homes	New
Western Australia	Duties Act 2008 – Div 3	First homeowners concession	Not specified
South Australia	Stamp Duties Act 1923	s71c concessional rates of duty in respect of purchase of first home	Not specified
Tasmania	Duties Act 2001	s36G exemptions and concessions	New from 2014

Source: Various State Duties / Stamp Duties Acts

The second part of Table 2.1 highlights the purpose and rationale for the property tax across the tiers of local and state government post-Federation. While local government had built a rapport with the tax paying public as the provider of local services, the re-entry of state government in the imposition of land tax during the 1950s resulted in the perception of the tax being a consolidated revenue tax. For the states, land tax was never a major source of revenue in contrast to local government, which is more reliant on rate revenue, as shown in Table 2.3.

Table 2.3 Local rates and state land tax as a percentage of total revenue, 2008–09

Own Source Revenue	*NSW*	*Vic*	*Qld*	*WA*	*SA*	*Tas*	*NT*	*Total*
Local Gov't Rates	33.6	43.7	27.0	41.3	55.2	32.7	17.1	35.6
State Land Tax	12	8.6	11	8.4	15	10.4	N/a	10.6

Source: 2008/09 Local Government National Reports cited by Comrie (2013)

Since WWII, local government across Australia has experienced increasing financial burden as services from the states are progressively assigned to local government. In response to the evolving commitments of local government, in 1972 the financial funding of Local Government changed significantly through a longstanding initiative of the Whitlam Government.

The Grants Commission Act 1973 (Cth) was introduced and designed to overhaul the Commonwealth Grants Commission (CGC) and, among other objectives, provide a more fluent conduit of Commonwealth grant funding to local government via the states. In the Parliamentary Debates, Hansards (1973) defined the words of Whitlam in the provision of Local Government grant revenue as, 'to provide a standard of service to their communities that will be comparable with that enjoyed by communities elsewhere' (para. 2304).

The peak in Commonwealth Government financial support for Local Government continued during the Fraser Government, which introduced the Local Government (Personal Income Tax Sharing) Act 1976 (Cth). The Oakes Inquiry (1990) highlights that between 1980/81 and 1985/86 the guaranteed share of income tax money granted to Local Government by the Commonwealth reached a high of 2 per cent. The Hawke/Keating Government abandoned this system, adopting a "macroeconomic policy of fiscal restraint" and by 1988/89 Commonwealth income tax receipts represented 1.32 per cent of grant revenue provided to Local Government across Australia and grant revenue having declined in real terms" (Oakes Inquiry 1990:58).

During the transition of the Whitlam to Fraser Government (Commonwealth), the state government in NSW introduced rate pegging in 1977. "Its introduction was seen as a response to the economic conditions of the time including spiralling cost-push inflation. However its use in NSW has no parallel in any other State" (Local Government Association of NSW 2003:3). When rate pegging was introduced in 1977, there was little resistance from Local Governments as Commonwealth Government grants were increasing during the Whitlam and Fraser period.

While the Commonwealth attempted to address the financial difficulties of local government, the states were experiencing their own financial challenges. Following the surrender of income taxes to the Commonwealth in 1942, as the need to fund WWII ceased and prosperity increased, government was presented with the need to reform a number of its taxes, of which death duties were high on the agenda. As homeownership increased post–WWII, the need to protect the inheritance of the next generation commenced in earnest as the states progressively abolished death duties, replacing this revenue with increases in conveyance stamp duty.

This move resulted from a tax being levied on which death was a deemed disposal, to a levy imposed on the acquisition of an asset at the point of purchase. While stamp duty existed at the time, death duties were progressively abolished, the rates in the dollar imposed on stamp duty from property transfers progressively increased during the years that followed the phasing out of death duties. Revenues from stamp duty, particularly on real property, progressively increased through bracket creep as property values continued to climb and more property tripped into higher tax brackets. Since the abolition of death duties, both property values have progressively increased, along with the rates in the dollar, and the states have become hooked on stamp duty revenue. This has worked on the premise that the tax is paid once while purchasers are in a euphoric frame of mind after they have purchased the property.

For state government, the transition to increasing tax revenue from a recurrent land tax while reducing conveyance stamp duty is a zero sum gain. Reducing revenue on acquisition taxes, imposed once on the purchase price of property, and replacing this revenue with an annual and highly visible tax imposed on land value is not a simple transition for the states. While in principle the states recognise the economics benefits of reform, these appear to be outweighed by an unpopular land tax imposed on the holding of land. The introduction of the GST from 1 July 2000 was to result in the abolition of wholesales sales tax and the tax revenue from the GST was to be paid to the states for the abolition of stamp duties.

The latter did not fully occur, with stamp duty being abolished on share transactions in July 2001, while being retained on housing. The rationale for not removing stamp duty on conveyances was attributed to the large concession of leaving food out of the GST tax base, which was reluctantly conceded in order for the GST to gain support in the Senate and ultimately succeed as an important landmark tax reform. In lieu of reforming conveyance stamp duty, both the states and Commonwealth provided exemptions to first homebuyers in the form of First Home Owners Grants (FOHG).

The First Home Owners Grant (FHOG) was introduced in July 2000 to offset the impact of the GST on new dwellings, and primarily focused on new homeowners who had not previously been in the market (Office of State Revenue 2009:1). This grant is administered by the states under their respective Duties legislation, which are referred to in Table 2.1, and highlights that in three states the grant is now applicable to new dwellings only, while in the other three states the legislation is silent on the type of housing to which the grant applies.

Given the additional costs of construction, resulting from the introduction of the GST, the focus of the FHOG scheme was in fact to offset impacts of the higher cost of housing and encourage developers to continue to build. The increased cost of housing would be offset by the FHOG for the first homebuyer segment and hence would not be impacted by the higher price resulting from the GST. However, at the time of introducing the FHOG scheme, it applied to first homeowners, regardless of whether the housing was new or existing. The grant was amended in New South Wales, Queensland and Victoria during 2009 when the FHOG only applied to new dwellings.

Shortly after the introduction of FHOG, Kupke and Murano (2002) studied the impact of FHOG and identified that it was the single largest factor contributing to first homeowners purchasing property. Subsequent to the transition of FHOG applying to new dwellings only, Irvine (2009), in citing surveys carried out by Bankwest and the Mortgage Finance Association of Australia, found that only 6.2 per cent of first homebuyers cited FHOG as the reason behind their home purchase. This is supported by Chancellor (2013:1), who states that currently only 20 per cent of first homebuyers purchase new property. The rationale for reforming the FHOG to only apply to new housing moved the focus from purely assisting first homebuyers, to stimulating the supply of new housing stock and maintaining the economy through support to the development and construction drivers of the economy.

Despite the fiscal challenges confronting sub-national government, Australia's Future Tax System (AFTS 2008) has identified that Australia – in contrast to the United States, United Kingdom, New Zealand and Canada – has capacity to increase revenue from land tax. This capacity was further identified by the Productivity Commission (2004), though it was not stated as to whether increase in revenue from land tax should be assigned to state or local government, or shared between these tiers of government. It is suggested that the states broaden their base of state land tax by including the principal place of residence, currently exempt from land tax in each state of Australia (AFTS 2010).

Despite capacity to increase land tax revenue, it is highlighted in Table 2.4 that in many OECD countries property taxes have decreased as a percentage of the total tax collected and also as a percentage of GDP. Since 1965, tax revenue sources have moved towards consumption-based taxation, which include the GST in Australia and Value Added Tax (VAT) in the United Kingdom and United States. The percentages used to measure taxes are defined by Bird and Slack (2004:7) as fiscal benchmarks for measuring the tax efforts of countries.

Table 2.4 shows that between 1965 and 2009, as a percentage of total tax collected, Australia's revenue from land tax reduced by 18.5 per cent; however, over the same period it has increased marginally by 1.1 per cent as a percentage of GDP. Unlike the United States, Canada and the United Kingdom, where the property tax is imposed and retained by local government, as highlighted in Australia, this tax is collected by states and local government. In the case of state land tax, the exemption of the principal place of residence and tax thresholds expended by each state result in less than 20 per cent of property owners in Australia who are subject to local government rating being liable to pay state land tax. To this end, the question arising is not whether land tax should be increased in Australia, but how this increase is to be achieved in a progressive and sustainable manner.

Table 2.4 Global trends in land tax revenues

	Percentage of total tax			*Percentage of GDP*			*Rank in OECD countries*
	1965	*2010*	*% change*	*1965*	*2010*	*% change*	
Portugal	0	1.9	. . .	0	0.6	. . .	20
Italy	1.7	1.5	–16.5%	0.44	0.62	40.4%	19
Finland	0	1.9	. . .	0	0.65	. . .	18
Netherlands	1.02	1.8	77.3%	0.334	0.7	109.6%	17
Korea	. . .	3.2	. . .	. . .	0.79	. . .	16
Sweden	0.025	1.7	–6868%	0.008	0.793	9812%	15
Ireland	12.2	3.2	–74.2%	3.05	0.87	–71.5%	14
Spain	0.45	2.7	511%	0.066	0.88	1235%	13
Poland	. . .	3.7	. . .	. . .	1.2	. . .	12
Belgium	0.027	2.8	10,363%	0.008	1.229	15,262%	11
Denmark	4.9	2.9	–41%	1.5	1.4	–6.2%	10
Australia	**6.8**	**5.5**	**–18.5%**	**1.4**	**1.42**	**1.1%**	**9**
Iceland	1.7	5.2	212%	0.4	1.9	320%	8
New Zealand	8.3	6.6	–20.9%	2.0	2.1	4.4%	7
Japan	5.2	7.7	49.3	0.9	2.1	131.6%	6
Israel	–	7.2	. . .	–	2.3	. . .	5
France	1.9	5.7	200%	0.7	2.5	268%	4
United States	13.7	12.2	–11%	3.4	3.0	–10.4%	3
Canada	11.9	10.1	–15.5%	3.0	3.1	2.1%	2
United Kingdom	11.2	9.8	–13%	3.4	3.4	–0.4%	1
Unweighted ave							
OECD–Total	**3.8**	**3.25**	**–15.4%**	**0.95**	**1.05**	**9.9%**	**Ranking**

Source: OECD 2011, Revenue Statistics 1965–2010

Land tax revenue trends and principles of good tax design

As highlighted in the previous section, stamp duty and land tax are impacted by a number of factors which influence their operation and acceptability in principle and in practice. This section reviews these points in more detail, and provides an overview of revenue trends from these taxes and a summary critic against the principles of good tax design. It further examines these factors within the parameters of the broader tax principles of ability-to-pay and benefits-received, which tie in with the earlier discussion on the perceived taxpayer rationale for the imposition of land tax and its evolution in Australia.

Revenue trends

The review of land and property tax revenue trends across Australia show that stamp duty is an important source of revenue for state government, as shown in Figure 2.2 and Table 2.6. Further noted from trends in stamp duty revenue is volatility compared with revenue from local government rates and state land tax across Australia. The volume of revenue generated from stamp duty is significant and is not replaceable with revenue from land tax in the short term. Such a transition, now underway in the ACT, will require a progressive phase out/phase in over a period of 5 to 10 years.

Figure 2.2 shows that state land tax produces the lowest tax revenue from all three sources; however, it is the narrowest in its application applying to less than 20 per cent of property owners in Australia. This is a criticism noted in a number of inquiries which impacts the principle of equity and efficiency. The narrow application of the tax is attributable to the exemption of the principal place of residence and the investment threshold applied in each state. The total state land tax revenue derived from residential property is approximately 25 per cent of the revenue collected from this source across Australia. Despite being the lowest tax revenue generated of the three taxes, the revenue is closely aligned to movements in land or site values of non-residential property of which land/site values are reassessed annually or bi-annually by the states.

Local government rates, in contrast to land tax, are paid by over 98 per cent of all property owners in Australia, rates have the broadest base and lowest exemption. Revenues from council rates are the least volatile of the three tax sources; while tied to value they are also impacted by rate pegging in New South Wales, where increases in revenue are largely aligned to the movement in the Local Government Cost Index (LGCI), which comprises the movement in cost of a basket of goods from one year to the next, similar to the Consumer Price Index (CPI). The rates applied to land, site or improved value across local government areas vary annually to ensure rate revenues remain steady and in line with expected budgeted costs for each tax year.

A further level of contrast is now made between state land tax and local government rates across the states in Table 2.6, which sets out the relative changes in revenue between state land tax and local rates at the beginning, middle and end of the 12-year period of 2002–12. It is noted that over this period in each state, with the exception of Western Australia, state land tax revenue has outpaced tax revenue collected from local government rates. Between 2001 and 2006 this trend was noted across all states, with exception of Western Australia and Victoria. The largest increases in revenue from land tax as a percentage of local rates across the 12 years are noted in the states of South Australia and New South Wales. Western Australia, in contrast, shows a steady similar tax revenue trend between state land tax and local government rates.

It is evident from this analysis that increases in revenue from recurrent land taxation across Australia over the past 12 years has been in favour of

Table 2.5 Recurrent land tax expenditure 2013

Property class	*Principle residence*		*Investment / Place of business*	
Recurrent tax	Local rates	Land tax	Local rates	Land tax
Concession	Age discount	Exempt	Nil	Threshold / exemption
Tax expended	Ave $250 p.a.	$5bn total	Nil	Not states
Tax rationale	Benefits-received	N/a	Benefits-received	Capacity-to-pay

Source: Author

state land taxation over local government rates, with the exception of Western Australia. This trend is likely to increase further in favour of the states if hypothecated taxes are applied by the states and collected by local government. Such tax includes the recently introduced fire service levy imposed in Victoria during 2013. Fire service levies are one way in which expanding the land tax net to include the place of residence may be aligned to specific services provided by the states. The rationale for the transition from collecting this levy on insurance policies to collecting a tax on every property is cited as being due to some property owners avoiding the levy, as they remain uninsured.

The application of the levy is simple and transparent, being an amalgam of a base amount of $100 per residence and a percentage of improved value, while on non-residences, $200 per property plus a percentage of improved value (NSW Treasury and the Ministry for Police and Emergency Services 2012).

While capacity exists for increased revenue from recurrent land taxes, what remains unresolved are four key points:

1. How is unmet revenue capacity to be measured in determining the progressive and incremental increases in recurrent land taxes by state and local government?
2. Which tier of sub-national government is to be the beneficiary of increases in recurrent land tax revenue, and can increased revenue be shared between state and local government?
3. In relation to point 2, can increases in state land tax revenue be offset against reductions in conveyance stamp duty, and what incentives can the Commonwealth extend to the states to incentivise this transition?
4. How can increases in unmet local government rate revenue be used to contribute to local infrastructure across the states?

Opportunity exists for land tax revenue to contribute to reform of the broader tax system in reducing the number of less efficient transaction taxes on property. As noted in Chapter 1, in bringing Australia in line with

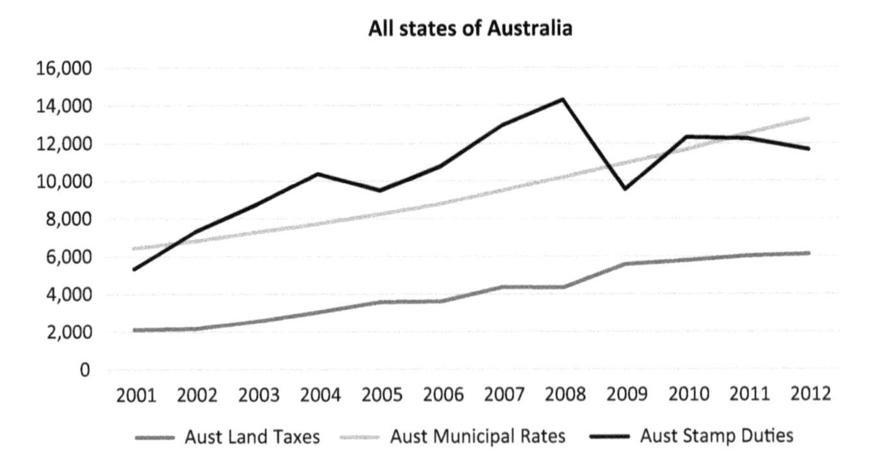

Figure 2.2 Land tax, council rate and conveyance stamp duty revenue comparison
Source: ABS 2013, Taxation Statistics 2001–12

advanced OECD economies, recurrent land tax revenues from state and local government combined firmly to assert this revenue as one of Australia's top five taxes by income.

Principles of good tax design

The review of tax systems, and in particular individual taxes, requires benchmarking against criteria, more commonly referred to as principles of *good tax design*. These principles, formerly known as the canons of taxation as defined by Adam Smith, comprise *simplicity*, *transparency*, *equity* and *efficiency*. In more recent reviews, a fifth principle of *robustness* or *buoyancy* has emerged which assesses the stability of tax revenue to government. These principles have both theoretical underpinnings and practical application of these taxes.

In the application of these principles to land tax and conveyance stamp duty, a number of state government tax reviews have provided context in ranking these taxes against these principles. In their seminal publication on public finance, Musgrave and Musgrave (1976:211) outlined the criteria that a good tax structure should exhibit, including:

1. The distribution of the tax burden should be equitable. Everyone should be made to pay his or her 'fair share'. (*Equity*)
2. That should be chosen so as to minimise interference with economic decisions in otherwise efficient markets. Imposition of 'excess burdens' should be minimised. (*Efficiency*)

3. At the same time, taxes may be used to correct inefficiencies in the private sector, provided they are a suitable instrument for doing so. (*Efficiency*)
4. The tax structure should facilitate the use of fiscal policy for stabilisation and growth objectives. (*Stabilisation and growth*)
5. The tax system should permit efficient and non-arbitrary administration and it should be understandable to the taxpayer. (*Transparency*)
6. Administration and compliance costs should be as low as is compatible with the other objectives (*Simplicity*)

In the most recent review of state taxes in New South Wales, Independent Pricing and Regulatory Tribunal (IPART 2008) and the Victoria State Business Tax Review Committee (2001) have examined and ranked land tax and conveyance stamp duty against the tax principles of good tax design. In each review, land tax ranks as the most efficient tax of all state taxes, with an overriding provision in the Victorian review. The Victorian tax review ranks land tax as the most efficient tax in theory, of which several matters impact its operational efficiency.

The New South Wales state tax review highlighted the weaknesses in the taxation of land under the principles of transparency and simplicity, as shown in Figure 2.3. Contributing to poor performance against the criteria of transparency and simplicity is the concern that taxpayers have poor information when a property exceeds the threshold and subsequently becomes liable for the land tax. Additionally, the three-year averaging of values has added a further level of complexity to this tax (IPART 2008). Impeding land tax further under these principles is the availability of information to taxpayers as to how land value is determined, being the base on which their tax is assessed (Walton 1999 and NSW Ombudsman 2005).

On the principles of equity and efficiency, the tax in principle ranks high; however, in practice and as per the NSW tax review, as shown in Figure 2.3,

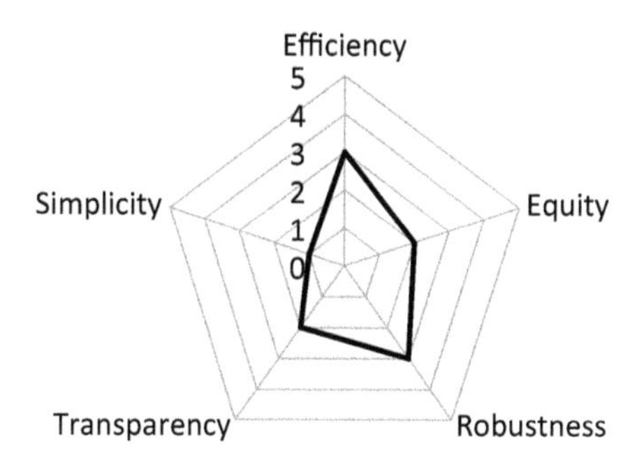

Figure 2.3 Land tax and tax design principles
Source: IPART NSW 2008

the tax scores are low. This is primarily due to the tax being applied to such a narrow number of property owners and the exemption of the principal place of residence applied in every state. This factor impacts the principle of equity, to which the tax-free threshold for investors further erodes the principle of equity. The Victorian tax review further adds the exemption for primary production land as a further ground impact equity of this tax. Another negative impact of land tax raised in the Victorian review, which sits outside the principles of good tax design, is the impact of cash flow difficulties on asset-rich and income-poor taxpayers.

In expanding on the points of equity and efficiency, Table 2.5 sets out two categories of property; the first is the principal place of residence and the second is investment property, of which the principal place of business is included within this category. In the case of the place of business, land used for primary production is land tax exempt in each state. In both the case of the principal place of residence and investment property, local rates apply with limited concession. As shown in Table 2.5, a concession of $250 per annum applies to pensioners for council rates on their principal place of residence. In the case of state land tax, there is significant tax expended, which equates to approximately $5 billion dollars per annum due to the absence of any form of land tax on the principal place of residence.

In contrast to land tax, conveyance stamp duty ranks opposite to land tax under the principles of good tax design, with simplicity and transparency ranking high, while efficiency, equity and, in particular, robustness ranking low. The low ranking attributed to robustness has been amply covered early in the analysis of tax trends under the volatility of revenue encountered in changing economic markets, of which revenue from this tax is shown to be volatile. Under efficiency, conveyance duty scores low, as this tax adds to the cost of real estate and deters business and people from purchasing property (IPART 2008). In addition, this tax creates a lock-in effect deterring the trading up and down of property, which creates greater emphasis on renovating or extending existing property, rather than moving to more appropriate housing (State Business Tax Review Committee 2001). The latter point emphasises the impact conveyance stamp duty has on mobility in freeing up housing and better accommodating the needs of occupants more suited to residential property.

On principles of transparency and simplicity, stamp duty scores high as the tax is transparently assessed on the price paid for the property, which is determined by the purchaser/taxpayer. As an upfront cost on the purchase of a property, the amount of tax is relatively simple to calculate. On the principle of equity, this tax scores low and has mixed reviews. In the absence of land tax on the principal place of residence, stamp duty is stated to be the second best solution (State Business Tax Review Committee 2001). It is further stated to be equitable from the perspective that the progressive rate and threshold structure of higher tax applied to more valuable property renders it a wealth turnover tax (State Business Tax Review Committee 2001).

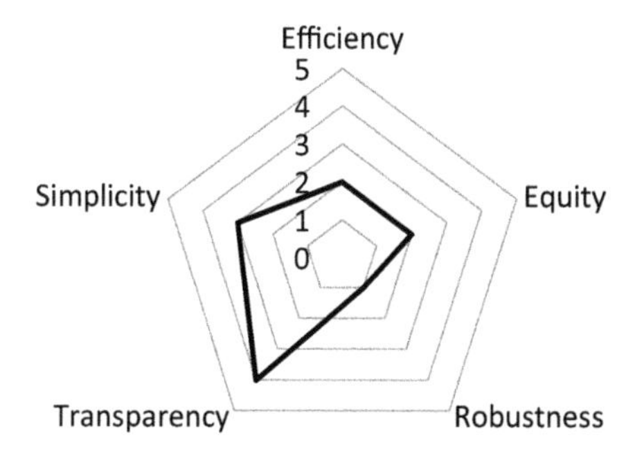

Figure 2.4 Conveyance stamp duty and tax design principles
Source: IPART NSW 2008

This point is countered by the issue of bracket creep and the progressive increase in value pushing more and more property purchases into higher tax brackets.

Benefits-received and capacity-to-pay

As discussed in the previous section of this chapter, Australia's recurrent land taxes are aligned with varying rationales which have emerged in the imposition of state land tax and local government rates. Musgrave and Musgrave (1976) define two strands of thought for defining equity in the application of a tax system as being the *benefits-received* and *capacity-to-pay* principles. As one of the more visual taxes imposed annually, debate remains as to whether recurrent land taxes are consumption/benefits-received or capital/capacity-to-pay taxes. These important principles will shape the reform of land tax into the future, influencing which level of government should collect the tax, which bases the tax may be assessed on, and what provisions and policies will be needed in managing the expansion in revenue from this tax.

Benefits-received

The application of the benefits-received approach concerns itself with both tax collection and allocation of tax revenue policy. Under a strict application of benefits-received, each taxpayer would be taxed in line with their demand for specific public services. This demand varies from taxpayer to taxpayer and, as highlighted by Musgrave and Musgrave (1976:212), 'For the benefits principle to be operational, expenditure benefits for particular taxpayers must be known'. As preference patterns change and use of different services varies over time, it is not a system that is easily applied and administered. Where specific benefits or uses of public services may be observed or measured, consideration may be given to the use of this approach.

Table 2.6 Change in land tax revenue as percentage of local government rate revenue 2001–2012

	2001	*2002*	*2003*	*2004*	*2005*	*2006*	*2007*	*2008*	*2009*	*2010*	*2011*	*2012*
Qld stamp duty	700	1,056	1,382	1,863	1,728	1,949	2,542	2,912	1,806	1,978	1,933	2,023
Qld land taxes	230	231	279	313	419	404	485	610	838	1,033	1,042	1,013
Qld municipal rates	1,210	1,281	1,369	1,461	1,559	1,736	1,925	2,096	2,285	2,438	2,666	2,805
% change in revenue	**19**					**23.3**						**36**
Vic stamp duty	1,284	1,885	2,116	2,446	2,337	2,671	2,961	3,706	2,801	3,604	3,910	3,379
Vic land taxes	525	515	655	837	848	780	989	865	1,238	1,178	1,398	1,401
Vic municipal rates	1,543	1,676	1,827	2,001	2,170	2,294	2,500	2,724	2,927	3,159	3,416	3,656
% change in revenue	**34**					**34**						**38.3**
NSW stamp duty	2,267	3,119	3,677	3,918	3,282	3,237	4,166	3,938	2,736	3,739	4,045	3,764
NSW land taxes	929	1,001	1,136	1,355	1,646	1,717	2,036	1,937	2,252	2,296	2,289	2,350
NSW municipal rates	2,168	2,236	2,347	2,424	2,521	2,638	2,776	2,935	3,030	3,166	3,303	3,445
% change in revenue	**43**					**65.1**						**68.2**
WA stamp duty	624	647	833	1,207	1,218	1,906	2,037	2,243	1,008	1,615	1,039	1,340
WA land tax	221	226	260	280	315	313	386	415	562	519	516	548
WA municipal rates	669	705	754	801	869	928	1,001	1,088	1,220	1,317	1,454	1,581
% change in revenue	**33**					**33.8**						**34.6**
SA stamp duty	295	354	428	578	561	600	721	909	721	787	784	683
SA land tax	140	140	157	198	256	291	332	375	510	553	576	588
SA municipal rates	545	589	641	683	738	785	834	886	958	1,019	1,086	1,161
% change in revenue	**26**					**37.1**						**50.6**
Aust stamp duties	5,340	7,283	8,745	10,388	9,472	10,788	12,923	14,289	9,526	12,294	12,229	11,657
Aust land taxes	2,103	2,172	2,553	3,059	3,583	3,613	4,358	4,346	5,565	5,767	6,005	6,103
Aust municipal rates	6,441	6,808	7,276	7,726	8,237	8,788	9,476	10,194	10,938	11,645	12,506	13,265
% change in revenue	**32.7**					**41**						**46**

Source: ABS 2013, Taxation Statistics

The benefits-received principle is respected in most tax systems; however, it is tended with difficulty as it attempts to rationalise a relationship between tax paid and services provided by government. It is even more tenuous when attempting to draw a relationship with rates against services actually used by ratepayers, of which there is little research to support a proportional connection. It is more commonly aligned and better correlated with user pay charges, in which a more direct link can be made between the two. In more recent years, local governments have used the benefits-received principle in charging for street parking.

In the application of this principle using value of the taxpayer's land or property, it may be further argued that whether or not someone uses the services of government is not the test for determining that a tax is or should be defined as a benefits-received tax. It is the option value to use services by virtue of the land's location to those services, that is the ultimate arbiter which impacts its value and hence captures the inbuilt value of the availability of those services. The same option exists for all residents to use the services provided by government, and it is this option which adds value to the property, regardless of whether the taxpayer uses these services or not. In this regard, it is important that value is not totally disregarded as the base on which land tax or rates are determined.

While a degree of perceived linkage exists between local government rates and local services, what is less clear in Australia is which services are perceived to be linked to council rates. In contrast, Mangioni (2014) suggests that rates should be seen as a general revenue tax and not directly linked to benefits-received. It was highlighted in Chapter 1 that many property owners are not accustomed to paying taxes and hence, when the tax is paid directly by property owners, the default to a benefits-received link is made, as paying taxes into consolidated revenue is not an evolved rationale for most residential property owners and novice investors.

In contrast to local government rating, the rationale for state land taxation is detached from any service provision; it is a consolidated revenue tax. State land tax is largely seen as a non-earmarked tax and is strongly opposed by many who pay it (Nile 1998). This opposition is founded on two bases, the first being the select and limited application of the tax, which is applicable to less than 20 per cent of property owners in Australia. This underpins the second reason, being that the tax is perceived to be targeted at the wealthy, rather than at all property owners (Nile 1998). It is at this juncture that the current structure of recurrent land taxation in Australia is faulted, because of its narrow application by state government and the reluctance to expand the tax to all property owners, as recommended by AFTS (2008).

Capacity-to-pay

Capacity-to-pay stands in contrast to the benefits-received principle and there is no clear definition as to what it means (AFTS 2008). Its measurability may be either determined on wealth or income and include a number of different factors in its application to tax-raising policy. These factors include whether all forms of income should be included, the role of assets including owner-occupied housing and choices about work or leisure (AFTS 2008). In its application to land tax, this principle may be applied differently in the case of state land tax in contrast to local government rates, a fact that emerges in Chapter 3 once the mechanics of determining these taxes is demonstrated. These demonstrated differences highlight the strengths and benefits of a two-tier land tax system and the opportunities it affords in delivering greater equity across the spectrum of property and taxpayers.

In summary, land tax and, to a lesser degree, council rates are determined on the value of land which reflects wealth, rather than the actual income of the taxpayer. It is highlighted that, over the lifetime of a taxpayer, the relativity between income and wealth may vary significantly, as highlighted in Figure 2.5. In this example, in the age bracket of 15–24 income is high relative to wealth, which changes in the mid-life age bracket of 45–54, where net wealth exceeds income for the first time. In contrast, in the later age brackets of 65–74 and 75+, income is low relative to net wealth, where mortgage debt on property has reduced or has been paid off and income reduces in retirement.

As will be observed in Part 2, local government has statutory provisions for discounting rates to address the high net wealth versus low income issue in the later years of the lifecycle. This is further extended for approved applicants on government pensions. While some correlation exists between income of ratepayers and the value of property, the determination of rates

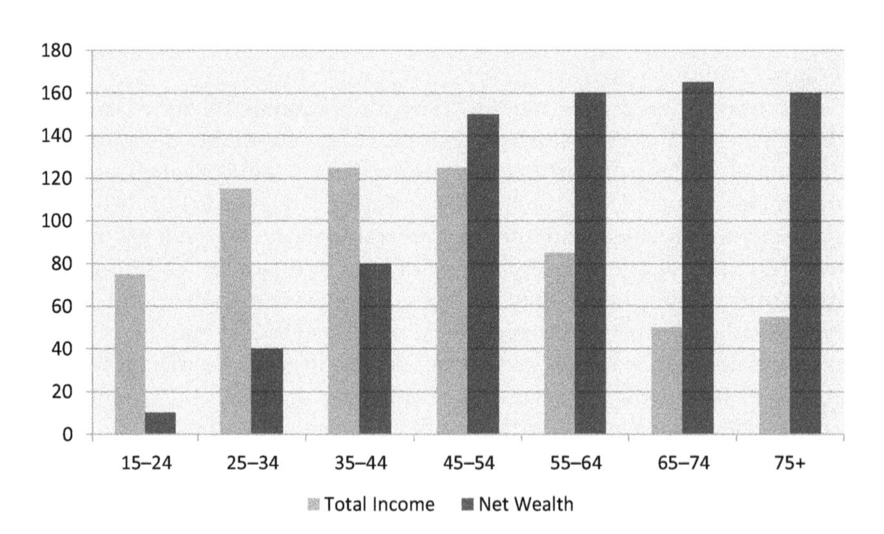

Figure 2.5 Relativity between income and net wealth over a taxpayer's lifecycle
Source: Kelly 2003 (cited by South Australian Centre for Economic Studies, 8)

on either value or income alone may be better addressed using a combination of these two measures. In progressing local government rating into the future and improving local government tax effort from property, formulation of a capacity-to-pay option determined on a combination of value and income of the owner warrants further consideration.

In Australia, Warren (2004) notes the progressive use of tax hypothecation, also known as earmarking, which has progressively emerged in a number of taxes over the past two decades in Australia. These include the Medicare Levy, Higher Education Contribution Scheme (HECS) and Superannuation Guarantee Charge (SGC). While tax hypothecation is not popular with tax economists, politicians are able to sell new taxes and tax reform when specifically aligned with services. This is particularly the case with local government rating of land where a higher degree of sensitivity exists in the imposition and payment of taxes. To this end, local governments are now focused on alternate revenue sources which can be directly linked to services, parking meters and charges for resident parking permits, which all constitute taxes paid for benefits-received and can be directly linked to local government's largest item of capital expenditure, road maintenance.

Summary

This section has examined several aspects of land tax, including its place among the top ten taxes by revenue raised in Australia and the contribution that may be made to reforming less efficient property taxes. The economics underpinning land tax, while robust, are met with a number of ideological and practical challenges in achieving reform and further expansion of revenue from this source. It was demonstrated that in highly urbanised locations, the traditional rationale for determining the supply of land may be expanded through vertical development and greater utilisation of airspace.

The discussion on the evolution of taxes on land and property highlighted that land was the most neutral base on which the tax has been assessed to date. In theory its economic efficiency is robust and, through the periods of time examined, the principles governing the underlying value have progressed from economic rent to the overriding factor of the land's location in an urban environment which ultimately determines its economic rent. Since the settlement of Australia, it was shown that the rationale for a tax on land has progressively evolved from a tax levied for the initial provision of infrastructure which was progressively capitalised into the value of land.

Identified among the challenges confronting recurrent land tax are the principles of good tax design. Land tax, while in principle and theory rates high, in practice rates poor against economic and operational efficiency, simplicity of assessment and transparency, of which the manufacture of value is central. The saleability of the land tax as a consolidated revenue tax aligned with the principle of capacity-to-pay, as opposed to its perceived understanding as a benefits-received tax, is one of the biggest challenges in driving reform and increasing tax revenue under the existing dual system.

The sections of this book which follow are important in defining some of the challenges confronting the reform of land tax. This important context underpins the final part of the book in articulating how reform may be managed and barriers overcome in the transition to a more efficient land tax system across Australia. It further paves the way for the states to contribute to the fiscal rebalance of own source revenue through a tax on land which removes distortions and impact on the mobility of housing and investment.

Part 2

Land tax assessment and administration in Australia

Introduction

Part 1 emphasised Australia's capacity to increase revenue from recurrent land tax while reforming less efficient taxes that apply to property at the point of purchase. As the need for tax reform exists at the macro level through tax revenue recalibration from less efficient to more efficient taxes, an operational review of land tax is also needed. Part 2 examines state land tax and local government rating, the component parts of these taxes, how they are determined and how the tax is applied by state and local government across Australia. An analysis of the principal place of residence exemption and threshold applied to business use and, in particular, investment property highlights the gross distortion that has emerged over the past two decades.

In Part 2, we review the way value is determined within valuation of land statutes and also within procedure manuals and guidelines designed to guide the valuer in the valuation of land process. We commence in Chapter 3 with an examination of the differences in the meaning of *value* within the Valuation of Land statutes and its manufacture through the valuation process. A study of the valuation process follows in Chapter 4, which maps the processes actually used by valuers to value land in highly urbanised locations and the challenges in undertaking this task in the absence of vacant land transactions. These two chapters bring forward questions about retaining land as the base of the tax and the frequency of its determination, and also paves the way for alternate forms of value.

Chapter 5 examines the differences between state land tax and local government rating in each state and the exemption of the principal place of residence from state land tax. Chapter 6 examines the application of land tax to land other than the principal place of residence, with particular reference to business use land and land used for primary production purposes. It further addresses a number of concessions and allowances embedded within land tax and valuation of land statutes. Chapter 7 reviews the process of objections and appeals to land tax, local government rates and the valuation of land used to assess these taxes. This part provides new research on the bases used to assess land tax and the evolving challenges of valuing land in highly urbanised locations.

Part 2 concludes by summarising the outdated operation and administration of recurrent land taxes and challenges that exist in the assessment of value. This leaves no doubt that, before revenue can be expanded from these taxes, a number of overdue reforms underpinning the integrity and acceptability by taxpayers are needed. This sets the tone for Part 3, which is a review of land and property tax in several international jurisdictions which define what works well and what does not work, and sets out potential reform options for Australia in modernising our recurrent land taxes.

In commencing Part 2, a summary of the component parts of state land tax and local government rating is provided. This is followed by a worked example of these two taxes as they apply in a hypothetical scenario. This provides a context and understanding of the assessment process, which in most states is eased through the availability of online land tax calculators provided by the various Offices of State Revenue. While these assist in the mechanical calculation of land tax liabilities on a primary assessment basis, the following chapters aim to assist in reviewing and explaining many of the factors that are yet to be articulated to the taxpayer in the assessment of their land tax and council rate liabilities.

The articulation of the component parts of these taxes further extends to explaining council rating and demonstrates the variability in differential rating across the various classes of land and property. As local government rating becomes an increasingly important conduit for the collection of recurrent land tax over the next decade across Australia, much work and reform will be needed across the spectrum of assessment processes that coexist across Australia. This will mean reforms to the base of this tax and the way it is applied within and across local government areas. In improving the economic and operational efficiency of land taxes, it is demonstrated that quantum improvement is needed in the simplicity and transparency in maintaining their integrity and acceptability with taxpayers.

The key components of recurrent land taxation in Australia are set out below, followed by a worked example which highlights the complexity of the assessment process and provides a context for Chapters 3 to 7, which examine the statutory definitions of value and the valuation processes on which the base of these taxes are assessed. While these taxes vary from state to state to some degree, the core principles of the aggregation of value, one threshold per entity and exemption of the principal place of residence are common across all states.

Component parts of state land tax and local government rates

Components	*Application summary*
Tax base	The bases on which these taxes apply vary from state to state and across local governments. (Land/Site Value, Capital Improved Value and Assessed Annual Value)

Components	*Application summary*
Rate	The rate is the factor applied to the base on which the tax is assessed and, like the base, varies between the states and also across local government.
Land/property category	Land and property categorisation apply in the rating of land and property by local government across Australia. The categorisation of land is the grouping of different land uses to which a different rate (differential rating) may apply.
Taxpaying entity	The taxpaying entity depicts the various ownership types in which land is held.
Tax-free threshold	The tax-free threshold is an exemption applied to state land tax and is distinguished by a number of factors, including the use of the property and the taxpaying entity, of which these factors differ from state to state.
Exemptions and concessions	These primarily apply to state land tax with the two main exemptions being the exemption for the principal place of residence and primary production land.

Taxpayer scenario: land tax and council rate example

The following example provides context for the operation of state land tax and local government rating as a prelude to Part 2. While the component parts of these taxes differ in each state of Australia, New South Wales is used to demonstrate the application of the tax using the 2014 rates and threshold applied to a single person taxpayer. In contrast to state land tax, the breadth of application and bases on which local government rates may be assessed across more than 600 local government areas of Australia cannot be simply demonstrated with one example. An example of how local rates are calculated is provided to demonstrate the fundamental differences between state land tax and local government rates in NSW; however, a number of the primary methods used to assess local rates will be discussed as they apply in each state throughout the following chapters.

Frances owns the house she lives in at 8 Brown Street and a retail shop in the nearby Main Street retail strip (15 Main Street). She operates her business packaging and distributing coffee beans from an industrial strata unit complex, in which she owns one strata unit (4/35 Green Street), and sells the coffee beans from the Main Street retail shop at 15 Main Street. What is Frances' land tax and council rate liability in the 2014 land tax and rating year?

Step 1: defining the land value

Land values for assessing land tax and council rates for 2014

Address	*Date of valuation (Base Date)*	*Land value*
8 Brown Street	1–7–2012	$600,000
	1–7–2013	$700,000
	1–7–2014	$800,000
	Average land value 2014	**$700,000**

(*Continued*)

Address	*Date of valuation (Base Date)*	*Land value*
15 Main Street	1–7–2012	$900,000
	1–7–2013	$1,000,000
	1–7–2014	$1,100,000
	Average land value 2014	**$1,000,000**
35 Green Street	1–7–2012	$6,000,000
	1–7–2013	$7,000,000
	1–7–2014	$8,000,000
	Average land value 2014	**$7,000,000**

Determining the land value of Unit 4 of the strata factory complex:

Seven million dollar average land value for the strata scheme complex. The complex comprises five industrial strata units. The unit entitlement of each of the units follows:

Unit / Lot	*Unit entitlement*
1	50
2	50
3	50
4	75
5	75
Aggregate unit entitlement	**300**

The average land value assigned to unit 4 is determined by reference to the allocation of the unit entitlement of unit 4 as a percentage of the aggregate unit entitlement. Calculation of the average land value of Unit 4/35 Green Street follows: 75/300 x 100 = 25 per cent. The average land value of Unit 4 is 25 per cent of $7,000,000 = $1,750,000.

Step 2: calculating the land tax

The following are the actual rates and thresholds that apply in New South Wales that were used to assess land tax liability for the 2014 land tax year.

NSW Land Tax Rates and Thresholds 2014

Rate in the dollar	*Threshold*
Nil	Up to $412,000
$100 + 1.6 %	$412,000 to $2,518,999
2%	$2,519,000

8 Brown Street

Principal place of residence: $700,000 average land value – **Exempt**

15 Main Street

Retail shop: $1,000,000 average land value

Unit 4/35 Green Street

Industrial Strata Unit: $1,750,000 average land value

Sum of assessable land value = $2,750,000 (Main Street plus Green Street)

$2,519,000 – $412,000 = $2,107,000 x 1.6 cents = $33,712
$2,750,000 – $2,519,000 = $231,000 x 2 cents = $4,620
$33,712 + $4,620 + $100 = **$38,432 – 2014 Land Tax Liability**

Council rate calculation in New South Wales

There are 152 local government areas in New South Wales, of which 41 are located within the Sydney Basin. Hypothetical rates in the dollar are used to demonstrate the calculation of council rates. The land values used are from the earlier example used to calculate land tax.

In contrast to using three-year average land values as per calculating land tax, council rates may be determined on individual annual land values, with land values being up to four years old. The primary method used to adjust rate revenue from one year to the next, where historic land values are used, is by adjusting for the rates in the dollar.

For the purposes of calculating council rates on the land owned by Frances, local government is using the 2012 land values, of which the rates are determined as follows:

Method 1: Ad valorem method / minimum rates

Category of land	*Rate in the dollar*	*Minimum rate*
Residential	0.003	$700
Business	0.006	$1,200
• Retail	0.0045	$900
• Industrial		
Farmland	N/a	N/a
Mining	N/a	N/a

N.B. There is no farmland or mining use land in this hypothetical local government area.

8 Brown Street

Principal place of residence: $600,000 land value x 0.003 = **$1,800** council rates for 2014

15 Main Street

Retail shop: $900,000 land value x 0.006 = **$5,400** council rates for 2014

Unit 4/35 Green Street

Factory Unit: $1,500,000 land value x 0.0045 = **$6,750** council rates for 2014

Land value of Unit 4/35 Green Street follows: 75/300 x 100 = 25 per cent. The land value of Unit 4 is 25 per cent of $6,000,000 = $1,500,000. As noted in this example, the rates for each property are above the minimum rate for each category of land.

Method 2: base amount plus ad valorem

Category of land	*Rate in the dollar*	*Base amount*
Residential	0.0015	$700
Business	0.004	$1,000
• Retail	0.003	$800
• Industrial		
Farmland	N/a	N/a
Mining	N/a	N/a

N.B. There is no farmland or mining use land in this hypothetical local government area.

8 Brown Street

Principal place of residence: $600,000 land value x 0.0015 = $900 + $700 base amount = **$1,600** council rates for 2014

15 Main Street

Retail shop: $900,000 land value x 0.004 = $3,600 + $1,000 base amount = **$4,600** council rates for 2014

Unit 4/35 Green Street

Factory Unit: $1,500,000 land value x 0.003 = $4,500 + $800 base amount = **$5,300** council rates for 2014

Land value of Unit 4/35 Green Street follows: 75/300 x 100 = 25 per cent. The land value of Unit 4 is 25 per cent of $6,000,000 = $1,500,000.

The primary difference between Method 1 and Method 2 is that under the latter method, rather than determining rates on pure ad valorem / land value, part of the council's revenue is determined on a base amount per property. In New South Wales, local government may collect up to 50 per cent of their rate revenue from a base amount per property. In contrast, Method 1 raises rates purely on the value of the land; however, it was noted that there is no averaging of land values in the calculation of council rates and further, in contrast to land tax in which the land values are updated annually, council rates are not necessarily determined on the most current land value.

The following table provides a summary of the land tax and council rates generated from the land holdings of Frances. This example demonstrates the differences in the mechanics between state land tax and local government rates in their application to raising tax revenues by two different tiers of government. At one end of the spectrum is state land tax with land value as the primary determinant of the tax, while at the other end, council rate revenue adjustment is driven by changes to rate in the dollar, of which increases are capped by state government. Differences in the determination of local government rates are further exemplified by local government having the choice of two methods of determining their rate revenue; the first being ad valorem method, while the second method is a mix of ad valorem and base amount per property.

Summary of land tax and local rates paid

Address	*Land tax*	*Rates ad valorem*	*Rates base + ad valorem*
8 Brown Street	Exempt	$1,800	$1,600
15 Main Street	$38,432	$5,400	$4,600
Unit 4/35 Green Street		$6,750	$5,300
Total	**$38,432**	**$13,950**	**$11,500**

3 Definitions and bases of value

Definitions of value

The definition of *value* within property valuation practice has evolved over time and is often interchanged with the term *market value*, which has been described by the High Court of Australia as reflecting highest and best use in which the reference is made to the most advantageous purpose to which the land was adapted. Rost and Collins (1993:36) refer to *Spencer v. The Commonwealth of Australia* (1907) 5 C.L.R. 418 at 432 and 441, in which the court set out what value constituted:

> In my judgment the test of value of land is to be determined not by inquiring what price a man desiring to sell could actually have obtained for it on a given day, i.e., whether there was, in fact, on that day a willing buyer, but not enquiring: What would a man desiring to buy the land have to pay for it on that day to a vendor willing to sell it for a fair price but not desirous to sell? . . . It is further stated that, 'regard must be paid for the most advantageous purpose for which the land was adopted.'

The High Court's description of market value resulted from a case in which value was evolved for the assessment of compensation for the compulsory acquisition of land in Fremantle, Australia, in which the determination of the value of the acquired land could not be agreed upon. Further to the High Court's description, in 2000 the Australian Property Institute (API) adopted a definition of market value from the International Asset Valuation Standards, as set out in the Australian Property Institute, *Professional Practice* (2000):

> the estimated amount for which a property should exchange on the date of valuation between a willing buyer and willing seller in an arm's length transaction after proper marketing wherein the parties have each acted knowledgeably, prudently and without compulsion.

While the courts have provided guidance as to the broad framework of market value, the word value itself is open to interpretation. A further level of complexity is added to the interpretation and understanding of value by virtue of its everyday use. The generic use and meaning of value has presented a number of challenges to its operational application in assessing the value of property. Its broad use as a word, in contrast to its use, particularly in the valuation of land for rating and taxing purposes, has resulted in disparity between common speak and its application in the manufacture of value for this purpose. The Australian Property Institute, *Professional Practice* (2004:39), makes this distinction in the following:

> Imprecision of language, particularly in an international community, can and does lead to misrepresentations and misunderstandings. This is particularly a problem when words commonly used in a language also have specific meaning within a given discipline. That is the case with the terms price, cost, market and value as they are used in the valuation discipline.

Other terms for value used within valuation practice include *insurable value, salvage value, liquidation* or *forced sale value, special value, plant and machinery value* and *going concern value* (Australian Property Institute 2007:26). The forms and purposes for which value is to be determined will impact on the process, method and considerations to be used in the task. The point of the valuation purpose is further extended to include 'problem settings and valuation purposes'. In this context, 1) purpose, 2) problem elements, 3) valuation considerations and 4) solution elements, are all parts of the valuation process as stated by R.T.M. Whipple, *Property Valuation and Analysis* (2nd ed, 2006:109), as follows:

> **Normative**: Tending to establish a standard of correctness by prescription of rules; evaluative rather than descriptive. The term '**positive**' has. . . the sense of that which is given or laid down, that which has to be accepted as we find it.

When considering the valuation task and its components, distinction is drawn between 'positive' and 'normative' definitions of value, which valuers and the courts have failed to distinguish between (Whipple 2006:87). In distinguishing these terms further, from a positivist perspective value is an outcome, while from a normative perspective value is the prescriptive process of attaining that outcome. In applying these concepts for rating and taxing purposes, in which equity is cited under the problem element, the solution element calls for a 'normative definition of value as prescribed' as the most appropriate of these two.

It is not the fact that a diversity of the meaning of value exists, or that more than one meaning may coexist in its determination for different purposes exist. It is the prescriptive precision in which the word value is given for the specific purpose to which it is applied that is central to its determination. This is particularly the case in the rating and taxing of land in which the principles of good tax design shape the definition of value in this case.

It may be argued that value without context or prescriptive determination is no more than a word open to interpretation by any person. The lack of framework in the case of land value taxation renders the opinion of the taxpayer on how value is determined to be a formidable view for those imposing the tax and, in particular, the valuer charged with assessing it.

The courts have interpreted the codified meaning of value under the various Valuation of Land statutes used to manufacture value for rating and taxing purposes across Australia. Codification within the context of economics and its application to tacit knowledge is defined by Balconi et al (2007:829) as 'the articulation of the rules on which skilful performance is made'. In the context of land taxation, the precipitating questions to be asked are: what is value? and, most importantly, how is it determined and applied in assessing land tax? Once this has been determined, the next question asked is: how is this skilful performance to be achieved when valuing land in highly urbanised locations, where most of this tax revenue is derived from?

Value is a tacit concept, in which the definitions assigned to it are little more than constructs of what land would have transacted for in the absence of an actual transaction over the land itself, as defined by the High Court of Australia. In its simplest form, this process is conducted by reference to other transactions of identical or similar land that have actually been transacted. In clarifying the intent of the codification of value, it is not to create its framework outside the existing law, but to consolidate the principles enunciated for its practical application in the valuation of land for rating and taxing purposes.

Determining the meaning and application of terms used within the environments in which land taxes operate is important when making comparisons of tax bases. Legislative measures, policy goals and administrative practice of tax from one jurisdiction to another demonstrate that value is not a catchall or universal term (Youngman & Malme 1994). Following the review of the bases of value internationally, the definitions of value within valuation of land legislation across six states and two territories of Australia are examined. The definition of value and detail of the valuation process is compared in the three adjoining states of Queensland, New South Wales and Victoria.

These states, and in particular their capital cities, are the most relevant, as they are the most highly urbanised in Australia, of which the absence of vacant land sales is most applicable. This review demonstrates that the highest and best use principle is implied within the definitions of value of the states and territories. It demonstrates that land value excludes all improvements, and hence existing uses, when it is valued. Its value indeed reflects its potential highest and best use. A review of the definition of *capital improved value* is also reviewed where it applies in the states of Victoria and South Australia.

The review of this tax commences with the valuation of land, how value is determined, the use of thresholds and rates in the dollar applied to the taxable value and the relevance of the tax paying entity. A summary of the various bases of value used to assess the land tax internationally is provided as a prelude to examining the bases of value in Australia. A review of concessions, allowances and exemptions that apply to the value of land and the assessment of the tax follows the review of its operation. This part further serves to identify provisions that work well, those that have become inefficient and impact revenues from land tax and those which no longer conform to the principles of good tax design. The objection and appeal process of the valuation of land and administration of the tax are also addressed under this part.

The legislation which governs the valuation of land and specific valuation practices used in each state includes the objection and appeal process. A review of the principal place of residence and the application of the 'Unity of Title Rule', concessions and allowances under the Valuation of Land Acts, Land Tax Management Acts and Local Government Acts across Australia are covered. Where relevant, examples are used to provide context to the rationale for reforms, as supported by inquiries and reviews as discussed in Chapter 2.

The examination of this tax raises questions as to whether greater harmonisation and efficiencies may be achieved across states and local government areas in the administration and application of the component parts of the tax. It is important to highlight at this point that the examination and review of these taxes is not for the centralisation of tax revenues. Its primary objective is to define the strengths and opportunities for refinement and improvement in facilitating increases in revenue at the expense of less efficient taxes on property.

Bases of value

The taxation of land commences with the basis of value on which the tax is assessed. As noted in Chapter 2, the basis on which land has been taxed in Australia has moved from unimproved capital value to either site or land value. The former excludes land improvements, while the latter now includes these improvements to land. Before examining the bases of value used in Australia and the valuation process, it is first apt to briefly review the various bases of value on which land tax is assessed internationally, which demonstrates that there is no one uniform base on which to assess this tax. A more specific and detailed account of how land tax operates in a number of these jurisdictions is covered in Part 3.

International bases of value

Value-based taxes internationally are divided into three broad categories. The first category is *capital improved value* (CIV), the second is income or *annual rental value* (ARV) and the third is *land value* (LV) or *site value* (SV). Value-based assessments are those determined by reference to the marketplace, 'being a price that would be struck between a willing buyer and willing seller in an arms-length transaction' (Bird and Slack 2004). In a number of Eastern European countries where land and property markets have been evolving over the past 20 years, land or building area is the basis on which land tax is assessed. RICS (2007) highlight that the use of area is the basis for assessing land tax in the Czech Republic, Poland, Hungary, Slovenia and Slovakia. McCluskey Bell and Lim (2010) also identify its use, to a limited extent, in parts of Africa and Asia. Internationally, area is used in one form or another in 42 countries.

The international survey undertaken by McCluskey, Bell and Lim (2010) between 2007 and 2010 across 122 countries summarises the various bases on which this tax operates by world region. The most common basis of value used is CIV, which is applied in 52 countries, followed by annual rental value used in 37 countries and land value (aka site value) used in 16 countries. Further to a regional review, Table 3.1 sets out a more detailed list of the base of the property tax in several countries, which shows that, since cessation from Russia in 1998, Estonia, Lithuania and Latvia have all adopted land value as the base of the tax (Tomson 2005), in line with neighbouring countries Finland and Denmark. In contrast, McCluskey et al (2010:125) highlight that 'for other countries Capital-Improved Value was the original system put into place (Canada, US, Brazil, Mexico, Japan, Sweden and Chile)'. Table 3.1 also indicates that in the majority of countries, state governments do not often impose a recurrent land tax, leaving the base to the lower tier of government. While this is the case, however, the administration of the tax and valuation of land is centralised at higher tiers of government, which improves consistency and reduces the impact of local influences opposed to the tax.

In other countries, including New Zealand and South Africa, land value as the base of the tax is in demise, while in Australia the trend towards improved value for assessing council rates in Victoria and South Australia moves their approach towards that common in the US, Canada, parts of New Zealand and parts of the United Kingdom. Foremost in the arguments in support of the move to CIV has been that it is more easily measured and better understood by the taxpayer. Despite such commentary, there is no empirical research which supports that any one basis of value is better understood than another. This is particularly the case where highest and best use applies and improved value no longer reflects that use or improvements are no longer maximally productive.

In the case of neighbouring New Zealand, it has a well-developed rating system in which local government has the option of adopting one of three

Table 3.1 Bases of value by level of government

Country	*Improved value*	*Land value*	*Building value*	*Income ARV*	*Area*	*Revenue ownership*
United States	X					Local
New Zealand	X	X				Local
Australia*	X	X				Local/State
United Kingdom	X			X		Local
Canada*	X					Local/Prov
Hong Kong	X			X		Local
Denmark*	X	X	X			Local/State
Finland*	X	X	X			Local/State
Czech Republic					Land Bldg	Local
Estonia		X	X			Local
Poland					Land Bldg	Local
Latvia		X	X			Local
Lithuania		X	X			Local
Hungary					Land Bldg	Local
Slovenia					Land Bldg	Local
Slovakia					Land Bldg	Local
Kenya		X				Local
South Africa*	X	X				Local

Sources: Ministries of Taxation Denmark & Finland and RICS 2007 (X = applicable)

*Denotes variable bases of value or that the tax may operate at different levels of government.

bases of value for the rating of property. Four of the main cities of New Zealand (Auckland, Wellington, Christchurch and Hamilton) utilise a capital or annual value rating system (McCluskey et al 2006:389). Further noted in the early 1990s was the removal of national land taxes in New Zealand at the time the GST was introduced. Subsequently, recurrent property taxation has existed at the local government level in the form of property rates assessed on improved value in the cities and land value in regional parts of New Zealand. The case for the move to CIV in New Zealand is not clear; among the reasons stated are the ability to pay, volatility of the property market and assessment difficulties and rating incidence (McCluskey et al 2006:382).

In the case of South Africa, following the end of Apartheid, local government evolved and the property tax continued to increase as a source of revenue during the 1990s, in which the tax still remains an important source of revenue of local government (Bahl 2002 cited in Bell and Bowman).

In the former system, which operated up until 1 July 2009, municipalities had a choice of three tax bases. As part of property tax reforms in South Africa, almost a century of land value taxation was replaced with market value (Franzsen in Dye & England 2009:39&41).

Denmark introduced land value as the base of the property tax in 1981, reformed it in 1992 and introduced a tax freeze in 2002 (Muller 2003). The meaning of land value is its market value, of which land values are re-determined every two years. Denmark has a unitary structure of government and the revenue raised from land tax is split between local and the county level of government of which the lower tier collects the tax but is centrally administered, which includes the valuation of land (Falk-Rassmussen & Muller 2010). The valuation of land used for taxation purposes in Denmark sets the standard worldwide and is the benchmark in terms of resourcing and quality assurance; however, despite its success in Denmark, a new system for taxing land is due to be introduced during 2015/16.

Finland taxes residential land and introduced legislation permitting this annual tax in 1993. Since 2010, a higher rate applies to the value of land that is undeveloped as a means of discouraging land banking. The rate applied to land values ranges between 0.6 to 1.35 per cent of the land value, of which the rate is determined by local government. The base of the tax, while referred to as land tax, is defined as the taxable value of land. It is not necessarily its market value, but cannot be above market value (J. Tiihomen & E. Narhi, directors of property taxation, Skat Ministry of Taxation, personal communication, 16 September 2010). Finland, like Denmark, is a unitary structure of government, in which the tax on land is centrally administered and collected with revenue channelled back to local government.

Estonia introduced land value taxation in 1993. Since the introduction of the tax, land values have been assessed in 1993, 1996 and 2001 (A. Tomson, Director Valuer, BPE Baltic Property Experts, Tallinn, Estonia, personal communication, 22 September 2010). Increases and changes in the relativity of values since 2001 have resulted in the reassessment of values being politically unpopular. Increases in property tax revenues since 2001 have been achieved through increases in the rate applied to the 2001 land values, as there is no statutory governance of the frequency of the valuation cycle (Tomson 2010, pers. comm). The rates applied to values, while originally centrally set, have been divested to local government, who are now responsible for the tax rate in their jurisdictions.

Up until 1985, land value was the preferred base on which to assess the property tax in New Zealand; however, by the 2006–2007 fiscal year, capital value had become the tax base for the majority of local authorities (Franzsen in Dye & England 2009:37). The rationale for the transition to CIV in the cities of New Zealand was due to the limited transaction of land. Despite the transition to CIV in the capital cities of New Zealand, the use of land

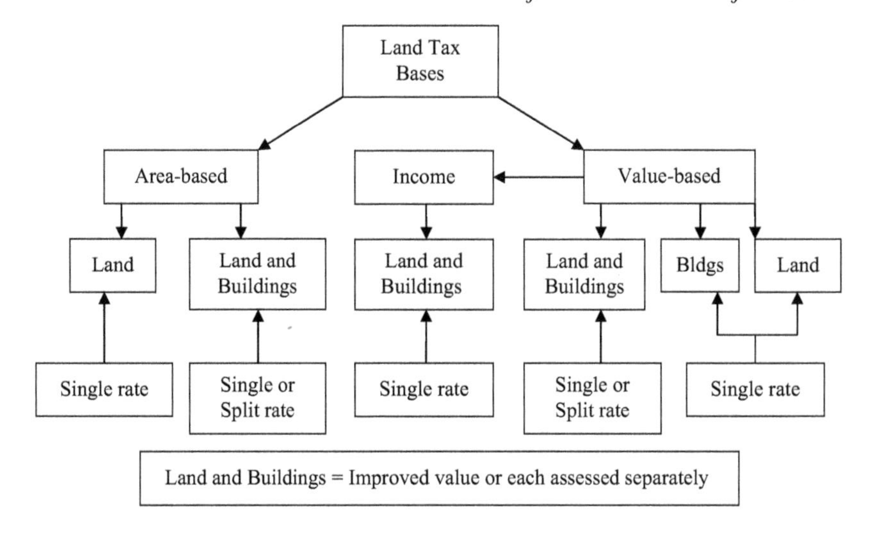

Figure 3.1 Land tax bases
Source: Author

value as the base of the property tax remains strong in regional New Zealand (Local Government Rates Inquiry Panel 2007).

The tax raised through local government property rates in New Zealand accounts for 56 per cent of local government revenue, which is estimated to increase to 60 per cent by 2016. In tempering revenue from this tax source, it has been suggested that total revenue from the property tax be reduced to 50 per cent. The sustainability of land value as the base of the property tax and the revenue raised is being questioned in New Zealand (Local Government Rates Inquiry Panel 2007). Land value is determined on the market value of land, which is reassessed at least every 3 years. The efficiency of land value used for the rating of land outside the capital cities of New Zealand is effective; however, the efficiency of land value based on highest and best use has come under scrutiny. In contrast to land value being determined on highest and best use, it is suggested that land value be determined on the existing use of land (Local Government Rates Inquiry Panel 2007).

Australian bases of value

It is noted in the earlier examples used to demonstrate the calculation of land tax and council rates in the introduction to this chapter, that land value was the first component to be determined in the calculation of these two taxes. Values must first be determined for each parcel of land in Australia by the relevant valuing authority. The centralisation of the valuation process is important in ensuring that values are determined consistently and transparently for the various land uses within and across local government areas and indeed the state. While contract valuers are engaged in a number of states, the valuations are centrally reviewed by the Valuers-General before being issued. This process is rigorously applied and valuations are scrutinised before they are issued to state and local government taxing authorities.

Each state of Australia has its own valuation of land legislation and definition of value as set out in Table 3.2. The definition of land and site value used to assess state land tax and local government rating vary despite being commonly referred to as land or site value. Within Australia, for the purposes of local government rating, a move away from land or site value is evident over the past 30 years. Since the 1980s, the States of Victoria and South Australia have moved to assessing local government rates on improved value, while Western Australia and Tasmania have moved to Gross Rental Value (GRV); however, Tasmania is reviewing their bases of value for rating purposes, with land value being considered an option. Local government rating in New South Wales and Queensland retain land and site value as the basis of their property tax.

Table 3.2 Bases of value across Australia

Land tax		
State / Country	*State gov't land tax*	*Local gov't council rates*
New South Wales	Land Value	Land Value
Queensland	Site Value	Site Value
Victoria	Site Value	Improved / Site / Annual Value
South Australia	Site Value	Improved / Site / Annual Value
Western Australia	Site/Unimproved Value	Gross Rental Value
Tasmania	Land Value	Gross Rental Value
Northern Territory	N/a	Unimproved Capital Value
Australian Capital Territory	Unimproved Value	Unimproved Value
New Zealand	N/a	Improved value in 4 main cities Land value in regional locations
Perceived objective/ purpose	General purpose or consolidated revenue tax	Quid pro quo tax for local services provided
Value premise	Land value includes land improvements as defined within various state valuation of land statutes, i.e. excavation, retention, filling and servicing of land.	
Valuation methods	Direct comparison where vacant land sales exist. Paired sales analysis and cost method with the use of improved sales.	

Source: Author

The bases of value used to assess land tax and local government rates in each state are now reviewed.

New South Wales (NSW)

Land tax was reintroduced in NSW in 1954, and was assessed on the unimproved value of land as set out under the now repealed provisions of section 6 of the Valuation of Land Act 1916 (NSW) (Smith 2005). In 1982, the Valuation of Land Act was amended to remove Unimproved Value and to introduce Land Value under section 6A of the Act. The former label, Capital Unimproved Value, assumed land to be in its 'virgin untouched' or 'en-globo' state, in which soil fertility, excavation, retention and improvements made to the land did not form part of the unimproved value, but assumed the existence of all extrinsic circumstances such as roads, railways and public services.

As difficulties with the determination of unimproved value evolved, unimproved value was replaced with land value in NSW in 1982. As highlighted in Figure 3.2, challenges to the assessment and the determination of unimproved value in urbanised locations required the notional removal of services and improvements to the land itself. Unimproved value became irrelevant within urbanised locations in which most land was fully serviced and included the improvements to land, in contrast to its en-globo state. The artificiality of unimproved value was beyond the understanding of taxpayers and some administrators in the intention of its meaning. Its intended meaning often collided with its perceived understanding and how it is applied in practice.

As urbanisation continued in the major cities of Australia and land became fully serviced, its en-globo definition as unimproved value was the antithesis of its use in domestic and business built environments. The ability to assess en-globo or unimproved value of land in highly urbanised and fully serviced locations eventually gave way to land value. Prest (1983:2) highlights that the difference in value between unimproved value and land value as of 1982 was in the range of 5 to 8 per cent. That is, the move to land value resulted in an increase in value on which land tax was assessed in the range of 5 to 8 per cent at the time it was introduced.

Following the change from unimproved value to land value in 1982, statutory codification of the definition of land value within the Act was made explicit by the courts. The courts crafted meaning where the black letter of the law was unable to articulate the intention of the Act and legislature. Included within the Act was an additional section which provided what was deemed to be included as part of the land value. Section 4(1) was added to the Act at the same time that Land Value was introduced under section 6A.

Under s6A(1) land improvements are deemed included as part of and merged with the land in the assessment of its value. Land improvements are set out under s4(1). In qualifying their application, any work that has been carried out to land, including retention, clearing and excavation, is deemed to be part of land and included in its value. Land improvements generally constitute improvements below the ground up to its surface; however, they also include the clearing of land. The economic challenge associated with the transition to a value which encompasses an element of improvement, even land improvements, is the disincentive in carrying out improvements which impact value and increase the tax burden.

In countering this economic disincentive of including land improvements in the definition of land value, provisions were introduced to negate, to some degree, the profitable expenditure made to land which impacted the increase in rates and tax payable at the time of its introduction. An allowance for profitable expenditure to land was introduced under the Valuation of Land Act 1916, which gave the owner of the land who actually carried out improvements for profitable expenditure to the land up to 15 years to deduct these costs from the land value for the purposes of assessing land tax and local government rates.

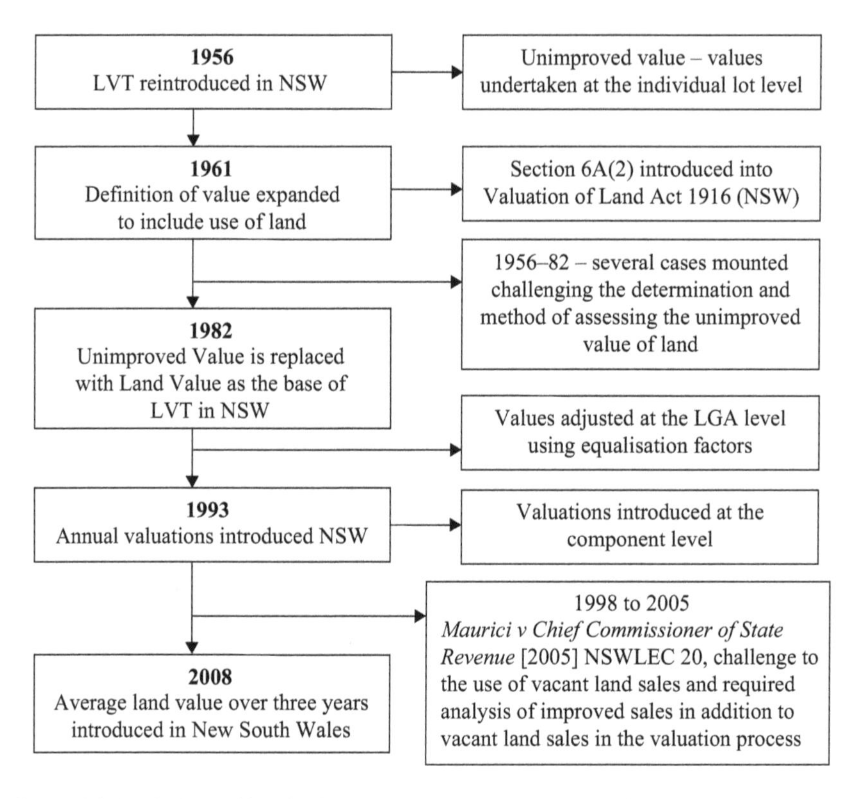

Figure 3.2 Evolution of land value as a base in New South Wales
Source: Author

Further, to statutory provisions which seek to codify the meaning of value, a number of court rulings have refined the statutory meaning of Land Value in NSW. Where *fee simple* land has impediments which might affect its value – such as easements, covenants or sub-interests – these are to be disregarded for the purposes of ascertaining the value of land under the Act. The term fee simple is taken to be unencumbered and subject to no conditions, regardless of what impediments actually exist on the land itself. Hyam (2004) refers to Lord Radcliffe, *Gollan v Randwick Municipal Council* [1961] AC 82 at 101, who states:

> The fee simple of the land does not refer to actual title vested in the owner at the relevant date but to an absolute or pure title such as constitutes full ownership in the eyes of the law. (p. 16)

In contrast to the title of land and interests therein, matters relating to the permitted use and town planning are to be accounted for in the value of land. In referring to a number of previous decisions, the distinction between the former matter of impediments to title and the latter matter of permitted use and planning is well-defined by J. Else-Mitchell in *Port Macquarie West Bowling Club v The Minister* [1972] 2 NSWLR 63 at 65, in referring to a number of previous cases states:

> The difficulties adverted to in these decisions ensue from the fact that each planning scheme ordinance must be, as the High Court said, characterized as 'a law operating over an area of country within the state . . . chosen independently of all questions of title or ownership' so that its effect has to be taken into account in determining value.

In further qualifying the use of land and, more specifically, specialised uses that require an additional layer of approval or consent over and above town planning consent, a number of rulings have been determined.

In *Tooheys Ltd v Valuer-General* [1925] AC 439, the Privy Council ruled that the license to operate a hotel was not attached to the land but to the building and hence is to be disregarded when assessing the value of land. In referring to the distinction between licensed premises and other uses of land that may also require licenses over and above town planning consent, the Queensland Court of Appeal in *Department of Lands v Webster* (1995) 89 LGERA 341 at 345–347 unanimously agreed: 'The nexus between a liquor licence and licenced premises is sufficiently strong to enable it to be regarded as an improvement, and as therefore notionally separable from the unimproved value' (*Tooheys Ltd v Valuer-General* [1925] AC 439). This makes an important distinction between land used as a business in its own right, and land which is used as part of another business of production, on which that use requires additional consent and approval.

In contrast to the above points relating to the title and interest in land, or the additional value of intangible improvements such as goodwill or licenses, the physical restrictions affecting land are not to be disregarded. These include such aspects as contamination and flooding that may relate to land and their impact on value. These factors are to be accounted for by deducing the land value for rating and taxing purposes. In summarising the disparity of the above points, Table 3.3 highlights the distinction between the valuation of land which is determined by the Valuation of Land Act, in contrast to the uncodified approach in the analysis of the sales evidence used to value land. Table 3.3 distinguishes that, while codification exists in the definition and meaning of value, there is no compulsion for valuers to select similar sales or analyze sales on the same basis on which the land is being valued.

While the valuation of land is definitive on what is to be included and not included to the land being valued, the same level of codification does not dictate the transactions to be used or how such transactions are analysed or adjusted for these factors. This will be discussed in more detail in the sales analysis process in Chapter 4.

Victoria

The base for state land tax in Victoria is the 'site value' of land. The definition of site value under s2 is explicit in qualifying that the fee simple is to exclude any lease, mortgage and charge over the land. In contrast to New

Table 3.3 Value v evidence of value

Land value (Valuation of Land Act 1916)	*Sales analysis (Valuation practice)*
Land value is to include: 1. Land improvements 2. Planning regulations Land value is to exclude: 1. Licenses which are deemed to be attached to the business. 2. Actual title of the land is assumed to be fee simple and any actual title impediments of the land are to be disregarded. 3. Leases over land are not specified as either included or excluded in the New South Wales Act; however, strict application of *Gollan v Randwick Municipal Council* [1961], requires leases to disregarded.	There is no structured codification for the sales analysis process as found by the NSW Ombudsman (see 3.1 above). The challenge to Land Value in New South Wales. Valuers may select and analyse sales on the following bases: 1. Land with agreements to lease versus land without agreements to lease. 2. Land with or without impediments to title. 3. Land with or without development consent. 4. Land with or without land improvements. 5. Land with variable planning or permitted uses.

South Wales, Victoria has a level of codification of the sales analysis process within its valuation of land legislation under section 5A. Of particular note are inclusions for comparability in time, circumstances and terms and conditions of sale of the evidence between the date of sale and the date of valuation. Victoria has recently incorporated 'Valuation Best Practice Specification Guidelines' into the Valuation of Land Act. These guidelines are reviewed prior to each biennial general valuation, in which feedback is invited on the operation of these guidelines up to two years prior to the issuing of site and capital improved and net annual values.

Victoria specifies that highest and best use of the land and suitability of the improvements on the land is to be included in the determination of value, as per the Valuation of Land Act. Further, while not prescriptive as to how value is to be determined, the broad points for consideration in the analysis and sales comparison process have been defined. This section provides the need for sales of comparability, the terms and conditions of those sales, planning and physical attributes of the land and the suitability of any improvements on the land. The Act sets out that any improvements on the land shall be accounted for by reference to the amount the improvements increase its value if offered for sale. Like the NSW legislation, land improvements are defined under the Act and encompass improvements to the land to be included within the definition of Site Value. Definitions of improvements are set out under s2 of the Act.

While highest and best use is a component of market value, within the statutory definition of Land or Site Value, its consistency and inclusion in the valuation of land may be distorted by the use and analysis of improved sales which are not highest and best use of land. The Act is limited to making reference in principle only to highest and best use of land. This point is further discussed in the valuation simulation results in Chapter 4. In addition to the Act, Victoria has two valuation guidelines for valuers undertaking rating and taxing valuations. The general provision guidelines refer back to the Act and its reference to highest and best use of land. It does not describe highest and best use or prescribe how this principle applies in the determination of value.

In contrast to state land tax which is confined to taxing land under the definition of 'site value', local government has the choice of rating land on one of three bases of value which include either site, improved or net annual value, s157 Local Government Act 1989. While a majority of local governments in Victoria determine rates on improved value, Melbourne City Council uses net annual value (NAV). The definition of NAV under s2 of the Valuation of Land Act 1960 is included in Annexure 1. It means the estimated annual value or 5 per cent, whichever is the higher; however, the term 'estimated annual value' is further defined under s2 of the Act, which sets out, to some degree, which outgoings qualify as expenses. Further, for land tax purposes (b) refers to the impost of land tax being on a single holding basis comprising the tax applicable to the subject property as if it were the only land owned.

Net annual value of any land means –

(a) except in the case of the lands described in paragraphs (b) and (c) (see full definition):
   (i) the estimated annual value of the land; or
   (ii) five per centum of the capital improved value of the land – (whichever is the greater).

Estimated annual value of any land means the rent at which the land might reasonably be expected to be let from year to year (free of all usual tenants' rates and taxes) less –

(a) the probable annual average cost of insurance and other expenses (if any) necessary to maintain the land in a state to command that rent (but not including the cost of rates and charges under the Local Government Act 1989); and
(b) the land tax that would be payable if that land was the only land its owner owned;

Queensland (Qld)

Following a number of court decisions circa 2007 against the government's taxing and valuing authority, the former Valuation of Land Act 1944 (Qld) was replaced, effective from the 2011 taxing year. This change was precipitated by the Qld Land Appeal Court, which discussed the level of complexity in deducing the unimproved value of land in the absence of clear guidelines for the determination of value. The Qld Land Appeal Court stressed the need and importance for the codification of the valuation process in its deliberation.

The review of the Valuation of Land Act 1944, was precipitated by a number of appeals and cases rendering both the legislation and valuation practices used outdated and unsuitable for the highly re-urbanising and expanding metropolitan areas of Queensland. In *PT Limited & Westfield Management Limited v The Department of Natural Resources and Mines [2007]* (Queensland Land Appeal Court, Brisbane, 17 October 2007), the court stated:

> The respondent carries out unimproved valuations of most land in Qld on an annual basis. If litigation of the cost and complexity of the present case is to be avoided in the future, we are of the view that it would be desirable that if the VLA was to provide for a mechanism by which the unimproved value of major commercial enterprises could

> be assessed without the need for the difficult, lengthy and complex evidence of the kind before the Court below. [para 109]

This case exemplified the difference in the definition of unimproved value and the prescriptive processes adopted in determining the unimproved value of land in Queensland, prior to the transition to Site Value. In contrast to Land Value in NSW and Site Value in Victoria, in 2011 Qld moved away from taxing Unimproved Value and adopted Site Value under the Land Valuation Act 2010. Despite adopting the same label of Site Value as Victoria, the Site Value in Qld under the Land Valuation Act 2010 differs in parts from the statutory definition in Victoria as set out in Annexure 1.

The 2010 Act defines site value as being on an unencumbered fee simple basis, in which any encumbrance or lease is to be disregarded, similar to the Victorian legislation. In contrast to Vic, the Qld legislation sets out what a bona fide sale is deemed to be under s18. It does not make specific reference to comparability of physical attributes between the sale and the land being valued, as does the Victorian Act. The legislation sets out the use of bond rates to be used for discounting the added value of improvements in the sales analysis process under s20. This additional element of codification does not exist in either NSW or Victoria.

In determining the site value of land, ss 26 and 28 of the Act allow land to be hypothetically developed to its maximum potential and stripped back to produce a land value. This allows land to be valued based on its highest and best use in cases where the existing improvements do not represent this use. Despite there being provision for such a process, no reference to highest and best use is used to define the basis of value. That is, the starting point for determining the value of land by reference to the land's highest and best use requires reference to actual use of land improved within the same location of the land being valued. This process, while prescribed and used in valuation practice, adds an additional layer of responsibility to the valuing authority in complying with the tax principle of transparency.

As was the case in New South Wales in the transition from unimproved value to land value in 1982, where land value included land improvements, the same resulted in Queensland in the transition of unimproved value to site value. A major factor negated the impact of any increase in value in the transition from unimproved value to site value in many parts of Queensland. This was the devastating impact of the 2010/11 floods, which impacted rural, regional and urban parts of Queensland, including the parts of Brisbane Central Business District. The impact of the floods required reassessments of site values due for release and applicable to the 2011 taxing year.

Tasmania

At present, the valuation of land in Tasmania is one of the least codified in Australia, both in the statutory provisions of the Valuation of Land Act and the level of prescription used in the valuation process. The volatility in local government rates and state land tax is currently impacted by the bases of value used to assess these taxes. Attempts to moderate this volatility through the use of rating differentials impacts the integrity and transparency, as it removes value as the differentiating factor between different land uses and the need to first revisit the baseline of value leads the reform agenda.

Tasmania has undertaken a comprehensive root and branch review of the valuation of land which underpins the rating and taxing of land across the state (Access Economics 2010 and Department of Premier and Cabinet 2013). A review of the valuation and local government rating was announced on behalf of the state government in December 2009 to address concerns about the volatility of local government rates and state land taxes that are linked to changes in property value. In April 2013, the Valuation and Local Government Rating Review Steering Committee submitted its Final Report, which made the following findings and recommendations in regard to future reform of Tasmania's valuation and rating systems.

The Final Report recommendations include that the state government:

- Discontinues valuations on assessed annual value (AAV) and assists councils to transition to capital value (CV) by 1 July 2016;
- Transitions to a valuation cycle of two years for Land Value (LV) and four years for CV;
- Maintains LV and AAV adjustment factors for each municipality until fresh valuations are completed;
- Seeks advice from local government on the preferred strategy for managing cost implications for councils associated with the transition; and
- Works with Local Government Association of Tasmania to improve the capacity of councils to manage differential rates resolutions.

(Department of Premier and Cabinet 2013)

South Australia

State land tax is assessed on the site value of property in South Australia. Like Victoria, local government may elect to impose local government rates on either site value, capital value or annual value of land under s151 Local Government Act 1999. While capital value is the most common basis of value used for rating purposes in South Australia, Adelaide City uses annual value. As set out in Annexure 1, the definition of annual value defines this to be three quarters of the gross annual rent, or 5 per cent of the capital value as per section 5 of the Valuation of Land Act 1971. There is no codification of the definition of any of the values used in South Australia within the Valuation of Land Act.

Adelaide City Council set out the rationale for the determination of the three values that may be used to assess rates under the Council's Operating Guidelines, *Rating Operations Guidelines.* The explanation provided in these guidelines as to how values are determined are well-hypothecated and provide a standard for elements of depreciation for the conversion from annual value with reference to gross rent. These explanations are stated to have been agreed upon in consultation with professional bodies and the Valuer-General.

Western Australia

State land tax is assessed on the unimproved value of land with capping of increases in assessments introduced from 2009/10. The unimproved value of land is its market value under normal sales conditions, assuming that no structural improvements have been made. Land within the Perth Metropolitan Region and townsites throughout Western Australia is assessed on the site value basis, which includes merged improvements with examples including draining, filling, excavation, grading and retaining walls. There is no codification of the definition of site value in the Act or how the value is determined.

Local government rates are determined on the gross rental value (GRV). These values are also used for the Water Corporation and water boards to determine residential sewerage rates and main drainage charges. GRVs are further used by the Department of Fire and Emergency Services to determine the Emergency Services Levy. The Act sets out under part (d) of s4, what is included and excluded in the determination of GRV. Landgate, WA set out an overview of the valuation methodology used for the determination of GRV in their online information services.

For land situated within a townsite, the UV is the site value of the land and, in general, this means the value of the land is as if it were vacant with no improvements except *merged improvements*. Merged improvements relate to improvements such as clearing, draining and filling. The UV of land outside a townsite is valued as if it had no improvements. In this case, the land is valued as though it remains in its original, natural state, although any land degradation is taken into account.

Bases of value in Australian Capital Territory (ACT) and Northern Territory (NT)

State land taxation does not apply in ACT or NT, as these are territories of the Commonwealth of Australia; however, local council rating does apply. ACT takes account of the leasehold interest applicable to the land and requires that an unexpired 99-year term be assumed at the date the unimproved value is determined.

Summary

In summarising the various Valuation of Land statutes across Australia, each has distinct labels and definitions of value which are set out in Table 3.2

with the full statutory definitions set out in Annexure 1. On balance, NSW is the least prescriptive and codified of the three eastern states, with the last major review of the Act in 2000. In contrast, Queensland, which is the most prescriptive and codified, introduced a new Act which took effect from the 2011 land tax year. The Queensland Act gives guidance as to how the added value of improvements are to be adjusted for in the analysis of improved sales in the value of land in highly urbanised locations.

In contrast to both NSW and Queensland, Victoria provides a more structured account of the physical attributes of land and timing of the sale and makes reference to values determined using the principle of highest and best use of land. Victoria provides a more detailed account of the factors which contribute to the sales selection process. In contrast, NSW is silent on all matters and provisions which are addressed in the Victoria and Queensland legislation. While articulation of the valuation process is generally limited in valuation statutes, the use of valuation procedure manuals has filled this gap, with the Victorian legislation recently introducing reference to best practice manuals in their valuation of land statute.

It is further noted that, across Australia, differences in the frequency of the valuation cycle vary from annual valuations up to six yearly valuations. While annual valuations are the most common frequency in the valuation of land for state land tax, greater variability exists in the frequency of the valuation cycle for local government rating. The valuation cycle used across Australia's local governments varies between two to four yearly. It was also shown that local government rating may be determined on a wide variety of value bases which include land or site value, capital improved value, net annual value and gross rental value.

Table 3.4 Valuation frequency and provider

State	*Frequency*	*Valuation provider*
New South Wales	Annually	Contract valuers
Queensland	Annually	Valuer-General
Victoria	Biennial	Contract or Valuer-General
South Australia	Annually (Max 5 yr)	Valuer-General (except Adelaide City)
Western Australia	Triennial in city and 3–5 yearly in regional WA	Government valuers
Tasmania	Up to 6 yearly	Valuer-General and External
Australian Capital Territory	Annually	Australian Valuation Office
Northern Territory	Three yearly	Australian Valuation Office

N.B. In the case of contract valuers, in some states government valuers are used to monitor the valuations and process.

4 Valuation of land and assessment of land tax

Introduction

While the last chapter reviewed the construct of value used to assess land tax across the states of Australia, this chapter examines the application of these constructs. It identifies a number of challenges confronting valuers for the rating and taxing of land in the absence of vacant land sales. In monitoring how valuers undertake valuations by deducing land value from improved sales, simulations were developed in a controlled environment to mirror the valuation processes and practices of valuers in the field. This chapter is a review of the valuation of land and examines the practices of valuers and consistency in the valuations of the same group of property, as undertaken by several different valuers.

The evolution of land taxation in Sydney (Australia) provides an insight into the challenges confronting all cities when imposing a land tax in increasingly urbanised locations. Figure 4.1 outlines the increasing lack of vacant land sales as the source of primary evidence for determining land value, which has resulted in greater reliance on improved property sales. This has resulted in additional layers of complexity which require accounting for the added value of improvements in the valuation of land (NSW Ombudsman 2005:7). With this has come a lack of transparency and simplicity and increasing pressure for the adoption of alternate bases of value for the assessment of recurrent land tax in Australia. The lack of vacant land transactions which impact simplicity and transparency are clearly defined as the rationale for the move to Capital Improved Value in some international jurisdictions highlighted in the previous chapter (Franzsen, 37 & 41, in Dye & England 2009).

The lack of consistency in accounting for the added value of improvements and the inability for valuers to articulate how land value has been determined from improved property transactions have raised questions as to whether or not land remains the most suitable base on which to assess land tax in highly urbanised locations. What is clear from the earlier review of the historical evolution of property taxation is that land tax has constantly

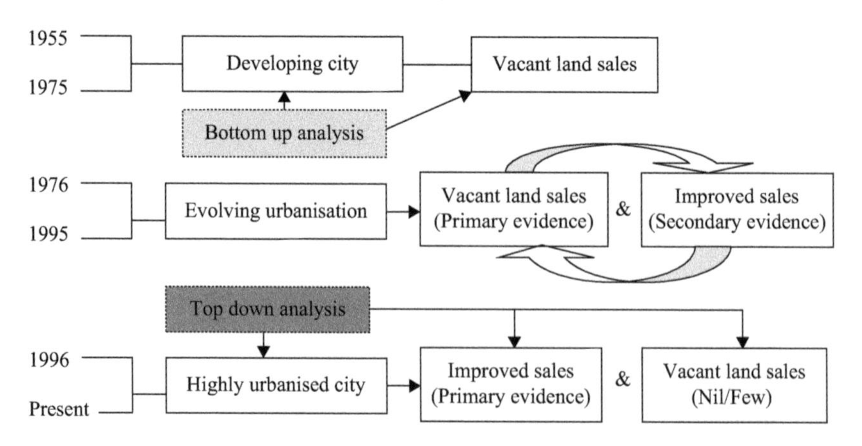

Figure 4.1 Evolution of value in the Sydney Basin

been challenged and is one of the more disliked, visual and least understood taxes imposed by governments.

The lack of sufficient vacant land transactions in a particular location has resulted in the practice of valuers being forced to use land transactions from adjoining locations (Bahl 2009:9). Another practice has been for land value to be determined by deducting the added value of improvements from improved property sales (NSW Ombudsman 2005). This emerging valuation process and, in particular, its use in the determination of the added value of improvements on land value have raised questions about its potential to compromise the economic efficiency, simplicity and transparency of land tax (Arnott & Petrova 2002:3).

While land has remained the dominant base of the tax across Australia, the greatest threats to its longevity and robustness of tax revenue generated are challenges to the determination of the base (land/site value) on which the tax is assessed. The safety net for both local and state government across Australia is the statutory requirement for the taxpayer to prove their case on any of the grounds permitted to challenge the correctness of a land or site value. Over 95 per cent of all objections to land tax and rating assessments are challenges to values and, more specifically, that the land or site value is too high.

Research method and rationale

While the challenges confronting valuers in valuing land in highly urbanised locations are apparent, what is not known is which specific elements of the valuation process valuers find most challenging and where they differ in practice in undertaking valuations of land in highly urbanised locations. This section sets out the research method used to examine the valuation process and factors that influence the practices that valuers use to value land. This is followed by a summary of the results which contribute to explaining the rationale of valuers and where they differ in their valuation approaches and the values used to assess land taxes.

Research method – simulations

In line with the researcher's approach, the API (NSW) Division invited members of the Institute to participate in valuation simulations, surveys and subsequent focus groups. This invitation was made directly to the chairperson of each valuer study group in NSW. The chair of each study group advised the group members of the research project. As set out in the letter of invitation to study groups, prospective participants were invited to contact the researcher direct. This resulted in communication between the chair, members of the valuer study groups and the researcher. The analyses of the simulations commenced with a review of the simulation response rates followed by respondent representativeness and explanation of the techniques used to analyze data. This was followed by a discussion of the results.

Simulation response rate

The procedures for undertaking the simulations and surveys were explained to participants, in which participant privacy and anonymity were ensured. The simulations were carried out using three means. The first means included meeting with participants at either their workplace or a location mutually convenient to both the participant and researcher; the second incorporated emailing simulations and surveys to participants with corresponding email responses; and the third was via surface mail. In the case of email and mail, the researcher communicated with a number of the participants by telephone post-completion of the simulation and survey to undertake discussions about the results and feedback.

Table 4.1 sets out the responses for both the simulations and surveys. As noted, the completed surveys of 25 valuers do not correspond with the number of valuer participants for either the retail or residential simulations. In total, 25 valuers participated in the simulation of which 21 of the

Table 4.1 Simulation and survey response rate

Response type	*Retail*	*Residential*
Gross simulation	40	40
Completed/Returned simulations	23	23
Total completed & returned surveys	25	
Non-returned	13	12
Returned & non-completed	3	4
Net responses after	23	23
Response rate completed & returned	57.5%	57.5%

valuers completed both the retail and residential simulations. Two additional valuers completed only the retail simulation and an additional 2 valuers completed only the residential simulation.

The completion rate was the same in each of the two simulations and was high at 57.5 per cent. Over half of the simulations were undertaken in the presence of the researcher. The completion rate of these simulations was significantly higher than the email participation rate. A review of the time taken to complete both the retail and residential simulations and the survey was between 90 and 100 minutes. An additional 20 to 30 minutes was spent with the participating valuers discussing and reviewing the results and outcomes of their simulation. General feedback from most participants was that the task appeared simple and straightforward at first. Upon completion, valuers reported the task to have been complex and challenging.

Respondents' representativeness

A basic comparison was made to gauge the representativeness of the participants, with specific reference to years of experience and gender, as set out in Table 4.2.

As noted in Table 4.2, the total number of participating valuers in the experience band of 1 to 7 years is 40 per cent higher in the residential simulation compared with the retail simulation population. Correspondingly, the number of valuers with over 15 years of experience is higher in the retail simulation by 13.5 per cent over the residential simulation. This observation is underlined by the fact that valuers will generally commence their valuation careers undertaking residential valuations and progressing onto commercial, retail and special uses property valuation later in their professional career.

The observation further highlights that the valuation profession has, until recently, comprised an older professional constituency. The void in the 7 to 15 years of experience across gender and both residential and retail property indicates a period of low entry into the profession. This trend seems to have changed during the past seven years, with an increase in valuers undertaking valuation work in the 1 to 7 year experience bracket. It is

Table 4.2 Participant representativeness of completed responses

Experience in years	*Retail*			*Residential*		
	Females	Males	Total	Females	Males	Total
1 to 7	1	4	5	1	6	7
8 to 15	2	1	3	2	1	3
15 & over	1	14	15	1	12	13
Total no.	**4**	**19**	**23**	**4**	**19**	**23**
Total percentage	**17.4**	**82.6**	**100**	**17.4**	**82.6**	**100**

further observed that the sample indicates the valuation profession to be a male-dominated profession at present.

Presentation of the data and explanation of analysis techniques

A broad overview of the results for each of the two simulations is set out with tables and figures at the beginning of each of the residential and retail simulations. This is followed by a more detailed account of results, in which the information is provided in a more detailed format.

An analysis of each of the sales, by each valuer in both the initial and revised simulation, is presented and discussed. This is then followed by an analysis of all valuers' responses for each of the three sales and an additional/control sale. The primary method of analysis for the responses of all valuers for each sale is the standard deviation across all valuers for each sale and a comparison of the standard deviation between the initial and revised analysis of each sale of both land and improvements. This is followed by an observation of the mean value of land across all valuers for each sale.

The analyses of the simulations have been conducted on the following bases:

1) Change in standard deviation in the land value between the initial and revised simulation results for each of the sales;
2) Change in standard deviation in the land value between the initial and revised simulations across all of the sales;
3) Change in the mean land value between the initial and revised simulations for each sale and across all sales;
4) Analysis of most relevant sale and least relevant sale as determined by all valuers in deducing the underlying value of land;
5) Analysis of most valuable location of land as determined by all valuers;
6) Variation between property nominated as most valuable location, but not assessed as the highest value property;
7) Analysis of the ratio between the sale price, land value determined and residual value of improvements for each sale;
8) Supporting discussions in the debriefing sessions after the simulations about the results from a number of valuers as to elements of process and judgment used in analysis process.

Simulation analyses

The residential simulation followed by the retail simulation results are presented and analyzed in three parts. The first part comprises a summary of the results of each simulation, which include the initial and revised simulations of each sale showing the standard deviation, change in standard deviation, mean value and change in mean value across all of the valuers. A graph of all responses in the before and after simulation has been included to provide a view of the changes in the overall results across all three sales and 23 valuers. A similar summary follows for the results in the deduced added value of improvements in the initial and revised simulations of all three sales across the 23 valuers.

The second part of the simulation analysis is a detailed sale-by-sale descriptive discussion of the results, commencing with the residential simulation and followed by the retail simulation. A discussion of the change in the standard deviation and mean values of the land is followed by a similar discussion on changes to the added value of improvements. These are supported by figures and graphs of both the deduced land value and added value of improvements highlighting these changes for each sale.

The third part of the analysis is a comparative critique across the three sales in each simulation, which establishes the sale that produces the most consistent and uniform result across all the valuers. The reduction in the standard deviation between the initial and revised simulations and the lowest standard deviation in the valuers' revised answers provides direction as to the nature and type of the most appropriate sale to be selected and analyzed in the absence of vacant land sales.

The result is then compared with the valuers' answers to the question recorded in the simulation as to which of the sales they nominated as the most relevant and least relevant in their analyses. This provided a review of the initial sales selection processes that valuers used as the first step in the analysis process. The last point for analysis is a review of each valuer's nomination of the most valuable location of the three sales in each of the simulations. The sale selected by valuers as the most valuable location is then compared against the valuers' deduced land values to establish the relationship between the value of the land and its location. This provides an insight as to the grading of values.

In conclusion, a cross comparison is made between the results of the residential and retail simulations which draws on the correlations and voids attributed to differences in the information provided and processes adopted by the valuers. Subsequent to completing the simulations, each valuer completed a survey which assisted in ranking elements of judgment, key information relied upon and the usefulness of the information provided in the simulations. A debriefing was conducted with the valuers following completion of the survey, in which additional feedback was gathered on the construct of the simulation, the valuers' answers and any additional information that valuers offered to assist in achieving greater uniformity and consistency of the values.

Residential simulation

A summary of the residential simulation results for the land is set out in Table 4.3 and Figures 4.2 and 4.3. Table 4.3 shows improvement in the standard deviation between the initial and revised simulations for two of the three sales. The additional sale of 11 Fiction Street, which was added in the revised simulation phase, shows the lowest standard deviation. No 11 Fiction Street is the proxy for improvements which are maximally productive, representing the highest and best use of land. The cost of new construction was deducted from the sale price of the improved property, resulting in the residual figure constituting the land value. From this residual figure,

Table 4.3 Residential land simulation summary

Residential simulation	*Initial*	*Revised*	*Initial*	*Revised*	*Initial*	*Revised*	*Add sale*
	10 Fiction	*10 Fiction*	*15 Fiction*	*15 Fiction*	*20 Fiction*	*20 Fiction*	*11 Fiction*
STDEV	$78,971	$46,715	$75,286	$56,363	$54,333	$60,311	$26,647
Mean	$465,676	$451,863	$459,480	$448,122	$486,241	$463,041	$477,326
STDEV	17.0%	10.3%	16.4%	12.6%	11.2%	13.0%	5.6%
Change in STDEV		39.4%		23.2%		(16.6%)	
Full Improvements v Light Improvements				79.4%		52.3%	
Mean percentage reduction in mean		-2.98%		-2.47%		-4.77%	

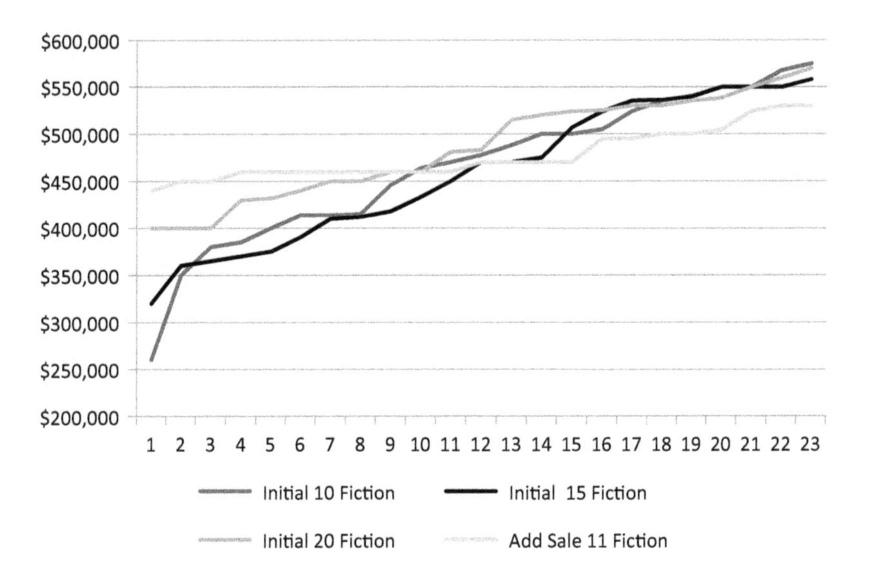

Figure 4.2 Initial residential simulation results

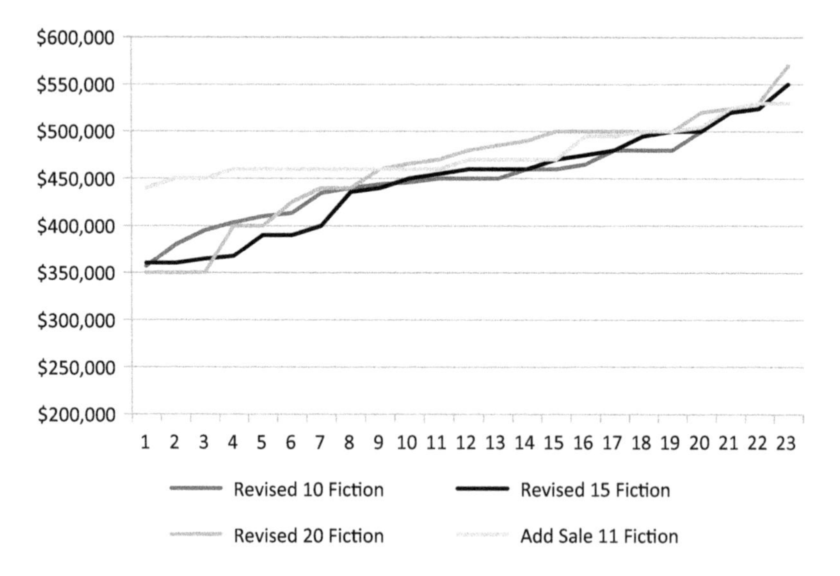

Figure 4.3 Revised residential simulation result

valuers were instructed to determine the land values of the other three sales in the revised simulation from the residual value of 11 Fiction Street.

The additional sale and the codified process which valuers were instructed to adopt in the revised simulation resulted in the improved standard deviations of 10 and 15 Fiction Street. In the revised simulation, the standard deviation of 20 Fiction Street increased in contrast to the other two sales. The revised simulation results for 20 Fiction Street show three valuers with lower values around $350,000 in their revised simulations, which were lower than the land values deduced in their initial simulations. During the debriefing and valuer interviews, it was found that these three valuers had also adopted a higher value for the improvements in the revised simulations for each of the sales. This indicated that the added value of the improvements of the additional sale impacted on their assessments.

In qualifying this result further, this occurrence was common among other valuers, but to a lesser degree. While it could not be fully determined, the limitations of the revised simulation, in which only one additional sale was provided, resulted in some level of caution being adopted in assessing the added value of improvements for 20 Fiction Street. In practice, the added value of new improvements would need to be analyzed a number of times to ensure that cost and value for new construction were similar. This factor is recognised as a limitation in using a simulated scenario, in which valuers do not have the option to look beyond the information provided.

In setting this aside, the simulation highlights that a number of valuers did not fully follow the instruction provided in using the deduced land value of 11 Fiction Street to redetermine the land values of the other three sales. Of particular note, which is elaborated on in the individual sale analysis, is that 15 of the 23 valuers, or 63.2 per cent, selected 20 Fiction Street as the most relevant sale of the three sales in the initial simulation. This sale also achieved the lowest standard deviation of the three sales in the initial simulation, which is in contrast to producing the highest standard deviation in the revised simulation.

A further point noted in discussions with valuers was the value of the location of 20 Fiction Street, being the second property from Henley Reserve in the residential simulation plan. Three valuers identified this as a detrimental value factor, due to the security factors of potential break-ins of living near a park. While not specifically quantified, this view was in the minority and was in stark contrast to the views of other valuers, who identified this as a positive influence on the value to 20 Fiction Street.

10 Fiction Street

Land: A mean value of $465,676 across the 23 valuers resulted in a standard deviation of 17.0 per cent in the initial simulation. In the revised simulation, this reduced to a mean value of $451,863 with a standard deviation of 10.3 per cent. The overall reduction in the standard deviation represents an improvement of 39.4 per cent as highlighted in Table 4.4. A reduction in the mean value of 2.98 per cent is noted between the initial and revised simulation result. Further noted later in Table 4.10, 5 valuers, representing 21.7 per cent, selected 10 Fiction Street as the most relevant of the three sales in contrast to 12 valuers, representing 52.2 per cent, who selected this sale as the least relevant of the three sales. The last point noted is that 3 valuers, representing 13 per cent, selected 10 Fiction Street as being the most valuable location, a point to be revisited in the residential simulation summary.

Improvements: A mean added value of improvements of $300,825 across the 23 valuers resulted in a standard deviation of 29 per cent in the initial simulation. In the revised simulation, the mean added value of improvements increased to $314,773 with a standard deviation of 15.8 per cent. This represents an improvement in the standard deviation of 46.7 per cent. The overall increased difference between the mean added value of improvements between the initial and revised simulation is 4.43 per cent, as shown in Table 4.5.

Comparative observation

Figures 4.4 and 4.5 set out the changes in each the land values and added value of improvements in the before and after simulations. Each graph represents the change in the standard deviation between the initial and revised simulation for this sale, in which improvement is noted in consistency at

Table 4.4 10 Fiction Street – change in standard deviation

	Land	*Improvements*
Initial	17	29
Revised	10.3	15.8
Percentage change	**39.4%**	**45.5%**

Table 4.5 10 Fiction Street – change in mean value

	Land	*Improvements*
Initial	$465,676	$300,825
Revised	$451,863	$314,773
Percentage change	**2.98%**	**4.43%**

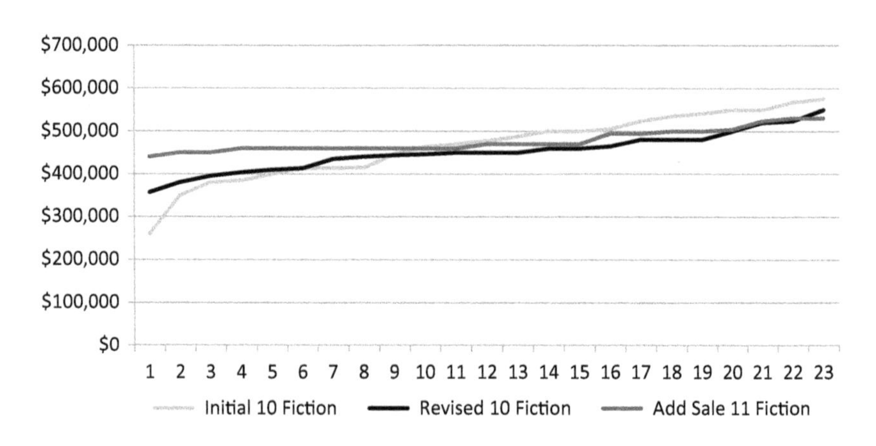

Figure 4.4 Land – initial v revised simulation mean comparison

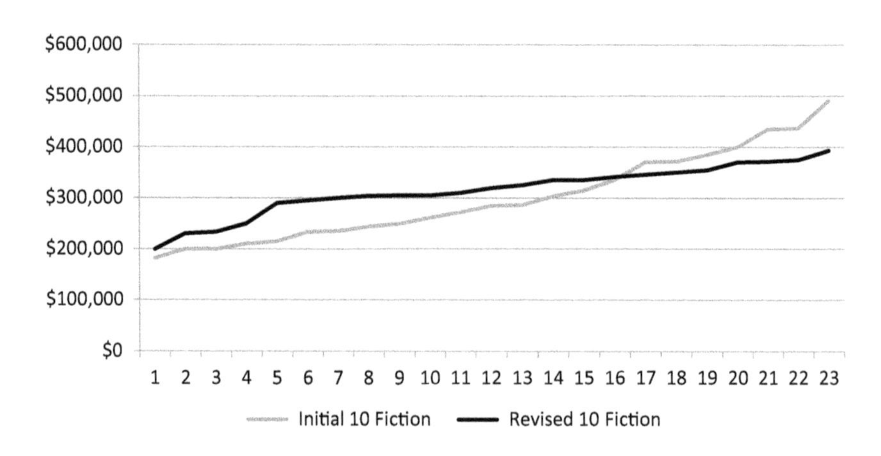

Figure 4.5 Improvements – initial v revised simulation mean comparison

both the upper and lower ends of the range. It may be concluded that the additional sale, 11 Fiction Street, and the codified method in dealing with the added value of improvements have increased consistency in the land value for this sale. It is further highlighted that 10 Fiction Street has the newest improvements of each of the three sales in the initial simulations, a point to be reviewed in the residential simulation summary.

15 Fiction Street

Land: A mean value of $459,480 across the 23 valuers resulted in a standard deviation of 16.4 per cent in the initial simulation. In the revised simulation, this reduced to a mean value of $448,122 with a standard deviation of 12.6 per cent. The overall reduction in the standard deviation represents an improvement of 23.2 per cent, as shown in Table 4.6. A reduction in the land mean value of 2.47 per cent is noted between the initial and revised simulation result. Further noted in Table 4.10, three valuers, representing 13 per cent, selected 15 Fiction Street as the most relevant of the three sales, in contrast to five valuers, representing 21.7 per cent, who selected this sale as the least relevant of the three sales. The last observation is that 47.8 per cent of valuers selected this sale as being the most valuable location, a point to be discussed in the residential simulation summary.

Improvements: The mean added value of improvements of $189,991 across the 23 valuers resulted in a standard deviation of 40.2 per cent in the initial simulation. In the revised simulation, the mean added value of improvements increased to $200,893 with a reduced standard deviation of 28.8 per cent. This represents an improvement in the standard deviation of 28.4 per cent. The overall increased difference between the mean added value of improvements in the initial and revised simulation is 5.7 per cent compared with the 2.47 per cent change in the mean value of the land, as per Table 4.7.

Table 4.6 15 Fiction Street – change in standard deviation

	Land	*Improvements*
Initial	16.4	40.2
Revised	12.6	28.8
Percentage change	**23.2%**	**28.4%**

Table 4.7 15 Fiction Street – change in mean value

	Land	*Improvements*
Initial	$459,480	$189,991
Revised	$448,122	$200,893
Percentage change	**2.47%**	**5.7%**

Comparative observation

Figures 4.6 and 4.7 are an overview of the differences in each of the land values and added value of improvements in the before and after simulations. Figure 4.6 shows improvement in consistency; however, at the upper end of the range of values of the land, only a small improvement is noted. Conversely, this corresponds with the results at the lower end of the added value of improvements, as shown in Figure 4.7, in which a small improvement is also noted. No 15 Fiction Street had a development application lodged at the time of sale for an extension of 50m^2 at the rear of the house, while the rest of the improvements are similar and in line with 10 Fiction Street. It is concluded that, due to the smaller improvements on the land compared with 10 Fiction Street, the improvements are not maximally productive and hence additional judgment in determining the added value of improvements resulted in a higher standard deviation in both the initial and revised simulation.

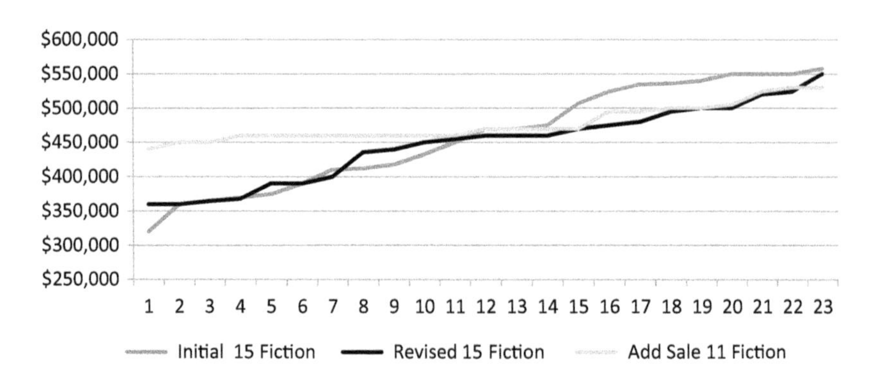

Figure 4.6 Land – initial v revised simulation mean comparison

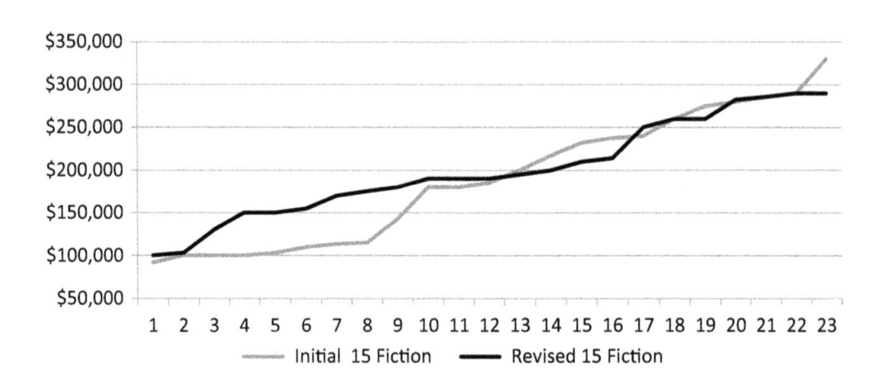

Figure 4.7 Improvements – initial v revised simulation mean comparison

20 Fiction Street

Land: A mean value of $486,241 across the 23 valuers resulted in a standard deviation of 11.2 per cent in the initial simulation. In the revised simulation, this decreased to a mean value of $463,041 with a standard deviation of 13 per cent. In contrast to the revised sales analysis of 10 and 15 Fiction Street, in which an improvement in the standard deviation was noted in each case, an increase in the standard deviation of 16.6 per cent resulted in 20 Fiction Street. A detailed analysis of the deduced values at the lower end of the range of each the initial and revised simulation shows that the increase in the standard deviation resulted from the three values in the revised simulation being below the lowest value in the initial simulation. Table 4.10 shows 15 valuers, representing 65.2 per cent, selected 20 Fiction Street as the most relevant of the three sales in contrast to 26.1 per cent of valuers who selected this sale as the least relevant of the three sales. The last point of note is that 47.8 per cent of valuers selected this sale as being the most valuable location.

Improvements: The mean added value of improvements of $63,843 across the 23 valuers resulted in a standard deviation of 83.8 per cent in the initial simulation. In the revised simulation, the mean added value of improvements increased to $86,659 with a reduced standard deviation of 69.6 per cent. This represents a reduction in the standard deviation of 17 per cent, as highlighted in Table 4.8. The overall increased difference between the mean added value of improvements in the initial and revised simulation is 35.7 per cent, as shown in Table 4.9. This represents a corresponding decrease in the change of the mean value of the land of 4.77 per cent.

Comparative observation

Figures 4.8 and 4.9 are an overview of the differences in each the land values and added value of improvements in the before and after simulations.

Table 4.8 20 Fiction Street – change in standard deviation

	Land	*Improvements*
Initial	11.2	83.8
Revised	13	69.6
Percentage change	**16.6%**	**17%**

Table 4.9 20 Fiction Street – change in mean value

	Land	*Improvements*
Initial	$486,241	$63,843
Revised	$463,041	$86,659
Percentage change	**4.77%**	**35.7%**

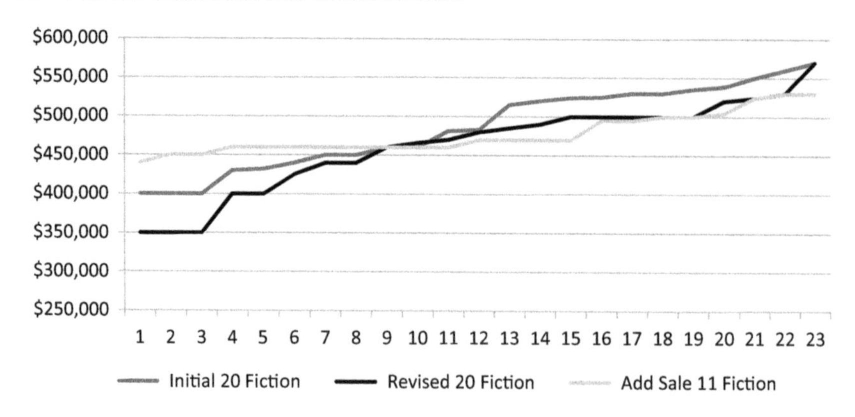

Figure 4.8 Land – initial v revised simulation mean comparison

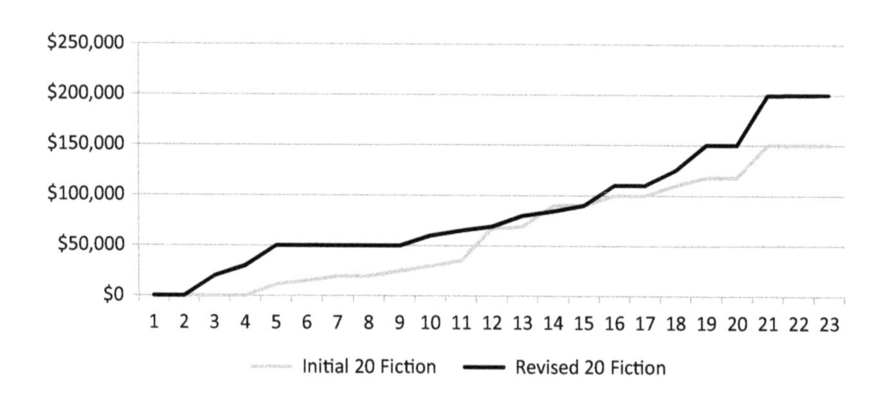

Figure 4.9 Improvements – initial v revised simulation mean comparison

As discussed above under Land, Figure 4.8 shows that the lower end of the range of land values in the revised simulation is below those of the initial simulation. Conversely, this corresponds with the results at the upper end of the added value of improvements, as shown in Figure 4.9, in which the added value of improvements are above those of the initial simulation. It is further noted that 20 Fiction Street had the most dilapidated improvements of the three sales and sold with a development application in place to demolish and rebuild the property.

Residential simulation summary

This summary deals with three important points resulting from the analysis of the residential simulation results. These points are the results and rationale for the selection of the most suitable sale, the selection of the most valuable location and the consistency of the deduced land values across the 23 valuers. In reviewing the collective results, it is apt to commence with the simulation task sheet, as this sets out the tasks that were addressed by each of the valuers. The first task was for the valuers to analyse the land value and added value of improvements of each of the three sales in the initial simulation. The second task instructed valuers to identify the most relevant to least relevant of the three sales. The last task instructed valuers to rank the location value of each of the three sales.

A discussion of the results of all the valuers for each sale has earlier been addressed on a sale-by-sale basis. Of more specific importance is the outcome and rationale for those outcomes which are now addressed collectively across all the sales analyses. This analysis accounts for, where possible, the distinction between the processes versus the elements of judgment that resulted in the answers derived by the valuers. It provides some explanation and plausibility for those points of difference.

It is clear from the results of the standard deviations in the initial and revised simulations that a mixed view may be formed as to the usefulness yielded from improved sales, particularly in circumstances where little or no guidance is provided in accounting for the added value of improvements. A review of the standard deviations of the three sales in the initial simulation shows that only one of the three sales, 20 Fiction Street, resulted in a standard deviation of 11.2 per cent, being within the acceptable margin of error of +/– 15 per cent (NSW Ombudsman 2005:iv). No 10 and 15 Fiction Street resulted in standard deviations of 17 per cent and 16.4 per cent respectively. These two results will be addressed again in the analysis of the sales selection task outcomes.

The specific question to be addressed, however, is, 'Where improved sales are used, which improved sales yielded the most consistent results and what processes could be adopted to achieve greater consistency?' In adopting a codified approach to the analysis in accounting for the added value of improvements, a different outcome resulted across all three sales. In the revised simulation, all three sales resulted in deduced land values within the +/– 15 per cent acceptable margin of error. No 10 Fiction Street produced the lowest standard deviation of 10.3 per cent and the largest improvement of 39.4 per cent of the three sales in the revised simulation. This was followed by 15 Fiction Street, resulting in a standard deviation of 12.6 per cent, an improvement of 23.2 per cent. In contrast, 20 Fiction Street resulted in a reduction in the standard deviation to 13 per cent, being a reduction of 16.6 per cent; however, it was still within the acceptable margin of error of +/– 15 per cent.

In articulating these results further, a review of the results from the second task of the initial simulation in conjunction with valuer interviews, qualifies the processes and judgment adopted by the valuers in the sales selection process. As set out in Table 4.10, a review of the sale rankings shows that valuers ranked 20 Fiction Street as the most relevant in deducing the underlying value of land. Fifteen of 23 valuers, representing 65.2 per cent, selected 20 Fiction Street as the most relevant, followed by 5 valuers, representing 21.7 per cent, who selected 10 Fiction Street as the most relevant sale. No 15 Fiction Street, with 3 valuers or 13 per cent, was selected as the most relevant sale.

The most consistent response by valuers who selected 20 Fiction Street in the initial simulation was that this property was closest to land value, having the oldest and most dilapidated improvements of the three sales. The primary approach in deducing the underlying value of land was by judgment in which the added value of improvements was deducted from the sale price. Three valuers who selected this sale as most relevant attempted to establish a rate of depreciation of the improvements by reference to the sale of 10 Fiction Street in conjunction with the cost new of improvements. The balance of the valuers used judgment by approximation in the determination of the added value of improvements, with reference to the new cost as an approximate guide.

In Table 4.10 it is noted in the colour-coded responses of valuers who selected 20 Fiction Street as the most relevant sale, that 10 of the 15 valuers

Table 4.10 Sales analysis and land value summary

Sale	*11 Fiction Street*		*10 Fiction Street*		*15 Fiction Street*		*20 Fiction Street*	
Sale price	$970,000		$785,000		$650,000		$550,000	
Land value mean revised	$477,326		$451,863		$448,122		$463,041	
Land value mean STDEV revised	5.6%		10.3%		12.6%		13%	
Land improved value ratio revised mean	49.2%		57.6%		68.9%		84.2%	
Age/Last upgrade of improvements	New		10 years		Not stated		Not stated	
Size m² of improvements	250		230		190		180	
	Void		*10 Fiction Street*		*15 Fiction Street*		*20 Fiction Street*	
			No	*%*	*No*	*%*	*No*	*%*
Most relevant sale	N/a		5	21.7	3	13	15	65.2
Least relevant sale	N/a		12	52.2	5	21.7	6	26.1
Most valuable location	N/a		3	13	11	47.8	11	47.8
Land value mean Initial simulation	N/a		$465,676		$459,480		$486,241	
Valuers who identified most valuable location but did not assign highest land value	**Total No** 11	**Total %** 47.8%	2	8.7%	5	21.7%	4	17.4%

were tightly clustered consecutively at the upper end of the land value range in the initial simulation, with a deduced land value range of between $515,000 and $560,000 being well above the mean value of $486,241. A further pertinent observation was that one valuer deduced the land value at the sale price of the property, attributing no added value to the improvements. Two additional valuers deduced a land value higher than the sale price. These latter two valuers accounted no value to the improvements, but also accounted for the added cost of demolition in removing the house, bringing it back to its value as vacant land. This observation highlights the breadth of interpretation of the added value of improvements to the extent that three valuers, or 13 per cent, viewed the added value of improvements as having no value and hence not being maximally productive or, more specifically, not utilising the land to its maximum potential.

In contrast to 20 Fiction Street, five valuers, representing 21.7 per cent, selected 10 Fiction Street as the most relevant sale. Their responses were at the mid and lower end of the valuer range in the initial simulation. Overall, this sale resulted in the highest standard deviation of all three sales in the initial simulation. An outlier land value of $260,000, being the lowest land value, resulted in a standard deviation of 17 per cent. Without this land value, the standard deviation for this sale in the initial simulation would have been 14 per cent. This valuer stated that he or she had adopted the added value of improvements as cost new less 5 per cent, resulting in a lower land value compared with the rest of the deduced land values. Of further note, no valuer in this sale attributed a zero value to the added value of improvements.

In the revised simulation, valuers were given additional information and a codified approach to the sales analysis process in deducing the land values of the three sales. In contrast to the initial simulation, 10 Fiction Street resulted in the lowest standard deviation of 10.3 per cent, with 20 Fiction Street resulting in the highest standard deviation of the three sales with 13 per cent. This is almost the opposite result to the initial simulation. The two key changes resulting from the codified approach were firstly, that all three sales in the revised simulation resulted in deduced land values within the acceptable margin of error; and secondly, that 10 Fiction Street, with the newest improvements of the three sales, resulted in the lowest standard deviation in the revised simulation of all three sales in both the initial and revised simulations. The standard deviation of all three sales in both the initial and revised simulations is set out next in Table 4.10.

The standard deviation of 20 Fiction Street in the initial simulation, which was largely deduced by valuers' judgment, compared with the standard deviation of 10 Fiction Street in the revised simulation, in which a codified process was used, resulted in an overall improvement of 9 per cent in these two standard deviations. This observation, in conjunction with all three deduced land values falling within the acceptable margin of error in the revised simulation results, notwithstanding the reduction in the standard deviation of 20 Fiction Street, is set out in Table 4.11.

Table 4.11 Standard deviation comparative analysis

	10 Fiction Street	*20 Fiction Street*
Initial simulation	17.0%	11.2%
Revised simulation	10.3%	13%

The third task valuers were instructed to undertake was to rank the most valuable to least valuable locations of the three sales. It is reiterated that all of the parcels of land are the same size and shape and sod within the same time period. Once the added value of improvements are accounted for and deducted from the sale price, the deduced land value ultimately reflects the value of the location of the land. It is noted in Table 4.10 that 11 valuers, representing 47.8 per cent, did not assign the highest land value to the property selected as the most valuable location in the initial simulation.

Following completion of the simulation and during the debriefing discussion it was identified that once valuers had deduced the land value of each sale, a majority of the 11 valuers who selected the most valuable location which was not the highest deduced land value did so in reference to the location of each of the three properties from the street plan. There was no direct correlation between the results and the relativity of the results of the land values deduced from the sale analysis task in ranking the value of the location of each of the three sales. In contrast, the valuers who did rank the value of the location in line with the results from the deduced land values did have regard to the results of their earlier analysis and the deduced underlying value of the land. It was indicated by the valuers who had regard to their analysis that, in ranking property based on the location, additional information would be needed to undertake this task more conclusively. This point will be discussed again after the analysis of the results from the retail simulation, in the conclusion and summary of this chapter.

Retail simulation

A summary of the retail simulation results for the land are set out in Table 4.12 and Figures 4.10 and 4.11. Table 4.12 shows improvement in the standard deviation between the initial and revised simulations for all of the three sales. The additional sale of 22 Main Street, which was added in the revised simulation, shows the lowest standard deviation. No 22 Main Street is the proxy for improvements which are maximally productive and highest and best use of the land, in which the cost new of construction was deducted from the sale price of the improved property resulting in the residual value constituting the land value. Valuers were instructed to redetermine the land values of the three sales in the revised simulation based on the land value determined from the analysis and valuation of 22 Main Street.

The additional sale and the codified process which valuers were instructed to adopt in the revised simulation resulted in the improved standard deviations of all three sales. In contrast to the standard deviation changes in the residential simulation, the results of the standard deviation in the retail simulation improved by between 27 and 56 per cent. A greater contrast between the most relevant sale nominated by a majority of all the valuers and the results that follows provides a further basis for a codified approach to the analysis of improved sales where they are to be used.

Table 4.12 Retail land simulation summary

Retail simulation	*Initial*	*Revised*	*Initial*	*Revised*	*Initial*	*Revised*	*Add sale*
	15 Main	***15 Main***	***5 Bank***	***5 Bank***	***20 Main***	***20 Main***	***22 Main***
STDEV	$57,671	$24,432	$57,188	$35,307	$46,448	$31,731	$10,261
Mean	$566,467	$549,890	$583,889	$541,939	$566,989	$531,439	$542,152
STDEV	10.18%	4.44%	9.79%	6.52%	8.19%	5.97%	1.89%
Change STDEV		56.4%		33.5%		27.2%	
Fully Imp v Light Imp				65.6%		59.9%	
Mean percentage reduction		-2.93%		-7.18%		-6.27%	

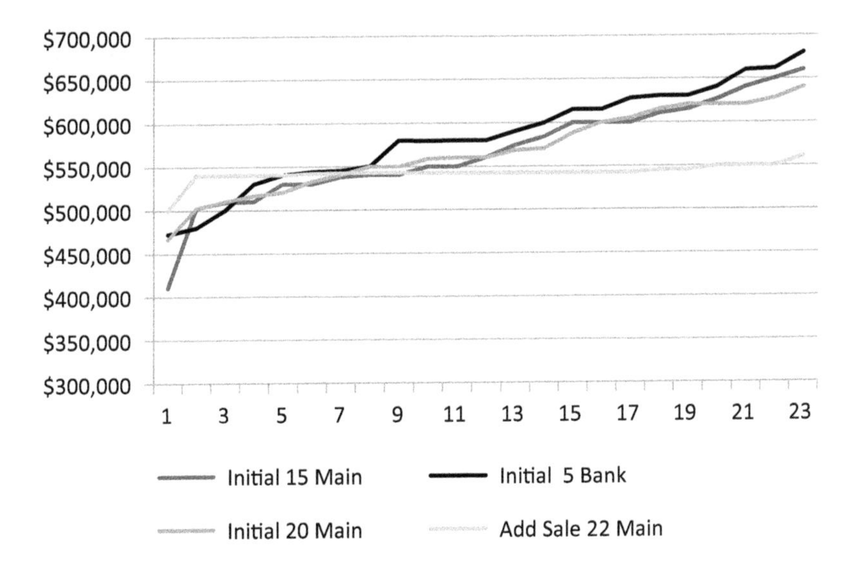

Figure 4.10 Initial simulation results

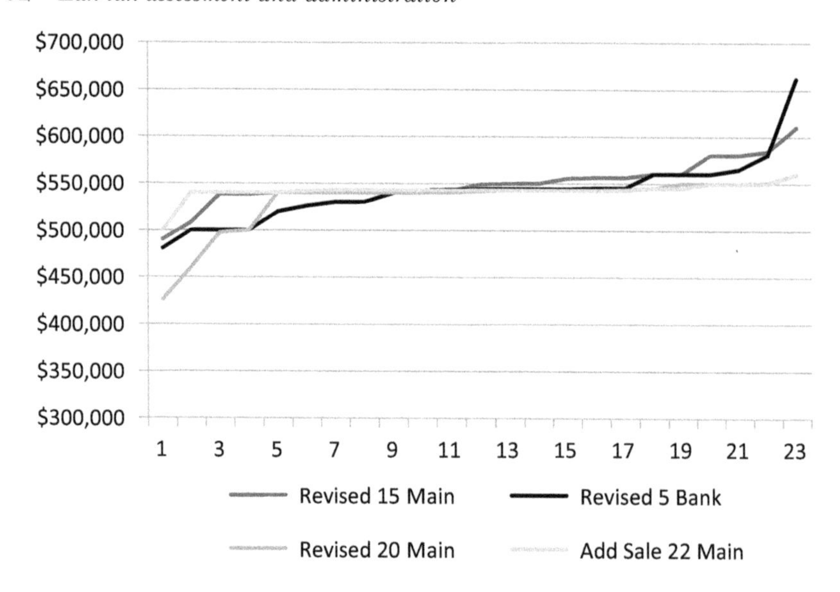

Figure 4.11 Revised simulation results

15 Main Street

Land: A mean value of $566,467 and standard deviation of 10.18 per cent was recorded across the 23 valuers in the initial simulation. In the revised simulation, the mean value reduced to $549,890 with the standard deviation reducing to 4.44 per cent. The overall reduction in the standard deviation as a percentage of the mean value between the initial and revised simulation represents an improvement of 56.4 per cent, as shown in Table 4.13. A reduction in the mean value of 2.93 per cent is noted between the initial and revised simulation result, as set out in Table 4.14.

Of further note in Table 4.19, six valuers representing 26.1 per cent selected 15 Main Street as the most relevant of the three sales in contrast to seven valuers, representing 30.4 per cent, who selected this sale as the least relevant of the three sales. The last point of note is that four valuers, representing 17.4 per cent, selected 15 Main Street as being the most valuable location.

Improvements: The mean added value of improvements of $292,446 and standard deviation of $56,565 represents 19.3 per cent across the 23 valuers in the initial simulation. In the revised simulation, the mean added value of improvements increased to $307,936 with a standard deviation $61,293, representing 19.9 per cent of the mean value. This represents a change in the standard deviation of 3.1 per cent. The changes in standard deviations and mean values are summarised in Table 4.14.

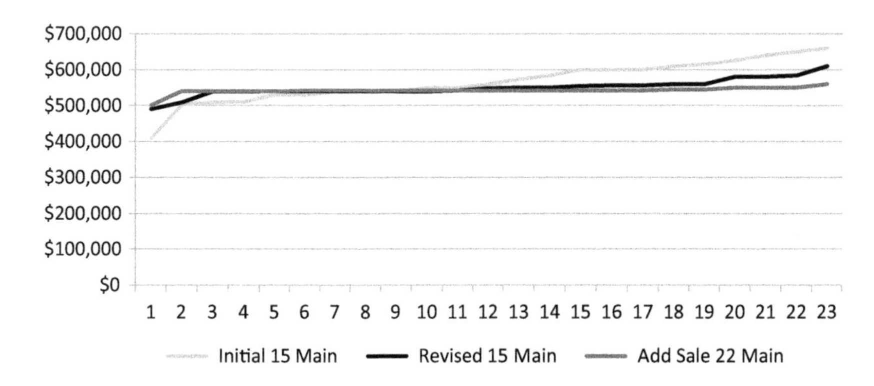

Figure 4.12 Land – initial v revised simulation mean comparison

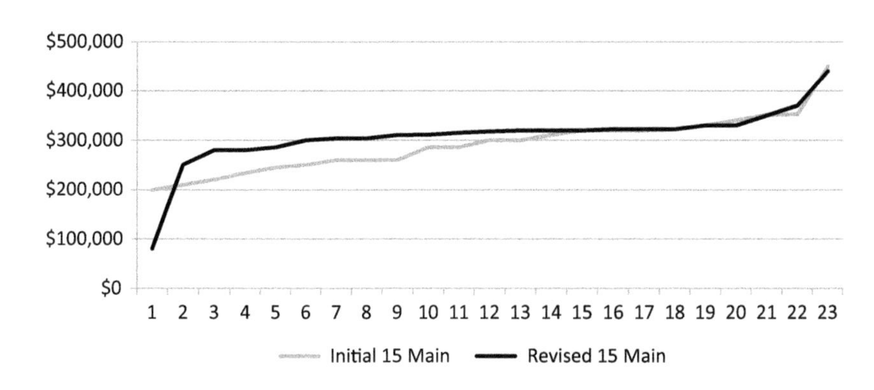

Figure 4.13 Improvements – initial v revised simulation mean comparison

Table 4.13 15 Main Street – change in standard deviation as percentage of mean values

	Land value	*Value of improvements*
Initial	10.18%	19.3%
Revised	4.44%	19.9%
Percentage change	**56.4%**	**3.1%**

Table 4.14 15 Main Street – percentage change in mean value

	Land value	*Value of improvements*
Initial	$566,467	$292,446
Revised	$549,890	$307,936
Percentage change	**2.93%**	**5.3%**

Comparative observation

Figures 4.12 and 4.13 are an overview of the differences in the standard deviation and mean values of each the land value and added value of improvements in both the before and after simulations. As shown in Figure 4.12, greater consistency is noted at both the upper and lower end of the land value range, resulting in a significant improvement in the standard deviation of the land value. In the case of the improvements, with the exception of one valuer's response, the lower end of the added value of improvements has improved, with the upper end of the range remaining constant in the revised simulation.

5 Bank Street

Land: A mean value of $583,889 and standard deviation of $57,188 was recorded across the 23 valuers. The standard deviation as a percentage of the mean value represents 9.79 per cent in the initial simulation. In the revised simulation, the mean value reduced to $541,939 with the standard deviation reducing to $35,307. The standard deviation as a percentage of the mean in the revised simulation represents 6.52 per cent. The overall reduction in the standard deviation as a percentage of the mean value between the initial and revised simulation represents an improvement of 33.5 per cent. A reduction in the mean value of 7.18 per cent is noted between the initial and revised simulation result.

Further noted in Table 4.19, no valuer, representing zero per cent, selected 5 Bank Road as the most relevant of the three sales in contrast to 12 valuers, representing 47.8 per cent, who selected this sale as the least relevant of the three sales. The last point of note is that eight valuers, representing 34.8 per cent, ranked 5 Bank Road as being the most valuable location.

Improvements: The mean added value of improvements of $246,763 and standard deviation of $56,141 across the 23 valuers resulted in the standard deviation as a percentage of the mean value at 22.8 per cent in the initial simulation. In the revised simulation, the mean added value of improvements increased to $280,940 with a standard deviation $66,677, resulting in a standard deviation representing 23.7 per cent of the mean value. This represents a change in the standard deviation of 3.9 per cent, as shown Table 4.15.

Comparative observation

Figures 4.14 and 4.15 are an overview of the differences in the standard deviation and mean values of each the land value and added value of improvements in both the before and after simulations. As set out in Figure 4.14, greater consistency is noted at both the upper and lower ends of the land value range, resulting in significant improvement in the standard deviation as a percentage of the mean value across the 23 valuers. The marginal increase in the standard deviation in the revised simulation, as shown in Table 4.15, results from one valuer recording a lower added value of improvements in the revised simulation compared with the initial simulation, as shown in Figure 4.15.

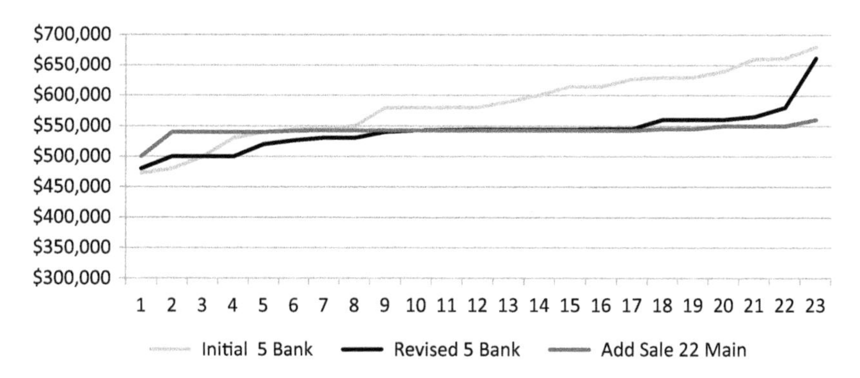

Figure 4.14 Land – initial and revised mean and standard deviation comparison

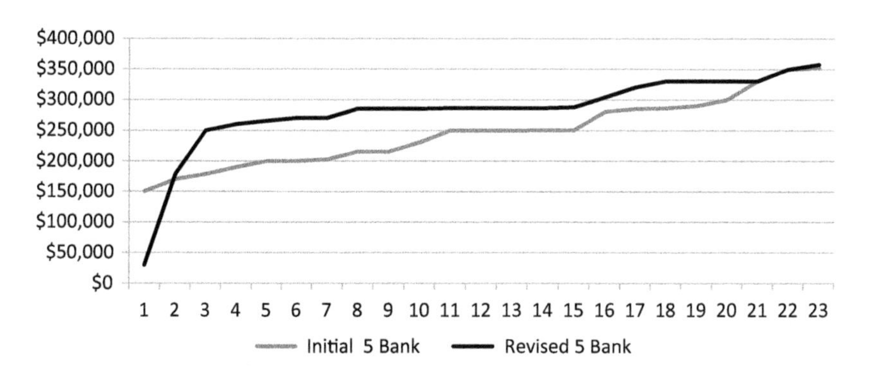

Figure 4.15 Improvements – initial and revised mean and standard deviation comparison

Table 4.15 5 Bank Road – change in standard deviation as a percentage of mean values

	Land value	*Value of improvements*
Initial	9.79	22.8
Revised	6.52	23.7
Percentage change	**33.5%**	**3.9%**

Table 4.16 5 Bank Road – change in mean value

	Land value	*Value of improvements*
Initial	$583,889	$246,763
Revised	$541,939	$280,940
Percentage change	**7.18%**	**13.9%**

20 Main Street

Land: A mean value of $566,989 and standard deviation of $46,448 were recorded across the 23 valuers. The standard deviation as a percentage of the mean value represents 8.19 per cent in the initial simulation, as per Table 4.12. In the revised simulation the mean value reduced to $531,439 with the standard deviation reducing to $31,731 being 5.97 per cent. The overall reduction in the standard deviation as a percentage of the mean value between the initial and revised simulation represents an improvement of 27.2 per cent. A reduction in the mean value of 6.27 per cent is noted between the initial and revised simulation result in Table 4.18.

Of further note in Table 4.19, 17 valuers, representing 73.9 per cent, selected 20 Main Street as the most relevant of the three sales in contrast to four valuers, representing 17.4 per cent, who selected this sale as the least relevant of the three sales. The last point of note is that 11 valuers, representing 47.8 per cent, selected 20 Main Street as being the most valuable location.

Improvements: The mean added value of improvements of $85,402 and standard deviation of $73,935 across the 23 valuers resulted in the standard deviation of 86.6 per cent in the initial simulation. In the revised simulation the mean added value of improvements increased to $115,517 with a standard deviation $49,931, resulting in the standard deviation representing 43.2 per cent of the mean value. This represents a change in the standard deviation of 50.1 per cent as shown in Table 4.17.

Comparative observation

Figures 4.16 and 4.17 are an overview of the differences in the standard deviation and mean values of each the land value and added value of improvements in both the before and after simulations. Figure 4.16 shows

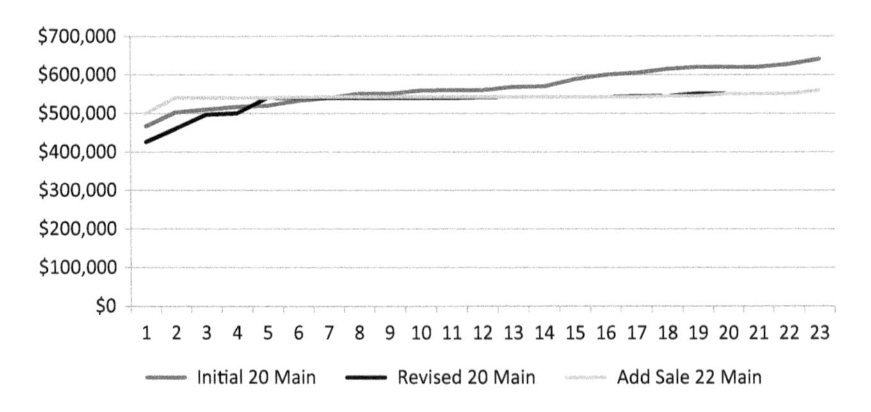

Figure 4.16 Land – initial and revised mean and standard deviation comparison

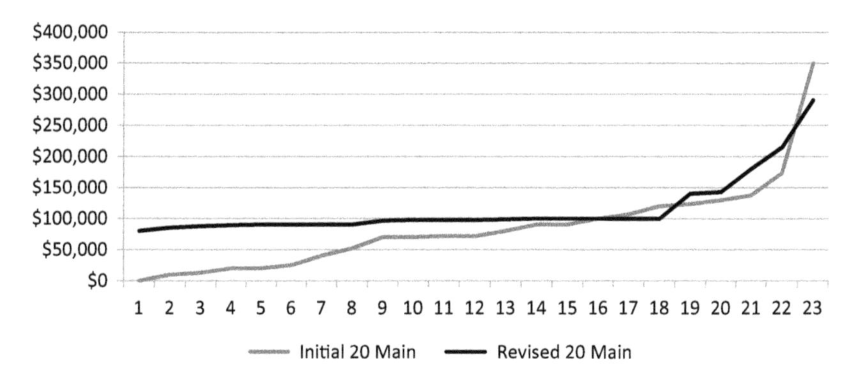

Figure 4.17 Improvements – initial and revised mean and standard deviation comparison

Table 4.17 20 Main Street – change in standard deviation as percentage of mean values

	Land value	*Value of improvements*
Initial	8.19%	86.6%
Revised	5.97%	43.2%
Percentage change	**27.2%**	**50.1%**

Table 4.18 20 Main Street – change in mean value

	Land value	*Value of improvements*
Initial	$566,989	$85,402
Revised	$531,439	$115,517
Percentage change	**6.27%**	**35.3%**

a significant improvement in the consistency at the middle and upper end of the land value range, resulting in an improvement in the standard deviation across the 23 valuers of 27.2 per cent. In the case of the improvements, Figure 4.17 shows a significant corresponding improvement at the middle and lower end of the values. A marginal improvement is noted at the upper end of the values.

Retail simulation summary

This summary deals with three important points resulting from the analysis of the retail simulation results. These points are the results and rationale for the selection of the most suitable sale, the selection of the most valuable location and the consistency of the deduced land values across the 23 valuers. In reviewing the collective results, it is apt to commence with the

Table 4.19 Retail sales analysis summary

	22 Main Street		*20 Main Street*		*5 Bank Road*		*15 Main Street*	
Sale price	$900,000		$640,000		$830,000		$860,000	
Land value mean revised	$542,152		$531,439		$541,939		$549,890	
Land value mean STDEV revised	1.89%		5.97%		6.52%		4.44%	
Land: improved value ratio revised mean	60%		83%		65.3%		63.9%	
Age/Last upgrade of improvements	1 month		50 years		15 years		7 years	
Size m² of improvements	130m²		130m²		130m²		**130m²**	
	Void		*20 Main Street*		*5 Bank Road*		*15 Main Street*	
			No	*%*	*No*	*%*	*No*	*%*
Most relevant sale	N/a		17	73.9%	0	0	6	26.1%
Least relevant sale	N/a		4	17.4%	12	47.8%	7	30.4%
Most valuable location	N/a		11	47.8	8	34.8	4	17.4
Land value mean Initial simulation	N/a		$566,989		$583,889		$566,467	
Valuers who identified most valuable location but did not assign highest land value	**Total No** 7	**Total %** 30.4%	4	17.4%	0	0%	3	13%

simulation task sheet, as this sets out the tasks that were addressed by each of the valuers. The first task was for the valuers to analyze the land value and added value of improvements of each of the three sales in the initial simulation. The second task instructed valuers to identify the most relevant to least relevant of the three sales in undertaking task one. The final task instructed valuers to rank the location value of each of the three sales.

The discussion on the numerical results across all the valuers for each sale has earlier been addressed on a sale-by-sale basis. Of more specific importance is the rationale for those outcomes which are now addressed collectively across all the sales analysis. This analysis accounts for, where possible, the distinction between the cognitive processes and the elements of judgment that resulted in the answers derived by the valuers and some explanation and plausibility for those points of difference.

In contrast to the residential simulation, the standard deviations across all three sales in both the initial and revised simulation results are within the acceptable margin of error +/– 15 per cent. The question to be asked resulting from this observation is whether changes need to be made given the standard deviation in the initial simulation is within this acceptable margin. The key difference between the residential and retail simulations is the level of information provided in each of the two simulations; the retail simulation included rental data which provided a second method of analyzing the sales. The second reason is that the difference between the initial and revised results of the retail simulation has resulted in an improvement of approximately 40 per cent in the consistency and standard deviations across all of the valuers.

Of relevance to the above change and improvement in the standard deviation are the questions of why and, more importantly, how these improvements were achieved. In following the analysis used in the residential simulation result, a review of the results from task two of the initial simulation, in which valuers were asked to rank sales from most to least relevant was analyzed. Table 4.19 shows that valuers ranked 20 Main Street as the most relevant sale. Seventeen valuers, representing 73.9 per cent of all valuers, selected 20 Main Street, followed by six valuers, representing 26.1 per cent of all valuers, who selected 15 Main Street as the most relevant sale. No 5 Bank Road was not selected by any valuers as the most relevant sale and was ranked as the least relevant sale by 12 valuers, representing 47.8 per cent.

Of the responses of valuers who selected 20 Main Street as the most relevant sale, 8 of the 17 valuers were tightly clustered at the upper end of the land value range in the initial simulation. The deduced land value range of between $600,000 and $640,000 is well above the mean value of $566,989. A further observation, not restricted to this group of valuers, was one valuer who deduced the land value at the sale price, attributing no added value to the improvements. This was the highest land value in the initial simulation results. At the opposite end of this range of responses, five valuers with a range of $510,000 to $532,750 were below the mean value for this deduced land value. Despite these two distinct groups, this sale resulted in the lowest standard deviation of the three sales in the initial simulation.

In the revised simulation, valuers were given additional information and specific instructions for the sales analysis process in deducing the land values of the three sales. In contrast to the initial simulation, in the revised simulation 15 Main Street recorded the lowest standard deviation of 4.44 per cent, with 20 Main Street recording a standard deviation of 5.97 per cent. As discussed earlier, the standard deviation of 8.19 per cent for 20 Main Street in the initial simulation, which was largely derived from valuers' judgments while within the acceptable margin of error, has improved considerably using a codified approach.

As noted in Table 4.20, the greatest improvement and lowest standard deviation resulted from the analysis of 15 Main Street, the sale that was the most maximally productive of the three initial sales, when using a codified approach. When using the information in the revised simulation information sheet, valuers were able to deduct the cost new of the additional sale to deduce the underlying value of the land. The additional sale of 22 Main Street recorded a standard deviation of 1.89 per cent. In the case of 15 Main Street, which was structurally refurbished seven years earlier, the amount of depreciation attributed to the improvements was more easily accounted for. This was also assisted via a depreciation schedule which was provided to the participants in the revised simulation. This confirms the commentary of O'Keefe (1974:88) that, when using improved sales, the most suitable are those closest to cost new where less depreciation is to be accounted for in assessing the added value of improvements.

The third task in the initial simulation was to rank the location value of the three sales. With all of the parcels of land being the same size and shape, once the added value of improvements is accounted for, the deduced land value ultimately reflects the value of the location of the land. It is noted in Table 4.19 that seven valuers, representing 30.4 per cent, did not select as the highest land value the sale that they deduced as the most valuable location.

During the debriefing, it was identified that, once valuers had deduced the land value of each sale, of the seven valuers who selected the most valuable location which was not determined the highest land value in their calculation, they selected the most valuable location by reference to the street plan only. There was no direct correlation between their results and relativity of the results of the land values deduced from the sale analysis task in ranking the value of the location of each of the three sales.

This was similar to the response for the residential simulation. In contrast, the valuers who ranked the value of the location in line with the results from the deduced land values did have regard to the results of their analysis and the deduced underlying value of the land. This observation highlights the disconnection of some valuers between the valuation mechanics used to calculate the land value and the judgment used in selecting the most valuable location, which did not correlate with their calculations.

A further observation from Table 4.19 in the analysis of the initial mean land values is that 11 valuers nominated 20 Main Street as the most

Table 4.20 Summary of standard deviation comparison

	15 Main Street	*5 Bank Road*	*20 Main Street*	*22 Main Street*
Initial simulation	10.18%	9.79%	8.19%	
Revised simulation	4.44%	6.52%	5.97%	1.89%
% Improvement	**56.4%**	**33.5%**	**27.2%**	

valuable location; however, it recorded a lower mean land value to 5 Bank Road, which 8 valuers nominated as the most valuable location. No 15 Main Street, with a similar mean land value to 20 Main Street, was nominated by only four valuers as the most valuable location. A review of the mean land value in the initial simulation compared with the revised simulation would further indicate a level of disparity between the valuers in accounting for the added value of improvements, which contributed to this difference. It may be further concluded that some valuers determined the most valuable location by reference solely to the geographic location of the property on the plan, with little regard to the answers achieved in the valuation process.

It was indicated by the valuers who had regard to their analysis and results in ranking property based on location that, while additional information would be needed to undertake this task, the rental information provided greater assistance in determining the value of the location. As highlighted in the residential simulation summary, this point will be discussed again in the conclusion and summary of this chapter. It does, however, point to the importance of information in drawing conclusions and establishing a sound basis on which qualified judgment can be made.

Residential/retail comparison

A comparison of the results between the residential and retail simulation results and conclusions provides a summary of differences that exist in the assessment of these two uses, as well as the factors contributing to those differences. Of particular note are the differences in the standard deviations in the initial simulations for both the residential and retail cases. In summary, only one of the two land values deduced from the sales in the initial residential simulation was within the acceptable margin of error of +/– 15 per cent. In contrast, all three of the land values deduced from the retail sales were within this acceptable margin. The rationale for this difference is of importance and is highlighted as follows:

1. The retail simulation included rental evidence which provided valuers with an additional anchor point on which to assess and deduce the use of improvements on the land. This was not provided in the residential simulation.
2. Valuers were able to consistently identify the maximal productivity of the improvements, by reference to their size, which was uniform across the retail use. The gross building area of the improvements in the retail simulation was consistent at 130m^2 for all of the sales and all of the retail improvements in the retail strip. The maximal productivity of the land, by reference to the improvements, is clearly identifiable in the retail simulation. This was absent in the initial residential simulation and hence, while there was an indication that the area of the improvements

on the land were expanding in the residential simulation, valuers were not able to clearly assess the maximal productivity of improvements.

3. The age of the improvements was the final key aspect in the assessment of the added value of improvements. In the retail simulation, both the age of construction and, most importantly, reference to the last date the improvements were structurally refurbished were clear. This was not as clear in the residential simulation, as the improvements on the land of the most recently refurbished improvements, being 10 years old, were the only indication of the age of improvements.

An analysis of the standard deviations of the three sales in the initial simulations of both the residential and retail land uses, shown in Table 4.21, highlights that the least developed of the three sales produced the lowest standard deviation. It is further highlighted that the majority of valuers chose 20 Fiction Street and 20 Main Street in the initial simulations as the most relevant sales.

Table 4.21 Residential and retail standard deviation comparison

	Residential simulations	*Initial Stdev*	*Revised Stdev*	*Revised Stdev*	*Initial Stdev*	*Initial retail simulation*
Codified	10 Fiction Street	17.0	10.3	4.44	10.18	15 Main Street
	15 Fiction Street	16.4	12.6	6.52	9.79	5 Bank Road
Uncodified	20 Fiction Street	11.2	13.0	5.97	8.19	20 Main Street

Summary

The results from these simulations demonstrate two important points. The first point is that improved sales can be used in the determination of land values in highly urbanised locations, in conjunction with vacant land sales when these are available. Secondly, when a codified approach is used to analyze improved property transactions and account for the added value of improvements, the consistency of results in deducing the underlying value of land is improved.

As highlighted in this section, the codification of value and its deduction is greatly assisted and improved with information relating to three key areas, which include income and rental information from property which is highest and best use, details of the size of improvements and finally the age of the improvements which are maximally productive. A review of the survey results next provides further clarification and strengthens these three key points.

The definition of value is a well-articulated concept within the property profession; however, within taxation the meaning of value is subsumed under the principles of good tax design, particularly in assessing land tax.

In summary, the value determined and used to assess recurrent property tax is an artificial construct resulting from a manufactured process in the absence of vacant land sales.

While requiring some resemblance to market value as defined in *Spencer v Commonwealth 1907*, the standard defined state of value and its manufacture is the key to an economically efficient recurrent land tax in Australia. This brings to the fore the importance that all bases of value are assessed on the same footing and, more specifically, that all land or property, in the case of capital improved value or assessed annual value, are assessed on the same footing, that is highest and best use and not existing use.

In the first instance, it has been observed that the success of taxing land on its highest and best use depends largely on the valuation practices adopted (Gaffney 1975; Hudson 2008; Oates & Schwab 1997). If land value is to remain the basis of recurrent land taxation, it will be necessary to ensure that valuers firstly define what constitutes the land's highest and best use. Once this has been determined and made transparent, the added value of improvements can be determined in a simple and transparent manner and, most importantly, improve the economic efficiency of the tax. A framework for determining the highest and best use of land, therefore, has the potential to facilitate the application and harmonisation of a recurrent tax on land within and across jurisdictions of Australia.

This study further highlights the fact that the highest and best use principle and, in particular, the determination of a structured and transparently codified valuation approach render any basis of value relevant in assessing the property tax, with one important proviso: that the approach adopted is consistently determined by all valuers within and across jurisdictions and is supported by market evidence of which all evidence is selected on the same basis and analysed in a consistent manner. Where land or site value is the basis of the tax, this then extends to a consistent basis of determining the added value of improvements which are maximally productive, of which reference to cost new is an acceptable proxy for value.

The primary rationale argued for land over other bases of value is that improvements are accounted for in the sales analysis process when valuing land. This is in contrast to including improvements in the tax base and hence attempting to communicate to the taxpayer that CIV is not what is on their land, but what should be on their land where improvements are not maximally productive. It may be further argued, however, that CIV is as economically efficient as land in highly urbanised locations where a Physically Defined Standard State (PDSS) of improvements is determined within a location.

It was shown in the simulations in this chapter that improved sales was the basis on which land value was determined. This is, in fact, not exceptional to find in many locations within the capital cities of Australia. What has not been apparent in the valuation process is that valuers deduce land values from sales with some form of improvements in more 50 per cent of cases.

It is further shown in these simulations that, when improved sales are used where improvements are new or near new and accounting for depreciation is nominal, greater consistency is achieved in the valuation process. This study raises the questions as to whether the economic efficiency of taxing land can be transposed to CIV as the base of the tax where improvements which are maximally productive (highest and best use) are used.

5 Principal place of residence

Introduction

This chapter focuses on the principal place of residence (PPR), its exemption from state land tax and factors impacting the fiscal reform of sub-national government directly resulting from this exemption. The revenue expended on the principal place of residence and the increasing need for the reform of local government rating are two of the greatest challenges confronting sub-national government in Australia. It was highlighted in Table 2.4 that revenue from recurrent land tax is low compared with advanced OECD countries of which the principal place of residence is the primary factor impacting revenue from this source in Australia.

Australia imposes recurrent land taxes on the principal place of residence at less than half of that applied to property across the United States and Canada. New Zealand does not impose a recurrent land tax beyond local government rating, with the collection from rates accounting for 2.1 per cent of GDP, which is well above the tax revenue collected across Australia from both state land tax and local government rating. Australia has much work to do in reforming its recurrent land taxes and ensuring that this tax source adequately contributes to tax revenue collected across the country. The principal place of residence is front and centre of tax reform in Australia; however, it requires careful and progressive transitioning as it is an emotive and sensitive reform.

While focus has centred on increases in utility charges as an alternate source of revenue from property, the taxation of land has languished and is often resigned to the too hard basket of government. Salience and saleability of land tax have impacted revenue from this source and now contribute to the rationale for reform. This is coupled with a clear vision as to how such reforms are to be operationalised and, more importantly, to what extent such reforms are to apply. The states are heavily dependent on revenue from conveyance stamp duty and while lending for housing increases and profits are capitalised into the banks' profit margins, little is done to capitalise profits into the community by properly taxing land.

Windfall gains are privatised by banks through interest rates rather than being channelled back to the public through fiscal revenues.

Following the fallout from the global financial crisis, to which governments provided statutory sovereignty and guarantees for financial institutions, government now has a duty to repay the taxpaying public by properly taxing the asset at the centre of this crisis, residential property. While much focus has centred on low doc loans, the underlying asset on which loans were written was housing. Governments of all persuasions should by now have learnt that privatising the profits through rouge lending and socialising the losses to shore up the banking and lending sector using taxpayer funds requires funds to be returned to the taxpaying public through fiscal reform. Tax policy is every bit as important as monetary policy in both moderating housing markets and raising revenue from the primary creation and store of wealth achieved from this undertaxed asset.

It would be ideal for states to continue to collect revenue from conveyance stamp duty and utility charges on property and for local government to increase their share of revenue from user pay charges such as parking fines, parking metre collections and charges imposed for issuing of resident parking stickers among other user charges. Rather than taxing property annually and allowing this to be factored into the price of housing, government allows this to be collected by the banking system through interest paid on loans used to service the purchase of property.

A taxation standoff across the three tiers of government in Australia exists for the tax revenue expended on the principal place of residence. Of all the potential opportunities for revenue realignment of state taxes, land tax is the preferred option and least distortive of the three taxes expended. While significant latitude exists for prolonging payment of capital gains tax on the principal place of residence if it became liable under the existing rules defining a deemed disposal, a recurrent land tax does not suffer the same distortion. Though the principal place of residence is a prime exempt tax source, neither the states or Commonwealth desire to occupy this important tax revenue void. For some local governments with capacity to increase rating revenue from property this is not the preferred option either.

It was noted in Chapter 1 that Australia has a highly centralised tax system, referred to as fiscal federalism. This allows tax revenue equalisation by the Commonwealth, which collects over 80 per cent of all taxes to be granted to sub-national government through the Grants Commission. While this is a perceived strength, the weakness of this approach is that sub-national government (particularly the states) would prefer the Commonwealth to continue the tax collection burden and pass additional tax revenue on to them. Some merit exists for this approach, as the states are highly focused on the provision of most services and infrastructure which the taxpaying public often does not identify with.

Sub-national government would prefer to see further expansion of consumption taxes in the form of increased GST revenue. Another alternative is for the Commonwealth to collect additional income tax revenue and pass this directly on to the states. The option of taxing consumption is taxing land or capital; however, if this revenue can be collected by the Commonwealth and granted to the states and local government, this would suit sub-national government rather than reforming their own limited tax bases, which collectively account for less than 20 per cent of the total tax collected in Australia. This subject will be addressed again under fiscal reform of sub-national government in Part 4.

A further factor impacting homeownership and residential property investment is Australia's aging population. As the retirement age of Australians moves further away by extending the qualifying age for the pension and the corresponding qualifying age for gaining access to structured superannuation funds, property is the default option for many Australians. Storing as much savings as possible in the home, both at the point of purchase and again through renovations and extensions over the holding period, will be the new sport for many Australians going forward. The purchase of a second property, either as an investment and rental return or as a place of retirement, will also become another savings option for many Australians.

Unstructured superannuation through homeownership, which is capital gains tax and land tax exempt, is a serious option. Further, by purchasing the retirement home and using this second property as a weekender/ holiday home is an increasingly important option and store of wealth. An emerging pattern of wealth creation through investment in the principal place of residence is noted across the capital cities of Australia, in particular Sydney, Melbourne and Brisbane, and has come to the attention of taxing authorities. This trend is defined further in the following example, which sets out the increasing trend of owning a second property primarily used as a second residence.

In demonstrating this approach in a hypothetical example, Couple A own their home in the city near their place of work, which they purchased 20 years ago and have paid off the mortgage. They also own a second home where they will eventually retire and currently use this property several times a year as a getaway; this property was purchased seven years ago and the debt is now low. The second property is not rented and hence does not have a rental bond lodged. All utilities are in the name of the owners, who are registered to vote in the electorate of their coastal property. The second property has a higher land value than their home in the city and attracts land tax and CGT on disposal. By moving into the property down the coast and commuting to work, their property in the city, with the lower land value, either avoids land tax or attracts a lower land tax burden.

This further raises the impact of the basis of value used across the states to assess state land tax in particular. The fast-emerging war in housing from a recurrent land tax perspective is units versus houses. This matter is addressed in Chapter 12 in more detail under the reform agenda. It warrants a brief mention here as taxpayers can clearly see that the land value component attributed to a unit is a fraction of that attributed to a house, which results in tax engineering where more than one property is owned. In a broader sense, it highlights the outdated two-tier government structure for imposing recurrent land tax in Australia and further questions the bases of value on which the tax is assessed.

The following sections of this chapter review the operation and administration of the principal place of residence exemption from state land tax in each state. They demonstrate the inordinate land tax framework needed to maintain the significant carve-out for the principal place of residence. Rather than taxing the principal place of residence, an extensive legislative framework and administrative oversight is needed to manage this inefficient exemption which adversely impacts revenue collected from this source.

Data matching or reform

The increasing benefits associated with home ownership and the principal place of residence exemption from any tax will continue to become more evident over time. Under the current regime, the need to police the status of this exemption has become a priority for taxing authorities across all tiers of government. Auditing this exemption has resulted in resources being allocated to determine the bona fides for both state land tax and capital gains tax exemptions. Auditing entails the cross referencing of data at both the macro and micro level. In one state, the severity of the problem associated with holiday homes being declared as the principal place of residence is expected to net $90 million dollars over the next three to four years (IPTI 2013a).

This audit will entail over 27,000 homeowners being contacted by a new tax audit unit at a cost of $10 million. The detection of owners declaring property as their principal place of residence was picked up during the 2006 and 2011 census, of which 11 of the 20 highest vacancy rates in Australia were located in one state during the 2011 census. Some beachside suburbs during the 2011 census recorded vacancies as high as 72 and 82 per cent. In the 2006 census, the vacancy rate of coastal and beachside towns was 42 per cent, with 24 towns having vacancy rates above 50 per cent.

While examples have been provided in the state of Victoria, other states with high vacancy rates at the census date are indeed also on the radar of taxing authorities. The level of sophistication in detecting these cases has improved and will need to improve further; however, other proactive measures may be adopted, which are used abroad and are addressed in Part 3 in the review of land tax practices in international jurisdictions. In addition to the use of census data, utility usage data is another form of monitoring household consumption. The monitoring of water and power usage is an important tool in detecting and profiling property occupation and usage.

There are a number of practices available and further data-matching measures which could become available in the detection of land tax avoidance. The relevant measures depend on the nature of the avoidance which may fall into a number of different categories, of which some would further breach Commonwealth income tax laws. The three main categories impacting tax laws are as follows:

1. Property owned and declared as the principal place of residence where that property owner does not own another property; however, the property owner resides elsewhere either within or outside the state or Australian boarders.
2. Property owned and declared as the principal place of residence where that property owner owns another property within Australia in which they reside.
3. As per 1 or 2, above also collects rental or occupancy income from the property which they declare as their principal place of residence, which sits outside the permitted income provisions of the relevant Land Tax Management statutes.

The relevant measures used to detect these cases will also depend on a number of different circumstances pertaining to the number of years the property has been held and the level of finance, if any, against the property. The primary audit measure for detecting whether a property is owner-occupied or rented is the auditing of rental bonds lodged on both a property and owner/renter basis. The first test is to determine if the property being audited is in fact rented and has a rental bond lodged. The second is to note whether any rental income has been declared from the property in the relevant taxing year(s).

The next test is the registered address of the owner, of which one source would include a rental bond lodgement with the owner as tenant in another property either within the state the property being audited is located in, or inter-state. In some circumstances, which are addressed in the review of each state in the sections which follow, this may be permitted and legitimate where the property owner's employment necessitates such arrangements and, in relevant states where provisions are made for such cases, the owner may state their case.

In no particular order, some of the measures which may be used to determine the principal place of residence include but are not limited to:

1. Drivers licence details.
2. Travel log details (through the use of travel smart cards used on public transport).
3. Electoral role addresses (including spouses or de facto partners).

4. Name and address registered for rates and particularly utility charges (of which fixed land line telephones are becoming less relevant with progressive transitioning to mobile usage as a primary source).
5. Average water and power usage (commonly used to determine under and over occupied property).
6. The delivery of mail or conversely the re-direction of mail.

The need for fiscal reform

The question arises as to why a property owner would attempt to declare a property they were not living in as their principal place of residence, particularly where the property is vacant and a rent is foregone. The answer to this question is income derived from capital gains, of which land tax captures this incremental benefit and is levied on an annual basis. The opportunity cost is the rent foregone for the residence; however, where the property has no or low mortgage debt and is not rented there is little incentive to rent the property where it is held as a lifestyle asset. Maintaining it as a lifestyle asset is the option of choice, particularly where the owner resides elsewhere and the subject property is the only property owned. Where two properties are owned, the election of the PPR is being determined by some taxpayers to be the property with the highest land value or fastest increasing land value in some cases.

The rationale for some owners comes down to return on investment, to which the net return of residential property may be as low as 3 per cent while the growth over a 10 year investment period is as high as 8 to 9 per cent per annum compound. The return is in the capital gain, rather than the rental return, of which the latter is further eroded once accounting for land tax, which is not deductible if the property is not rented, available for rent or income producing. This demonstrates the archaic state land tax rationale which provides an exemption for the principal place of residence while rate pegging and capping the rates of local government authorities. The outdated land tax rationale constitutes a fiscal void, costing subnational government billions of dollars in lost recurrent land tax revenue. The structural flaw of raising billions of dollars in revenue from inefficient mobility taxes (conveyance stamp duty) while stifling local government rate revenue has no place in a modern Australian tax system.

This chapter now examines the technical provisions governing the administration of the rating and taxing of land as they apply to the principal place of residence. The earlier part of this chapter identified the challenges and emerging erosion of revenue from this source as well as the response from administrators in defining the principal place of residence. The following sections review the statutory mechanism used to define what constitutes the principal place of residence in each state and in what circumstances this exemption applies. The objective of this review is to show the broad alignment of this exemption across the states and taxing jurisdictions, and it is not an exhaustive account of the application of these provisions. It does, however, define the extent of the intention through the diversity of provisions to detach the principal place of residence from the impacts of a tax assessed on value.

It is apt at this point to articulate the origins of the exemption of the principal place of residence and the differential status afforded in the taxation of this asset by state and local government. It is afforded no exemption from council rating, with the exception of pensioner concessions, while for state land tax purposes it is the largest exemption with the highest tax expended by state government from this revenue source. It is important, however, to put this distinction into perspective and revisit the rationale for the origins of this exemption. When the states relinquished income tax to the Commonwealth, they reintroduced land tax progressively during the late 1940s and 50s, a tax that the Commonwealth vacated. Residential land was initially to be excluded from state land tax altogether at the request of local government; however, a compromise was reached and the states provided an exemption for the PPR and soon after progressively introduced a threshold for holders of property that sat outside the PPR exemption.

The introduction to this chapter impresses that significant tax revenue is expended through the principal place of residence exemption from state land tax. The more accurate distinction is that recurrent land tax across all tiers of sub-national government is low and that significantly more revenue could and indeed should be raised from residential property. The fundamental questions that remain unanswered are: how this is to be achieved, from which tier of government additional revenue should be collected and which tier of government should be the beneficiary of increased revenue from this source. In New South Wales, local government is either politically or statutorily restrained from maximising revenue from this source and further shelters many residents by applying lower rates to residential property over business use property. What has hindered sustainable reform of recurrent land tax and revenues is the ability to measure the property owner's capacity to pay, of which this capacity is determined on the price paid for their land and its subsequent movement in value.

New South Wales – defining the principal place of residence

The principal place of residence (PPR) has undergone a number of changes in relation to land tax with its inclusion and removal from the states' land tax net over the past two decades. Prior to the 1998 land tax year, the principal place of residence was largely exempt from land tax. Prior to 1998, a 2,100m^2 area threshold existed to which land over and above that land area was subject to land tax. However, there were further provisions that exempted property used as a PPR with a land larger than 2,100m^2, where the land was not subdivisible under a planning instrument, in which case the whole property remained land tax exempt.

In 1998, the land tax net was expanded to include a tax on the land of the principal place of residence based on the land's value, which commenced with land values above $1 million. The Premium Property Tax Act (1998) was introduced to achieve this objective. The Valuer-General had the role of determining the number of dwellings to be assessed for the premium property tax (PPT) at no more than 0.2 per cent of occupied private dwellings under sections 10–12 Premium Property Tax Act (1998). This was undertaken by adjusting the threshold accordingly to ensure no more than 0.2 per cent of private occupied dwellings attracted the tax.

At the time the PPT was introduced in 1998, the threshold for the land value was significantly higher than the threshold for investment and other property, which was $160,000. The 'millionaire tax' was born to the dissatisfaction of those it affected. Having been caught out by the amendments to land tax, those affected found that, on parity, it was not actually a millionaire's tax, as unlike land tax paid on investment property, the tax paid on the principal place of residence was not a deductible expense. Therefore, the land tax on the principal place of residence was paid with post-taxed dollars, as opposed to investment property, in which land tax was paid with pre-taxed dollars.

> For example, for the 1998 land tax year, those affected by the PPT paying income tax at 48.5 cents in the dollar, their equivalent threshold for the PPT was in fact $515,000. That is: [$1,000,000 x (100 cents – 48.5 cents) / 100].

The premium property tax netted a small percentage of the total revenue raised from land tax. In 2003, the total revenue raised from land tax in NSW was $1,136 million, of which approximately $13 million was raised from the premium property tax, approximately 1.15% of total land tax revenue in that year. In the 2004 Mini Budget, the Premium Property Tax Act, was abolished onwards from the 2005 land tax year. Table 5.1 sets out the history of the principal place of residence since the reintroduction of land tax by the NSW in the early 1950s.

The removal of the land area threshold for the principal place of residence in NSW and its absence in other states has largely resulted in land banking on the fringes of the capital cities of Australia. This is particularly the case in (Sydney) New South Wales and (Melbourne) Victoria, where an absence of a land area threshold has allowed land ripe for amalgamation and subdivision to be used for less efficient purposes which no longer reflect highest and best use. Table 5.1 sets out the evolution of the area and value thresholds applicable to the principal place of residence since its reintroduction in 1956.

Unity of title (four unities) rule

The removal of the principal place of residence land area and value thresholds has elevated the status of the principal place of residence, a status

Table 5.1 Evolution of land tax and the PPR

Date	*Change*
1956	Introduction of the Land Tax Management Act in NSW. Principal place of residence exempt, no area or value threshold.
1973	1,200m^2 threshold imposed on the PPR to encourage subdivision of larger estates as demand on land increases.
1987	Area threshold expanded to 2,100m^2 subject to land not being sub-divisible under a planning instrument.
1998	Premium property tax introduced and the 2,100m^2 threshold replaced with $1,000,000 land value threshold.
2003	High Court of Australia establishes the recognition of a Scarcity Factor and orders Land & Environment Court to consider additional sales evidence in *Maurici v. Office of State Revenue* [2003] HCA 8. Land & Environment Court assesses Scarcity Factor at nil in that case.
2005	Premium property tax abolished with no value or land area threshold reinstated or applicable up to and including the present time.
2009	Unity of title and use of land rule was used in conjunction with provisions in the Valuation of Land Act 1916 to demonstrate that parcels of adjoining land to a principal place of residence owned and used are considered to constitute the one principal place of residence.

enjoyed across all states of Australia. Subject to the prevailing statutory provisions within the relevant state's Valuation of Land legislation and Land Tax Management legislation, it is the use of land rather than the strict nature of the residence on the land that dictates what qualifies as the PPR. The *unity of title rule* defines the principal place of residence status by defining the meaning of lots and parcels in the use of residential land. This was defined in *Triguboff v Valuer General* [2009] NSWLEC 9 (13 February 2009) in which the court ruled in favour of the applicant in confirming the appeal against the Valuer-General, of which extracts from this case follow:

> In this case, the applicants, are owners as joint tenants of the property comprising two adjoining lots with a combined area of 5,217 square metres. Immediately prior to its consolidation under one title by a plan of consolidation registered in May 2007, the Property comprised adjoining blocks. These blocks were approximately 3,952 square metres for House number 62 and 1,265 square metres for house number 64. The applicants have owned and lived in the house at No 62 (**House 62**) since 1984. They have owned and used the house at No 64 (**House 64**) since 1998.
>
> The respondent in this case, the Valuer-General assessed the land value of the Property as at the base date 1 July 2007 on the basis that it comprised two parcels. Records of the valuation notices tendered in evidence described one parcel as No 62 and its land value was assessed at $15,000,000. The other was described as No 64 and its land value was

assessed at $9,700,000. The applicants lodged objections stating as the reason for objection that the 'Land Should Be Valued Together'. On 13 August 2008 the objections were disallowed. The applicants appealed to the Court under s 37 of the Act, contending that they were dissatisfied with the Valuer-General's decision to disallow their objection to the Valuer General's determination that the adjoining lands at No 62 and No 64 be included in one valuation.

The agreed issues are whether the Property comprises one parcel; and, if it comprises two parcels, whether s 26(1) of the Act applies to require one valuation. If either of those questions is answered in the affirmative, the appeal must be allowed. The scheme of the Act imposes under s 14A(1) a primary obligation on the Valuer General to value each parcel of land each year. In the present case the Valuer General has valued the Property as if it comprised two parcels.

Unification of use and occupation

Immediately following their purchase of No 64 in 1998, the applicants undertook extensive renovation works to the existing house located there (House 64) and to the grounds. The renovations were completed by about the beginning of 1999. The renovations to House 64 were for the applicants' specific purposes and configured to suit their needs in their use and occupation of the Property as a whole. The boundary wall dividing No 62 and No 64 was removed – but the retaining wall along about 60 per cent of the former boundary remains in place because the ground level is higher on one side than the other – and sandstone steps linking paving leading to House 62 and House 64 were constructed. The gardens of No 62 and No 64 were landscaped to form one continuous garden for the Property and also to incorporate a putting green built on the site of the tennis court on No 64.

The linkages between House 62 and House 64 comprise common landscaped gardens (shrubs, lawn areas and trees) and three interconnected pathways for easy access from one house to the other. At least one of the pathways is wide enough to accommodate two persons walking side by side. The main driveway to, and the car garaging facilities for, the Property are located at House 62. The double garage located on No 64 is used for the storage of garden tools and to house a large bore-water tank which feeds the watering system installed on the Property.

In the upstairs level of House 64 are two bedrooms, one with an en-suite. These are used from time to time to accommodate friends and family. A large walk-in wardrobe in the second bedroom is used for storage of tables, chairs and other equipment which the applicants use for functions downstairs. The third upstairs room was converted into a studio or study which the applicant uses on the weekends. The applicant also has a small study in House 62 that is usually uses on weekday evenings when he returns from work. However, he prefers to use the study at House 64, especially during daylight hours, because it is lighter and has a better outlook than the study at House 62. He has a television set in his study at House 64 and spends a lot of time there, especially on weekends. There is no television set in the House 62 study. The applicant exercises in the indoor swimming pool which the applicants installed in House 64, usually two to three times a week. House 62 has no indoor swimming pool.

The applicants entertain on a regular basis a large number of family and friends. At many of the festive occasions that they celebrate annually they can have up to 40 family and friends present for a seated function. House 62 does not accommodate these types of functions. Since at least 1999 the entertainment area in House 64 has been used for their annual large family celebrations.

The question whether the Property is one parcel or two parcels is a question of fact. No 62 and No 64 are united in title, physically united and united by use and occupation. There are no distinguishing structural or geographical features that would indicate that there has been any degree of separation effected by the owners in using the Property. To the contrary, the owners have taken steps to physically unify No 62 and No 64. The whole Property is fenced as a single entity. The owners have removed the fence between No 62 and No 64 and have connected them by pathways, steps and a common garden area. Neither by established usage nor by regard to whether there is a reasonably well defined area of land nor by use of any current plan or map can it be said that there is more than one parcel. Because there is no dividing fence or other physical severance of No 62 and No 64, and paths and, in substantial part, the garden, flow across the former boundary, there is difficulty in determining with reasonable precision whether there is any boundary between them and, if so, where it is. That uncertainty is reflected in the Valuer General's submission that the boundary is either the outside of the path running along the southern side of House 64 or the retaining wall between No 62 and No 64 extended by a notional line running to the Harbour.

The whole Property was and still is used for domestic purposes. At No 64 the applicants converted the ground floor of House 64 into a large open entertaining area with related facilities; built an indoor swimming pool and putting green; and installed a large bore-water tank in what used to be the garage to feed the Property's watering system. The garage on No 62 is used as the garage for No 64. House 64 is regularly used for entertaining and guest accommodation and its indoor pool and study are also regularly used. It is a small point, but no privacy is maintained between House 62 and House 64 in that the windows in House 64 facing House 62 have no curtains or blinds, in contrast to the windows on the other side of House 64 facing a neighbouring property.

The use of No 64 in various ways, whether viewed singly or as a totality, is not a separate and independent use, occupation or enjoyment divorced, or for a purpose that differs, from the use of No 62. No 64 and its facilities, including those created and changed by the applicants, are an integral part of the domestic use and enjoyment of the Property as a whole for domestic purposes by the applicants, their family, friends and guests. They are incidental to the use of No 62. It makes no difference if the Property had all originally been purchased as one or whether (as occurred) No 64 was subsequently added to the adjoining No 62 for the purpose of enjoyment of its amenities.

In satisfying the 'four unities test' established in *Ryan & anor v Commissioner of Land Tax* [1982] 1 NSWLR 305, adjoining property must meet the test of four unities comprising, physical, use, occupation and title. It was determined by the court, that the appeal must be allowed in the above case as the property was considered to constitute one parcel under s26(1) of the Valuation of Land Act 1916.

Principal place of residence

The abolition of the Premium Property Tax Act from the 2005 land tax year would infer that the principal place of residence has regained its land tax exempt status; however, this is not necessarily the case. For some owners, land tax may still apply. Before discussing this further, it is important for all owners of their principal place of residence in New South Wales to assume that their property may attract land tax unless they qualify for a Principal Place of Residence Exemption. Whilst many owners will qualify for this exemption, some may not and may not learn of their obligation for some time. More detail on this exemption will follow; however, if in doubt about the Principal Place of Residence Exemption, property owners should consult their taxation advisor.

A number of provisions are covered under Schedule 1A – principal place of residence exemption (Section 10(1)(r)) of the Land Tax Management Act 1956, which sets out the application of the principal place of residence exemption. The following sections of this chapter cover some of the provisions of the exemptions and qualifications that apply.

Part 2 – principal place of residence exemption

SCHEDULE 1A PART 2 – LAND TAX MANAGEMENT ACT 1956

In order for a property to qualify as a principal place of residence, it must be a parcel of residential land, a lot under the Strata Schemes (Freehold Development) Act 1973 or a lot under the Strata Schemes (Leasehold Development) Act 1986.

The residence must be occupied continuously by the owner from 1 July in the preceding year to which the land tax is levied. Occupation of one's residence as of midnight on 31 December in a given year is not the test for determining the status of the property as a principal place of residence.

No other land is to be continuously used and occupied by the person for residential or other purposes from 1 July in the year preceding the tax year in which land tax is levied.

Part 3 – concessions in application of a principal place of residence exemption

CONCESSION FOR LAND USED FOR INCIDENTAL BUSINESS PURPOSES

For the purposes of the principal place of residence exemption, where land owned by a person is used and occupied by the owner primarily for residential purposes and no more than one room is used primarily for business purposes, the use of the land for business purposes may be disregarded if the business is primarily conducted elsewhere.

In simple terms, if a business is run from the PPR and does not operate from any other place, the principal place of residence exemption will only apply to the portion of the residence used as the PPR. Parties need to consider the viability of running a business from their home against the land tax liability it may raise.

The impact of taxing or partially taxing the PPR provision may be significant for the self-employed and some tradespeople, contractors and service providers who primarily run a business from their home and from no other place. A registered business name, company or Australian Business Number (ABN) at a principal place of residence may give rise to the first limb of a test. An owner using their PPR for business purposes who is challenged may need to prove that either an ABN, company or business name registered at their residence is incidental to the business operating elsewhere.

In accordance with Revenue Ruling No LT 020 dated 8–11–1989, where land tax is payable the tax payable is then reduced by the proportion of the building used as the owner's principal place of residence.

Table 5.2 Application of principal place of residence exemption

Circumstance	*Is land tax applicable?*
The ABN or business name is held by the owner, is not in use and there is no business being operated or conducted at their residence or other place.	No
There is only one room in the residence primarily used for business purposes and the business is primarily conducted elsewhere.	No
A business is being conducted from one room in a residence and from no other place.	Yes
A business is being operated from more than one room in a residence and is primarily operated elsewhere.	Yes
A business is being operated from more than one room of a person's residence and from no other place.	Yes

CONCESSION FOR UNOCCUPIED LAND INTENDED TO BE AN OWNER'S PRINCIPAL PLACE OF RESIDENCE

A concession may be granted where the Chief Commissioner is satisfied that the owner of unoccupied land intends for that land to be the used as their principal place of residence. The application of this concession applies where either:

1) two tax years immediately following the year the owner became the owner of the land; or
2) if the land is used by a person other than the owner two years immediately following the tax year in which the building or other works necessary to facilitate the owner's intended use and occupation are physically commenced on the land.

The Chief Commissioner may extend this period in the case of delay for reasons beyond the control of the landowner. This concession does not apply unless the Chief Commissioner is satisfied that the land is currently unoccupied because the owner intends to carry out building works to facilitate their occupation of the land as their PPR and that no income is derived from the use or occupation of the land since that commencement.

The PPR exemption is revoked in the event that the owner does not continue to use and occupy the land as their principal place of residence for at least six months immediately after the two-year period in which the exemption was claimed. In these cases, land tax will be assessed or reassessed accordingly. This concession does not apply if the person or any member of their family (as defined in the Land Tax Management Act) is entitled to have their actual use and occupation of other land taken into account. The exclusion of this concession also applies where the person or any joint owner of the land owns land outside NSW that is their principal place of residence.

CONCESSION FOR SALE OF A FORMER PRINCIPAL PLACE OF RESIDENCE

Relief has been provided to owners who have been caught buying and selling their principal place of residence where they hold both properties at midnight on 31 December in a given year. For an exemption to apply, there are two key criteria:

1) The owner must dispose of the former residence within the first six months after 31 December of the relevant year; and
2) There cannot be a lease over either of the two properties.

CONCESSION FOR ABSENCE FROM RESIDENCE

A person may live away from their principal place of residence for up to six years and claim a principal place of residence exemption based on the following criteria:

1) They occupied their residence for six months prior to the six-year exemption period; and
2) The owner must reoccupy their residence immediately after the six-year exemption for a minimum period of six months.

Under this provision, income may be derived from the use or occupation of the residence if:

1) The income is derived from a lease, license or any other arrangement under which a person has a right of occupation that does not exceed any more than six months in the tax year; or
2) The income is derived from any arrangement under which a person occupies a former residence, but the income is no more than is reasonably required to cover water and energy rates and maintenance costs of the owner in keeping the residence.

This clause is subject to the provisions of the Members of the Same Family rule discussed later.

CONCESSION ON DEATH OF AN OWNER

In the event of the death of a person, their principal place of residence will not attract land tax until the first of the following two events:

1) Twelve months after their death; or
2) The deceased's interest in the land vests in a person other than the deceased's personal representative.

Part 4 – restrictions

EXEMPTION DOES NOT APPLY TO LAND OWNED BY COMPANIES AND TRUSTEES

Land is not exempt from taxation under the principal place of residence exemption if:

> the land is owned by a company, unless the land is owned by a trustee company acting in its representative capacity or a company acting in its capacity as trustee of a concessional trust or the owner of the land, or each of the joint owners, who use and occupy the land as a principal place of residence is an owner only by reason of being a trustee, or the land is owned, or jointly owned, by a person who is a trustee acting in the person's capacity as trustee of a special trust.
>
> Land that is owned by a company acting in its capacity as trustee of a concessional trust is taken to be used and occupied as the principal place of residence of the owner of the land only if the person, or one of the persons, who used and occupied the land is a person who is a beneficiary of the trust.

ONE RESIDENCE ONLY FOR MEMBERS OF THE SAME FAMILY

Under the principal place of residence exemption, only one place of residence may be treated as the PPR of all members of the same family. Where the one family owns more than one residence jointly or separately, they may elect which property is to be treated as their PPR. An election may be made up until the last date for an objection. In the event a property is not elected, the Commissioner is to treat the residence with the highest land value as the PPR.

Within the meaning of the Property (Relationships) Act 1984, the capacity for spouses and de facto couples to each claim a different residence as a principal place of residence no longer exists, with only one residence claimable as the principal place of residence by members of the same family, which include dependents under the age of 18 and not legally married.

In *McNally and Anor v Commissioner of State Revenue* [2003] NSWSC, the Supreme Court heard a case whereby a retired person living four days per week with their spouse in Sydney and three days per week in a property they also owned on the central coast did not discharge the onus of proof that they had relinquished their principal place of residence in Sydney. The court commented that, had the party spent four days per week at the Central Coast property and three days at the Sydney property, the result might have well been different. With the tightening of the 'same family rule', only one residence per family will now receive the Principal Place of Residence Exemption, avoiding cases like this from slipping through the net.

If the Chief Commissioner is satisfied that a person is legally married but not cohabitating with their spouse and has no intention of resuming cohabitation with their husband or wife, the person is not regarded as the spouse of that other person.

New South Wales – council rating

A concession for local government rating is available to owners qualifying as eligible pensioners, as defined under the Act, of which the definition follows:

> **'eligible pensioner'**, in Division 1 of Part 8 of Chapter 15, in relation to a rate or charge levied on land on which a dwelling is situated, means a person:
>
> (a) who is a member of a class of persons prescribed by the regulations; and
> (b) who occupies that dwelling as his or her sole or principal place of living.

A number of concessions exist under Part 8, Division 1 – concessions for pensioners of the Local Government Act 1993, sections 575–584 cover matters of reductions in rates with the maximum amount rates may be reduced being set out in section 575. Further applications for postponement of rates exist where for that component of rates which are still payable. Eligibility for concession and postponement are further set out under these sections.

Queensland – defining the principal place of residence

Queensland undertook a substantial revision of its land tax statutes, which were introduced in 2010 and replaced the outdated provisions of the Land Tax Act 1915. The Land Tax Act 2010 is well-structured and is set out logically under Parts and Divisions. It has a comprehensive Dictionary and Schedules that set out the rates for the various holding structures of land. *Part 6 – Exempt land, Division 1 – Home* addresses the principal place of residence exemption under sections 35 to 45. The provisions for this exemption are further set out under the LTA000.1.2 Public Ruling, which has a structured means test. A précis of the key provisions applicable to land which qualifies for this exemption follows.

Section 35: provides an explanation of the operation of the overriding provisions for the application of this exemption which broadly addresses the land owning entity, residency and occupation of the owner and an overview of what constitutes an exempt purpose.

Section 36: sets out what constitutes the use of the home of a person and the Land Tax Ruling LTA000.1.2 further defines this section under three tests. The first is the main test, the second is a deeming test and the third is a residual test, to which the Act and the Ruling define that land is considered to be a home if one of these three tests are satisfied.

Main test

The main test in section 36(1)(a) of the Land Tax Act applies to where only one parcel of land is used continuously as a person's residence for a six month period, to which two conditions apply for the land to qualify as the person's home. The first condition is that the land must have been used for six months, ending on 30 June when the liability arises for the financial year, and that no other land has been used continuously for residential purposes for the same period; this period is also known as the residency period. This test is premised on the notion that 'most people have one residential property occupied as their home'.

Deeming test

The second test is the deeming test and it applies to *section 37 – Land taken to be used as a home – person who receives care*, and *section 38, Land taken to be used as a home – demolition or renovations.* the person being absent during the six month residency period. These sections of the Act are defined in LTA000.1.2 as follows:

> This test deals with the circumstances, outlined in ss.37 and 38 of the Land Tax Act, where land is taken to be a person's home despite the person being absent during the six month residency period. Section 37 of the Land Tax Act provides that land is taken to be used as a person's home in the circumstances where the person is not in occupation because of illness or having to reside elsewhere to receive care. Section 38 of the Land Tax Act provides that land may be taken to be a person's home when the person is not in occupation temporarily because of renovations to the residence or the existing residence has been demolished and a new residence is being constructed on the land.

Section 37 – land taken to be used as a home – person who receives care

(1) This section applies to land, for a financial year, if –
    (a) the person who owns the land received care for all or part of the six-month residency period; and
    (b) the person used the land for a qualifying residential use before the person started to receive care; and
    (c) the person has used the land for a qualifying residential use continuously for a period of at least six consecutive months; and
    (d) subsection (6) does not prevent the person from being taken to use the land as the person's home under this section.
(2) For this section, the person receives care if the person –
    (a) resides at a hospital as an inpatient; or
    (b) receives residential care at a residential care service; or
    (c) resides on other land that they do not own and is under the care of someone else.
(3) The land is taken to be used as the person's home for the financial year.
(4) However, subsection (3) does not apply if income was derived from use of the land during the one-year period ending when the liability for land tax arises.
(5) Despite subsection (4), income may be derived from a lease, licence or other arrangement under which a person has a right to occupy the land, if –
    (a) the right of occupation is for no more than six months in the one-year period; or
    (b) the income is not more than is reasonably required to cover the following –
        (i) rates and other charges levied on the land by the local government for the land; and
        (ii) maintenance expenses for the land.
(6) The maximum period for which the person may be taken to use the land as their home under this section is six years from the end of the

last period of at least six consecutive months during which the land was used by the person for a qualifying residential use.

Section 38 – land taken to be used as a home – demolition or renovations

Land is taken to be used as a person's home for a financial year if –

(a) the Commissioner is satisfied that the person is temporarily residing elsewhere, when a liability for land tax arises for the financial year, because –
  (i) a residence on the land has been or is being demolished and a new residence is being or will be constructed; or
  (ii) a residence on the land is being renovated to an extent requiring it to be vacated; and
(b) the land was used as the principal place of residence of the person at some time during the six-month residency period; and
(c) the person intends to resume using the land as their principal place of residence before a liability for land tax arises for the next financial year.

Residual test

Where the main or deeming tests do not apply, in order for the land to be considered for use as a person's home, the Commissioner must be satisfied that the land is used as the person's principal place of residence (whether alone or with other persons) at the relevant 30 June date.

In deciding whether land is used as a person's principal place of residence, section 36(2) of the Land Tax Act provides that the Commissioner may have regard to the following –

(a) the length of time the person has occupied a residence on the land
(b) the place of residence of the person's family
(c) whether the person has moved his or her personal belongings into a residence on the land
(d) the person's address on the electoral roll
(e) whether services such as telephone, electricity and gas are connected to the land
(f) whether the person acquired the land with an intention to occupy a residence on the land as his or her principal place of residence
(g) any other relevant matter.

Unlike the main test, the residual test depends on the Commissioner being satisfied that the facts support the conclusion that the land is used as the person's principal place of residence. This test must be applied by the Commissioner on a case-by-case basis after considering all the facts and circumstances.

Sections 42A and 42B – exemption for old home after transitioning to current home and Exemption for new home before transitioning from current home

These sections are designed to provide temporary exemption from land tax for the transition phase from one residence to another and are subject to sections 41 and 42. Section 42A provides a temporary exemption for the new residence while the existing residence remains the home in that financial year; conversely, 42B provides a temporary exemption for the old residence while new residence is the home in that financial year. A recent proposed amendment requiring that both premises be located in Queensland was flagged by OSR from 30 June 2014 for this transitional exemption to apply.

The expression 'principal place of residence' is not defined in the Land Tax Act. However, section 36(2) of the Land Tax Act provides a list of factors that the Commissioner will consider in determining whether a residence is a person's principal place of residence. These factors are listed above under the residual test. Where a person resides in more than one place during a year, the question of which place is the person's principal place of residence is one of fact and degree. In these cases, it is necessary to examine the history and circumstances, the purpose for which each property is used, the duration of ownership and the amount of time spent during the course of the year in each.

Land also used for purposes other than a home

Land Tax Ruling LTA000.1.2 also provides guidance as to exemption, partial exemption and no exemption for land used for purposes other than a home as follows.

Where land that is used as a person's home is also used for another purpose, a full home exemption may still be available depending on the extent of the other use. The Commissioner must decide whether the land is being used for a non-exempt purpose. The Commissioner will not consider land as used for a non-exempt purpose if:

(a) the land is used as the home of a person (principal resident) for a financial year; and
(b) when liability for land tax arises for the financial year, either or both of the following apply –
    (i) there is a permitted number of allowable lettings for the land; and/or
    (ii) a person has a work-from-home arrangement; and
(c) the Commissioner is satisfied that land is used only for the purposes mentioned in paragraphs (a) and (b) when a land tax liability arises for the financial year.

Allowable lettings

An allowable letting is one where a person (the occupant) has been given the right to occupy a residential area on the land (the leased area) under a tenancy agreement. The following conditions apply:

(a) the leased area is not more than 50 per cent of the total floor area of all residential areas on the land; and
(b) the leased area is not a residential area that –
  (i) is one of three or more flats in a building; and
  (ii) is not used for residential purposes by the principal resident; and
(c) the leased area is used by the occupant for residential purposes and the occupant has not given the right to occupy any part of the area to another person under a tenancy agreement; and
(d) the rent payable for the leased area is not more than the market rent for the area.

An allowable letting where the occupant is a member of the principal resident's family is called a family letting.

A member of a person's family means each of the following:

(a) the person's spouse
(b) the parents of the person or the person's spouse
(c) the grandparents of the person or the person's spouse
(d) a brother, sister, nephew or niece of the person or the person's spouse
(e) a child, stepchild or grandchild of the person
(f) the spouse of anyone mentioned in paragraph (d) or (e).

The permitted number of allowable lettings for the land is:

(a) one allowable letting; or
(b) two allowable lettings if –
  (i) at least one of the lettings is a family letting; and
  (ii) the total floor area of the leased areas for the lettings is not more than 50 per cent of the total floor area of all residential areas on the land.

Work-from-home arrangements

A work-from-home arrangement, referred to in paragraph 43(b)(ii), must meet the following conditions:

(a) the person doing the work must live on the land
(b) the person must be working from home under an arrangement with their employer
(c) the work must not involve using the land for a purpose or in a manner in which residential land is not ordinarily used.

Deciding if land is used for a non-exempt purpose

Where land is used as both a home and for another purpose the Commissioner may decide that the land is being used for a non-exempt purpose if:

(a) the Commissioner is satisfied that the use of the land for the other purpose is substantial; and
(b) the allowable lettings and work-from-home arrangements do not apply.

In deciding whether the use of land for other purposes is substantial, the Commissioner must have regard to the following factors:

(a) whether a person other than the principal resident has been given a right to occupy any part of the land under a tenancy agreement
(b) whether a person, other than the principal resident or a member of the principal resident's family who uses the land as his or her home, works on the land as an employee or contractor (disregarding work on the land itself or a building situated on the land, such as repairs, renovations and landscaping)
(c) the extent to which a person uses the land or has set the land aside for use, for purposes other than as the home of the principal resident
(d) whether the gross income generated during the financial year immediately before the relevant financial year from business or an income producing activity on the land is more than $30,000
(e) any other relevant matter.

Depending on the circumstances, any one or more of these factors alone may be sufficient for the Commissioner to determine that use of the land for a purpose other than as a home is a non-exempt purpose.

Apportioning the taxable value of the land for a partial home exemption

Where land is used both as a home and for a non-exempt purpose, a partial home exemption is available.

The Commissioner apportions the taxable value of the land between the use of the land as a home and the use of the land for non-exempt purposes, having regard to:

(a) the proportion of the land used for each purpose; and
(b) the extent to which each portion is used for the purpose.

The apportionment method will vary depending on the circumstances. For example, in cases where part of the land is set aside exclusively for the non-exempt purpose, it may be appropriate to apportion on a floor area basis. However, this may not be an appropriate basis where the non-exempt purpose use is not physically separated from the home use. In the latter case, it may be appropriate to consider both floor area used for the non-exempt purpose and the time and extent of use for that purpose.

Victoria – defining the principal place of residence

Exemption for the principal place of residence is found under *Part 4 — Exemptions and Concessions, Division 1* of the Land Tax Act 2005. The definition of what constitutes a principal place of residence, the right to reside and the extent of this exemption are each addressed under sections 53–54 of the Act.

Section 53 – what is a principal place of residence?

A principal place of residence requires the building to be affixed to the land (including a home unit) that, in the Commissioner's opinion:

(a) is designed and constructed primarily for residential purposes; and
(b) may lawfully be used as a place of residence.

In determining whether land is used or occupied as a person's principal place of residence, account must be taken of all of their places of residence, whether in Victoria or elsewhere.

For the purposes of Division 1, if land on which home units are situated is owned by a body corporate, the land is deemed to be owned by the shareholders of the body corporate, who are entitled to exclusive occupancy of the home units.

Section 53A – what is a right to reside?

The Act qualifies that a person has a right to reside on land if –

(a) the right was granted on the death of the person previously occupying the land; and
(b) the right was granted in writing under a will or testamentary instrument; and
(c) the right was not granted or acquired for monetary consideration.

It further qualifies that a right to reside on land does not include –

(a) a right to occupy land as a lessee; or
(b) a right to occupy land as a beneficiary of a discretionary trust or as a unit holder in a unit trust scheme.

Section 54 – principal place of residence exemption

(1) The Act sets out what is defined as exempt land as follows –

(a) land owned by a natural person that is used and occupied as their principal place of residence;
(b) land owned by a natural person who has a right to reside on that land that is used and occupied as the principal place of residence;
(c) land owned by a trustee of a trust that is used and occupied as the principal place of residence of a vested beneficiary in relation to the land;

(1A) despite the above, land referred to in that subsection is not exempt land unless –

(a) immediately before the natural person, who has a right to reside on the land, was granted that right, the land was exempt land under section 54(1)(a)or (b); and
(b) that person is not entitled to –
  (i) an exemption under this Division in respect of any land; or
  (ii) an exemption from land tax under a law of any other State or Territory that corresponds to this Division.

(2) Subject to section 55, subsection (1)(a) or (b) only applies if the land has been used and occupied as the principal place of residence of the owner or vested beneficiary –

(a) since 1 July in the year preceding the tax year; or
(b) if the owner or trustee became the owner of the land on or after 1 July in the year preceding the tax year, since a later date during that year.

(3) In addition to land of an owner that is used and occupied as a person's principal place of residence (the PPR land), land is also exempt land if it is owned by that owner and –

(a) is contiguous with the PPR land or separated from the PPR land only by a road or railway or other similar area across or around which movement is reasonably possible; and
(b) enhances the PPR land; and
(c) is used solely for the private benefit and enjoyment of the person who uses and occupies the PPR land and has been so used –
  (i) since 1 July in the year preceding the tax year; or
  (ii) if the owner or trustee became the owner of the land on or after 1 July in the year preceding the tax year, so used and occupied since a later date during that year; and
(d) does not contain a separate residence.

Section 56 – absence from principal place of residence

(1) For the purposes of this Division, land is taken to be used and occupied as the principal place of residence of a person despite their absence from the land if the Commissioner is satisfied –

(a) that the absence is temporary in nature; and
(b) that the person intends to resume use or occupation of the land as his or her principal place of residence after the absence; and
(c) that, in respect of the period of absence, no other land is exempt land under this Division or under a law of another jurisdiction (whether in or outside Australia) as the principal place of residence.

(1A) For the purposes of this Division, land is taken to be used and occupied as the principal place of residence of a person despite the person's absence from the land if –
(a) the person has lost the ability to live independently; and
(b) the person resides –
(i) at a hospital as a patient of the hospital; or
(ii) at a residential care facility or supported residential service within the meaning of section 76(4), or a residential service within the meaning of the Disability Act 2006, and receives residential or respite care at the facility or service; or
(iii) with another person who provides personal support to the person on a daily basis. Land will not be taken to be used and occupied as the principal place of residence of a person under this section unless the conditions set out.

(4) The condition in this subsection is that the land –
(a) was exempt land under section 54 immediately before the person's absence; or
(b) was used and occupied –
(i) in the case of an owner – by the owner as his or her principal place of residence for a period of at least six consecutive months immediately before the owner's absence; or
(ii) in the case of a trustee – by a vested beneficiary in relation to the land as his or her principal place of residence for a period of at least six consecutive months immediately before the vested beneficiary's absence.

(5) The condition in this subsection is that the owner or trustee has not rented out the land for a period which is six months or more, or for periods which total six months or more in the year preceding the tax year.

(6) The maximum period for which land can be taken to be used and occupied as the principal place of residence of a person despite the person's absence from the land under subsection (1) is six years from the date of the person's absence.

Section 57 – exemption continues on death of resident

(1) If land is used and occupied as the principal place of residence of a person and the person dies, liability for tax is to be assessed as if the person had not died but had continued to use and occupy the land as his or her principal place of residence.

(2) Subsection (1) operates only until the earlier of the following dates –
    (a) the third anniversary of the person's death or the expiry of the further period approved by the Commissioner under subsection (3); or
    (b) the day on which the person's interest in the land vests in another person under a trust; or
    (c) the day on which the person's interest in the land vests in a person (other than the person's personal representative) under the administration of the person's estate.

Section 58 – exemption continues if land becomes unfit for occupation

(1) If land that is exempt land under section 54(1) becomes unfit for occupation as the principal place of residence of a person because of damage or destruction caused by an event such as fire, earthquake, storm, accident or malicious damage, the land continues to be exempt land while the owner continues to own the land as if it had continued to be used and occupied as the person's principal place of residence.
(2) Subsection (1) operates until the second anniversary of the day on which the land became unfit for occupation as the person's principal place of residence.
(3) The Commissioner may extend the period of operation of subsection (1) beyond the period referred to in subsection (2) for a further period of not more than two years in any particular case if the Commissioner is satisfied that there has been an acceptable delay in that case.

Section 59 – purchase of new principal residence

(1) Land is exempt land in respect of a year if –
    (a) a person becomes the owner of the land in the preceding year for use and occupation as his or her principal place of residence; and
    (b) as at 31 December in the preceding year, the person is the owner of other land that he or she uses and occupies as his or her principal place of residence.

Section 60 – sale of old principal residence

(1) Land owned by a person is exempt land in respect of a year if –
    (a) either –
        (i) the land was exempt land under section 54(1) for the preceding year; or
        (ii) the person used and occupied the land as his or her principal place of residence for a period of at least six months during the preceding year; and

(b) as at 31 December in the preceding year, the person was the owner of other land that he or she used and occupied as his or her principal place of residence.

Section 61 – unoccupied land subsequently used as principal residence

(1) An owner who was assessed for and paid land tax in respect of a year in respect of land that is not occupied as his or her principal place of residence is entitled to a refund of that land tax if –
   (a) the owner was unable to occupy that land as his or her principal place of residence as at 31 December in the preceding year because a residence was being constructed or renovated on it; and (ab) once construction or renovation of that residence has been completed, the owner continuously uses and occupies the land as his or her principal place of residence for at least six months commencing in that year; and
   (b) an application for a refund is made before the end of the next following year.

Section 62 – partial exemption if land used for business activities

(1) Despite anything to the contrary in this Division, if land that would be exempt land under another provision of this Division, but for this section, is used by any person to carry on a substantial business activity, the exemption applies only to the extent that the land is used and occupied for residential purposes by the owner, a vested beneficiary in relation to the land, or a natural person who has a right to reside on the land (as the case requires).
(2) In determining whether land is used by a person to carry on a substantial business activity, account must be taken of –
   (a) whether paid employees or contractors (other than employees or contractors who are relatives of, and who ordinarily reside with, the person who uses and occupies the land as his or her principal place of residence) work on the land; and
   (b) whether any part of the land is used or allocated solely for business purposes; and
   (c) if part of the land is used or allocated (whether solely or partly) for business purposes, the proportion of the area of the land (or of the floor space of buildings on the land) that is so used or allocated; and
   (d) the amount of income (if any) and the proportion of the person's total income that is derived from business activities conducted on the land; and
   (e) any other matters the Commissioner considers relevant.

Section 62A – partial exemption if separate residence is leased for residential purposes

If land that would be exempt under this Division contains a separate residence that is leased for residential purposes –

(a) land tax is assessable on the part of the land that is leased for residential purposes; and
(b) Section 22 applies, if necessary, for that purpose.

South Australia – defining the principal place of residence

The principal place of residence exemption is addressed under section 5 of the Land Tax Act 1936. *Subsection 10 of section 5 – Proper grounds for exempting land from land tax under this section* also covers land used for primary production purposes, which is addressed in the following chapter.

Section 5 (10) Proper grounds for exempting land from land tax under this section exist as follows:

(a) a full exemption is available where:
  (i) the land is owned by a natural person (whether or not he or she is the sole owner of the land);
  (ii) the buildings on the land have a predominately residential character; and
  (iii) less than 25 per cent of the total floor area of all buildings on the land are used for any business or commercial purpose (other than the business of primary production).
(ab) land may be wholly exempted from land tax if –
  (i) the land is owned by a natural person (whether or not he or she is the sole owner of the land); and
  (ii) any buildings on the land of a predominantly residential character are uninhabitable; and
  (iii) the Commissioner is satisfied –
    (A) that the person has ceased to occupy any building on the land of a predominantly residential character because it has been destroyed or rendered uninhabitable by an occurrence for which the person is not responsible (whether directly or indirectly) or which resulted from an accident; and
    (B) that any such building constituted the person's principal place of residence immediately before the date on which the building was destroyed or rendered uninhabitable; and
    (C) that the person intends to repair or rebuild the building within a period of three years from the date on which the building was destroyed or rendered uninhabitable; and

(D) that the buildings on the land will, after the completion of building work, have a predominantly residential character; and

(E) that the person intends to occupy the land as his or her principal place of residence after the completion of the building work; and

(iv) the person is not receiving an exemption from land tax under another provision of this subsection in relation to other land that constitutes the person's principal place of residence;

(b) land may be partially exempted from land tax by reducing its taxable value in accordance with the scale prescribed in subsection (12) if –

(i) the land is owned by a natural person and constitutes his or her principal place of residence (whether or not he or she is the sole owner of the land); and

(ii) the buildings on the land have a predominantly residential character; and

(iii) a part of the land of 25 per cent or more but not more than 75 per cent of the total floor area of all buildings on the land is used for a business or commercial purpose;

(ba) land may be wholly exempted from land tax if –

(i) the land is owned by a natural person and constitutes his or her principal place of residence (whether or not he or she is the sole owner of the land); and

(ii) the buildings on the land are used for the purposes of a hotel, motel, set of serviced holiday apartments or other similar accommodation; and

(iii) more than 75 per cent of the total floor area of all buildings on the land is used for the person's principal place of residence;

(bb) land may be partially exempted from land tax by reducing its taxable value in accordance with the scale prescribed in subsection (12) if –

(i) the land is owned by a natural person and constitutes his or her principal place of residence (whether or not he or she is the sole owner of the land); and

(ii) the buildings on the land are used for the purposes of a hotel, motel, set of serviced holiday apartments or other similar accommodation; and

(iii) 25 per cent or more of the total floor area of all buildings on the land is used for the person's principal place of residence, (and for the purposes of the scale prescribed in subsection (12), the area used for the hotel, motel, set of serviced holiday apartments or other similar accommodation will be taken to be the area used for business or commercial purposes);

A sliding scale for partial exemption from land tax exists under subsection (10) (b) or (bb), which provides a range of exemption for more than 75 per cent of the premises used for business, given nil exemption, to less than 25 per cent, given full exemption.

Tasmania – defining the principal place of residence

The principal place of residence exemption in Tasmania is addressed under section 6 of the Land Tax Act 2000, of which the key subsections follow:

Section 6 – principal residence land

(1) Principal residence land is land on which the principal residence of an owner of at least a 50 per cent interest in the land or a related person of such an owner is situated.

(2) The Commissioner is to determine that adjoining land is principal residence land, if satisfied that –

  (a) the land is on a separate title held by the owner of the principal residence land; and

  (ab) there is no dwelling on the land that is used as a place of residence; and

  (b) the land is used by that owner solely in conjunction with the principal residence land; and

  (c) the owner does not receive any income from the use of that land; and

  (d) the owner of at least a 50 per cent interest in the principal residence land is also the owner of at least a 50 per cent interest in the adjoining land.

(3) The Commissioner, on application by a trustee of a trust, is to determine that land is principal residence land for a financial year if –

  (a) the land is held by –

    (i) a registered trustee company; or

    (ii) an executor, administrator, guardian, committee, receiver or liquidator; or

    (iia) the trustee of a special disability trust; or

    (iii) a trustee appointed by a court; or

    (iv) the trustee of a fixed trust in which all of the beneficiaries are individually named or are descendants of individually named beneficiaries; and

  (b) the principal residence of a beneficiary of the trust is situated on the land as at 1 July in that financial year; and

  (c) the Commissioner is satisfied that the beneficiary does not own any other principal residence land.

(3A) For the purposes of subsection (3)(b), a person is taken to be a beneficiary of a fixed trust, referred to in subsection (3)(a)(iv), only if the person would be entitled, on the winding up of the trust, to 50 per cent or more of the value of the income and capital of the trust.

(4) The Commissioner, on application by a company, is to determine that land is principal residence land for a financial year, if –
  (a) the land is beneficially owned by the company; and
  (b) the principal residence of a person who owns 50 per cent or more of shares in the company is situated on the land as at 1 July in that financial year; and
  (c) the Commissioner is satisfied that the person does not own any other principal residence land; and
  (d) the Commissioner is satisfied that the person, by reason of his or her ownership of 50 per cent or more shares in another company, does not have another principal residence situated on other land which –
    (i) is beneficially owned by that other company; and
    (ii) has been determined under this subsection to be principal residence land.

(5) The Commissioner is to determine that land owned by a home-unit company is principal residence land if any flat on that land is the principal residence of a person owning shares in the home-unit company.

(6) . . . .

(7) The Commissioner is to determine that a part of land owned by a cooperative housing society is principal residence land if that part is used for residential purposes.

(8) If a person occupies residential premises in a retirement village as his or her principal place of residence, any other land owned by the person is not that person's principal place of residence.

Western Australia – defining the principal place of residence

Exempt purposes for the principal place of residence in Western Australia are addressed under *Division 2 – Private residential property* of *Part 3 – Exemptions, concessions and rebates* of the Land Tax Assessment Act 2002. Sections 21 to 28 detail the various circumstances under which the exemption applies, which include but are not limited to beneficiaries, construction and refurbishment of a residence or second residence and residences of disabled persons held in trust.

Section 21 sets out the broad circumstances for the occupation requirements of the residence for it to qualify for the principal place of residence exemption as follows:

(1) Private residential property (except property held in trust) is exempt for an assessment year if, at midnight on 30 June in the financial year before the assessment year, it is owned –
  (a) by an individual who uses it as his or her primary residence; or
  (b) by a husband and wife, at least one of whom uses it as his or her primary residence; or

(c) by persons who have lived in a de facto relationship with each other for at least two years, whether or not they still live on that basis, at least one of whom uses it as his or her primary residence.

(2) However, if the property is also owned by another person or persons, it is exempt if each owner who does not use it for that purpose is an owner only because of a requirement by a financial institution for a guarantee of money advanced on the security of the property.

Section 22 – residence owned by executor etc., exemption for if beneficiary in will exercising right to reside

Private residential property is exempt for an assessment year if, at midnight on 30 June in the previous financial year –

(a) it is owned by an executor or administrator of a will as trustee; and
(b) an individual identified in the will –
  (i) is entitled under the will to the property as a tenant for life; or
  (ii) has a right under the will to use the property as a place of residence –
    (I) for as long as he or she wishes; or
    (II) for a fixed or ascertainable period,

whether or not the individual is or may become entitled under the will to ownership of all or part of the property at some future time; and

(c) the individual uses the property as his or her primary residence.

Further provisions exist for cases where the residence is owned by the executor and the beneficiary in the will has the right to future ownership and is resident, see s23A.

Section 23 – continued exemption after death of resident

(1) Private residential property owned by the executor or administrator of an individual's estate is exempt, but only for the assessment year following the financial year in which the individual died, if –
  (a) the individual's ownership and use of the property as his or her primary residence gave rise to an exemption under section 21 for the financial year in which he or she died, or would have given rise to such an exemption if he or she had owned the property and had been using it for that purpose on 30 June before his or her death; and
  (b) the executor or administrator is the owner of the property at midnight on 30 June in the financial year in which the individual died; and

(c) the individual's estate does not derive any rent or other income from the property between the date of the individual's death and the end of the assessment year.

Further provisions for exemptions in advance are available under section 23(2) and (3).

Exemptions for construction or refurbishment of a private residence, property of a disables person moving between two residences and construction of a second residence are highlighted in the following sections of the Land Tax Management Act 2002.

Section 24: provides a land tax exemption for one assessment year for the construction of a private residence, subject to the provisions of this section.

Section 24A: provides a land tax exemption for two consecutive assessment years for private residential property if the time taken to complete the construction of the private residence is two years or more.

Section 25: provides a land tax exemption for one assessment year for the refurbishment of a private residence, subject to the provisions of this section.

Section 25A: provides a land tax exemption for two consecutive assessment years for private residential property if the time taken to complete the refurbishment of the private residence is two years or more.

Sections 26 and 26A: provide an exemption for residence of a disabled person held in trust and an exemption for a residence of a disabled person owned by a relative.

Section 27: moving between two private residences.

Section 27A: construction or refurbishment of a second private residence.

Prescriptive detail of the application of these sections of the Act can be downloaded from the Department of Finance, Office of State Revenue at: www.osr.wa.gov.au/landtax.

Australia Capital Territory (ACT)

The Australian Capital Territory Legislative Assembly performs the roles of both a city council and territory government. The assembly consists of 17 members, elected from three districts using proportional representation. The three districts are Molonglo, Ginninderra and Brindabella, which elect seven, five and five members, respectively. The Chief Minister is elected by the Members of the Legislative Assembly (MLA) and selects colleagues to serve as ministers alongside him or her in the Executive, known informally as the cabinet (ACT Government 2013).

The ACT's taxation effort is higher than its assessed capacity to raise taxes, as measured by the Commonwealth Grants Commission. The territory's expenditure on services has been higher than the national average, funded by above average taxation. These assessments, however, are gauged against the national average capacity to raise taxes and deliver services. The ACT collects the lowest tax revenue per capita as a proportion of mean disposable income compared with the national average. Taxation per employed person is also below the national average.

The ACT has a higher than average capacity to tax transactions and relatively lower capacity to tax land. By contrast, actual taxation effort is greater for land-based sources and lower for transactions. Overall, taxation effort in the ACT, based on the Commonwealth Grants Commission's assessments, is more efficient relative to the national average (ACT Government, Treasury Directorate 2012).

In response to the recommendations of Australia's Future Tax System (AFTS) Review of 2009, the ACT government commissioned a review of the territories' tax system during 2012 in aligning the objectives with those of AFTS where appropriate. This review was the first undertaken by the ACT since being granted self-governance (ACT Government, Treasury Directorate 2012). In summary, the AFTS Review suggested revenue-raising should be concentrated on four robust and efficient broad-based taxes.

1. Personal income
2. Business income, designed to support economic growth
3. Rent on natural resources and land and
4. Private consumption.

The recommendations resulting from the review included reforms to specific taxes, including property under Recommendation 2, which follows:

> With regards to long-term structural reform, over a period of time that is adequate for appropriate transition:
>
> - abolish duty on conveyances
> - retain a form of tax on payroll to maintain a diversified tax system
> - abolish duty on general insurance and life insurance and
> - adopt a broad-based land tax as a base for revenue replacement.
>
> (ACT Government, Treasury Directorate 2012)

The ACT government is working through its potential land tax and general rate reform policy resulting from the recommendations of the Tax Review 2012. Establishing the nature and degree of reform, followed by an operational plan for transition with commitment to reducing less efficient tax on conveyance stamp duty, will test the resolve of government in the ACT. The reform sets an example for sub-national government in Australia to observe corresponding dollar reduction in revenue from stamp duty.

Land Tax Act 2004 — imposition of land tax section 10

Land tax exempted from section 9

(1) Land tax at the appropriate rate is imposed for a quarter on each parcel of rateable land that is –
  (a) rented residential land; or
  (b) residential land owned by a corporation or trustee.
(2) The appropriate rate of land tax for a parcel of land is the amount worked out for the parcel as follows:
  determined rate × average unimproved value
(3) However, land tax is not imposed on a parcel of land that is exempt under section 10 or section 11.
(4) In this section: average unimproved value means the average unimproved value of the parcel of land under the Rates Act 2004. Determined rate means the rate determined under the Taxation Administration Act, section 139.

Note: The power to determine a rate under the Taxation Administration Act includes the power to determine a different rate for different matters or different classes of matters (see Legislation Act, section 48).

Land exempted from section 9 generally:

(1) The following parcels of land are exempt from land tax imposed under section 9:
  (a) a parcel of residential land owned by an individual if the parcel is exempted under section 13 (decision on compassionate application) in relation to the parcel;
  Note: An exemption under section 13 is for 1 year or less.
  (b) a parcel of rural land;
  (c) a parcel of land owned by the housing commissioner under the Housing Assistance Act 2007;
  (d) a parcel of land owned or leased by an entity declared under the Duties Act 1999, section 73A (transfers etc. to entities for community housing);
  (e) a parcel of land leased for a retirement village;
  (f) a parcel of land leased for a nursing home;
  (g) a parcel of land leased for a nursing home and a retirement village;
  (h) a parcel of land leased by a religious institution or order to provide residential accommodation to a member of the institution or order and allow the member to perform his or her duties as a member of the institution or order;
  (i) a parcel of land, other than a parcel of residential land, leased to a corporation or trustee, being used for a purpose prescribed under the regulations.

Local government rating – exemptions and concessions across Australia exemptions

New South Wales: Land owned by the Crown, held in trust or subject to a conservation agreement; owned by a state water corporation or used for water supply works; used in connection with a religious purpose, public place, mines rescue stations, school or rail infrastructure facilities; public, benevolent or charitable institutions; the Sydney Cricket and Sports Ground Trust; Zoological Parks Trust; and land vested in Aboriginal land councils (Local Government Act 1993, s. 555).

Victoria: Land owned by the Crown, a Minister, a council or public statutory body; land used for public, charitable, religious or mining purposes; or land held in trust for memorial of war veterans (Local Government Act 1989, s. 154).

Queensland: Land owned by the state or a government entity (other than non-exempt government-owned corporations); land in a state forest or timber reserve; Aboriginal land; land used to facilitate specific transport infrastructure; and land used for religious, charitable, educational or public purposes (Local Government Act 1993, s. 957).

South Australia: Land owned by the Crown, occupied by councils, universities or emergency services organisations; land exempt under the Recreation Ground Rates and Taxes Exemption Act 1981; land subject to a mining lease; and land subject to division under the Community Titles Act 1996 (s. 147).

Western Australia: Land owned by the Crown or a local government; land used for the public, religious, charitable or agricultural purposes; and non-government schools (Local Government Act 1995, s. 6.26).

Tasmania: Land owned by the Commonwealth; land owned by the Crown and used for conservation and nature recreation purposes; the Hydro-Electric Corporation; Aboriginal land or land used for charitable purposes (Local Government Act 1993, s. 87).

Northern Territory: Crown land occupied by the Territory or the Commonwealth; public land; land used for religious, educational or charitable purposes; and public hospitals (Local Government Act 2005, s. 58). Councils can exempt classes of land or persons (s. 98), including Indigenous landholders.

ACT: Land used for public parks and reserves; cemeteries, public hospitals and benevolent institutions; land used for religious purposes; public libraries; land leased by the Commonwealth which is occupied by schools; and Commonwealth-unoccupied land (s. 8).

Rate concessions

Concessions on local government rates and service charges are granted to persons on the grounds of financial hardship. Eligible persons generally include pensioners and persons in receipt of particular allowances, such as veteran's allowances or social welfare payments. Concessions are usually granted as a partial reduction in the rates payable by a landholder. Concessions can also be granted to specified land or buildings for the purposes of preserving buildings or places of historical importance or encouraging proper development. The provisions relating to rate concessions are contained in each of the Local Government Acts (table B.3).

State or Territory Rate (and other) waivers, concessions, discounts and rebates specified in state legislation:

New South Wales Local Government Act 1993 (s. 565, 582, 601, 610E)
Victoria Local Government Act 1989 (s. 142, 171, 171A, 243)
Queensland Local Government Act 1993 (1031–1035, 1035A)
South Australia Local Government Act 1995 (s. 166, 181–2, 188)
Western Australia Local Government Act 1995 (s. 6.12, 6.47–48, 6.50)
Tasmania Local Government Act 1993 (s. 92, 99, 106A)
Northern Territory Local Government Act 1993 (s. 81, 86–87, 89)

'Other' indicates that some waivers, concessions, discounts and rebates relate to fees and charges (state government legislation various & Productivity Commission 2008). The majority of the states compensate local governments in part or in full for mandatory concessions. The NSW government is only required to reimburse 55 per cent of the value of concessions granted by local governments to eligible pensioners (Local Government Act 1993 (s. 581)).

Restrictions on rates increases

New South Wales is the only jurisdiction in which the state government currently enforces formal restrictions on the percentage by which local governments may increase rates. The Local Government Act 1993 (NSW) provides for the Minister to determine the maximum percentage by which general income from rates and charges may vary from the previous year (section 509). 'General income' is defined in section 505(a) of the Act to mean income from ordinary rates, special rates and annual charges other than:

- special rates for water supply and sewerage services charges for water supply and sewerage services
- annual charges for storm water and waste management services
- annual charges referred to in section 611.

The statutory limit on general income effectively caps or pegs increases in councils' rates revenue. The limit for 2006–07 was set at 3.4 per cent (DLG 2007d). Councils can apply to the Minister for a variation to exceed this limit (section 508(2) and section 508A).

In Victoria and South Australia, temporary rate-capping policies were introduced in the 1990s during amalgamation processes, with the intention of ensuring that any costs savings that resulted were passed on to ratepayers. Part 8A sections 185A–185C was added to the Local Government Act 1989, Victoria, which gives the Minister power to limit or regulate the increase in local government revenue.

6 Business use, investment and development land

Introduction

Investment and business use land comprises a broad range of land uses and classifications that attract exemptions and concessions from state land tax, which do not necessarily apply to local government rating. These concessions and allowances add to the tax revenue expended on the principal place of residence and further impact the reform of this tax across Australia. This chapter examines the concessions and exemptions, including the investment threshold, the exemption for primary production land and a brief highlight on the exemption of charitable organisations. After accounting for the principal place of residence exemption, investment threshold, the exemption of primary production land and land held by not-for-profits, less than 20 per cent of land in Australia is subject to state land tax. This is in contrast to over 98 per cent of all rateable land which attracts local government council rates.

Within the broad category of investment use land exists several sub-categories in which there is no distinction in the threshold granted. In the case of investment property, this may be sub-categorised into investment rental property and investment owner-occupied property, of which the latter constitutes property owned and used by the owner to operate their business. In the case of primary production land, this category of land holdings may be further broken down into a number of sub-categories. The first is rural land which remains unoccupied or unused, or unoccupied and used as a hobby farm from time to time. The second sub-category is where the land is simply leased or used by others under licence agreements. The third is where the land is used and operated as a business by the landowners themselves.

The third sub-category may be further broken down into land used by primary producers which is locally owned and whose produce is provided to local domestic markets in Australia versus those that are predominantly foreign-owned. This category may be extended to primary production land which is foreign-owned and from which the produce is exported for international consumption. In the case of taxing primary production land, the lines have blurred on land claimed to be used for primary production purposes which, on closer analysis, is not primary production land, but is either leased for that purpose or lay fallow on the fringes of Australia's capital cities as a land bank. Some of these properties sit across two forms of use, both as primary production land and as a principal place of residence where they are dually occupied by the owner(s).

The absence of a land area threshold for land tax purposes has adversely impacted the supply of land ripe for subdivision and urban use purposes, which has encouraged land banking and contributed to the escalation of land prices and speculation. In highly urbanised cities, where economic growth and development is contingent on highest and best use of land, the removal of the land area threshold, particularly on the fringes of Australia's capital cities, has led to both a tax expenditure and the mechanism used for freeing up of land ripe for residential subdivision. While counterargument exists for better use of infill land, both of these strategies have a place in Australia's fast-emerging international cities.

This chapter examines the investment threshold, how it is determined and the rationale for its existence, as well as primary production land and the exemption afforded for this use. It further provides an overview of land designated for charitable uses, which attracts exemption from state land tax and, in a number of cases, from local government rating. While examining both the technical details of these matters, the relevance of their existence and the format in which they operate are of equal importance in the fiscal reform and reorientation of sub-national government tax revenue across Australia. These factors are now reviewed and presented on a State-by-State basis.

New South Wales (NSW)

Land tax rates and thresholds

New South Wales was one of the first states to introduce a tax-free threshold which was set at $55,000 in the 1970s. By 1987 the tax-free threshold was $94,000 with a tax rate of 2 per cent. The initial rationale for introducing the land tax-free threshold was to incentivise residential property investment and for government to encourage more rental housing stock, a rationale which further underpinned the introduction of negative gearing. By the late 1980s and following the introduction of the first mass appraisal valuation of land in New South Wales, the government came under pressure to increase the land tax threshold, which increased to $125,000 in 1988, $135,000 in 1989 and $160,000 in 1990, where it remained at this rate for eight years. Table 6.1 sets out the land tax threshold and rates in the dollar over the past two decades.

In 1991 and 1992, the official rate in the dollar was 1.5 per cent; however, a rate in the dollar of 1.85 per cent was passed by the then state government to shore up falling conveyance stamp duty revenue from land tax, following the 1988/89 property boom and 1990/91 property market correction. It was at this point in time that state governments fully became aware of the volatility of revenue generated from conveyance stamp duty.

Pre-2005 land tax year

The mechanics for reviewing the land tax threshold pre-2005, as set out in Land Tax Management Act 1956, section 62TB – Tax threshold pre-2005 land tax year, follows:

(1) The tax threshold for the 1998 land tax year is $160,000.
(2) The tax threshold for a land tax year subsequent to the 1998 land tax year and before the 2005 land tax year is to be determined in accordance with the following formula, subject to subsection (3):

> $160,000 x (100 per cent + the percentage change in land values determined under section 62TA(1) during the month of September preceding the land tax year)

(3) The tax threshold for a succeeding land tax year is to remain the same as the previous land tax year if the tax threshold determined in accordance with the formula in subsection (2) for the succeeding land tax year is equal to or less than the tax threshold for the previous year.

Section 62TA – Determination of change in NSW property values pre-2004

(1) During the month of September in each year before 2004, the Valuer-General is to determine the percentage by which average land values of land within residential, commercial, business and industrial zones in New South Wales have changed between 1 July 1997 and 1 July last preceding the making of the determination.

Removal of the land tax threshold 2005 land tax year

The abolition of a land value threshold in the assessment of land tax reduces any incentive of tax structuring or planning by purchasing investment property in different names or spreading ownership across different family members. The NSW Office of State Revenue, like its Commonwealth counterpart the Australian Taxation Office, has look-through provisions to determine the purpose of any structure and the orderly arrangement of assets for the purposes of minimising land tax or other revenue. The impact of removing the land value threshold in 2005 dampened any such planning, as most urban property, except the principal place of residence, attracted some level of land tax in the 2005 land tax year.

The implications for investors with investment properties which were below the land tax threshold before the changes in 2005 are an incurrence of up to $1,268 increase per annum. However, for investors land tax is a deductible income tax expense in the running of income-producing property. The upside to the 2005 changes to land tax was the reduction in the rate from 1.7 cents to 1.4 cents being the top rate in the dollar applied to land values in calculating land tax. The breakeven point between the 2004 and 2005 land tax year was for aggregate land values at approximately $405,000.

There is one caveat to the above position on rates in the dollar, which is the ability for the rate in the dollar to be adjusted, just as it has been reduced from 1.7 cents in the dollar to 1.4 cents with these most recent changes to land tax. Since 1992, the rate in the dollar for calculating land tax liability has fluctuated between 1.4 cents to 1.85 cents in the dollar. The changes in the thresholds and rate in the dollar applicable to assessable land values over the past 12 years are shown in the following table.

Table 6.1 Land tax thresholds and rates in the $1.00

Land tax year	*Land value threshold: investment*	*Land value threshold: principal residence*	*$100 + percentage / cents in $1.00*
1996	$160,000	Land above 2,100m^2	1.5 cents
1997	$160,000	Land above 2,100m^2	1.65 cents
1998	$160,000	$1,000,000	1.85 cents
1999	$176,000	$1,116,000	1.85 cents
2000	$192,000	$1,234,000	1.85 cents
2001	$205,000	$1,319,000	1.7 cents
2002	$220,000	$1,414,000	1.7 cents
2003	$261,000	$1,680,000	1.7 cents
2004	$317,000	$1,970,000	1.7 cents
2005	Nil	No threshold (value or area) reinstated	< $399,999: 0.4 cents $400,000–$499,999: 0.6 cents > $500,000: 1.4 cents
2006	$352,000	No threshold (value or area)	1.7 per cent
2007	$352,000	No threshold (value or area)	1.7 per cent
2008	$352,000	No threshold (value or area)	1.6 per cent
2009	$368,000	No threshold (value or area)	1.6 per cent
2010	$376,000 and premium $2,299,000	No threshold (value or area)	1.6 per cent $30,868 for first $2,299,000 then 2 per cent above that
2011	$387,000 and premium $2,366,000	No threshold (value or area)	1.6 per cent $31,764 for first $2,366,000 then 2 per cent above that

Table 6.1 (continued)

Land tax year	*Land value threshold: investment*	*Land value threshold: principal residence*	*$100 + percentage / cents in $1.00*
2012	$396,000 and premium $2,421,000	No threshold (value or area)	1.6 per cent $32,500 for first $2,421,000 then 2 per cent above that
2013	$406,000 and premium $2,482,000	No threshold (value or area)	1.6 per cent $33,316 for first $2,482,000 then 2 per cent above that
2014	$42,000 and premium $2,519,000	No threshold (value or area)	1.6 per cent $33,812 for first $2,519,000 then 2 per cent above that
2015	$432,000 and premium $2,641,000	No threshold (value or area)	1.6 per cent $35,444 for first $2,641,000 then 2 per cent above that

N.B. the absence of a value or land area threshold on the PPR keeps NSW in line with other states.

Tax threshold – 2006 land tax year and subsequent land tax years

Section 62TBA Tax threshold – 2006 land tax year and subsequent land tax years

(1) The tax threshold for the 2006 land tax year is $352,000.
(2) The tax threshold for the 2007 land tax year and any subsequent land tax year is the average of the indexed amounts or the tax threshold for the previous land tax year, whichever is the greater.
(3) The *'average of the indexed amounts'* is the average of the following three amounts:
   (a) the indexed amount for the land tax year,
   (b) the indexed amount for the two preceding land tax years.
(4) For the purposes of this section, the *'indexed amount'* for a land tax year is the following:
   (a) in the case of the 2005 land tax year – $342,000,
   (b) in the case of the 2006 land tax year – $352,000,
   (c) in the case of the 2007 land tax year or any subsequent land tax year, the amount is determined as provided for by subsection (5).
(5) The indexed amount for the 2007 land tax year and any subsequent land tax year is to be determined in accordance with the following formula:
   'N' is the indexed amount for the previous land tax year.
   'I' is the indexation factor for the land tax year, determined in accordance with section 62TBB.

Accordingly, the tax threshold for the 2007 land tax year will be the average of the following three amounts or $352,000, whichever is the greater:

(a) $342,000,
(b) $352,000,
(c) $352,000 × (100% + I).

Premium rate threshold – 2009 land tax year and subsequent land tax years

Section 62TBC – premium rate threshold – 2009 land tax year and subsequent land tax years

(1) The premium rate threshold for the 2009 land tax year is $2,250,000.
(2) The premium rate threshold for the 2010 land tax year and any subsequent land tax year is to be calculated in accordance with the following formula:
"T" is the tax threshold for the land tax year for which the premium rate threshold is being calculated, as determined under section 62TBA.
"B" is the tax threshold for the land tax year preceding the land tax year for which the premium rate threshold is being calculated, as determined under section 62TBA.
"P" is the premium rate threshold for the land tax year preceding the land tax year for which the premium rate threshold is being calculated.

Land exempted from land tax

This section sets out land exempt from land tax. Parts 2 through 5 of Section 10 of the Land Tax Management Act 1956 provide further qualification as to the application of this section.

Part 3 – Section 10 – land exempted from tax

(1) Except where otherwise expressly provided in this Act the following lands shall, subject to sections 10B, 10D, 10E, 10G and 10P, be exempted from taxation under this Act:
(b) land owned by any marketing board constituted under the Marketing of Primary Products Act 1983, an agricultural industry services committee constituted by the Agricultural Industry Services Act 1998 or Local Land Services,
(c) land owned by or in trust for a public health organisation within the meaning of the Health Services Act 1997,
(d) land owned by or in trust for a charitable body,
(e) land owned by or in trust for a religious society if the society, however formed or constituted, is carried on solely for religious, charitable or educational purposes, including the support of the aged

or infirm clergy or ministers of the society, or their wives or widows or children, and not for pecuniary profit,

(f) land owned by or in trust for, and used and occupied solely by:
   (i) an association of employers or employees registered as an organisation under Part IX of the Industrial Relations Act 1988 of the Commonwealth,
   (iii) an industrial organisation of employers or employees within the meaning of the Industrial Relations Act 1996,

(f1) land owned by the New South Wales Aboriginal Land Council, a Regional Aboriginal Land Council or a Local Aboriginal Land Council constituted under the Aboriginal Land Rights Act 1983,

(g) land owned by or in trust for any person or society and used or occupied by that person or society solely as a site for:
   (i) a place of worship for a religious society, or a place of residence for any clergy or ministers or order of a religious society,
   (ii) a school registered under the Education Act 1990,
   (iii) a building (not being a building of which any part is used for the purpose of a commercial activity open to members of the public) owned and solely occupied by a society, club or association not carried on for pecuniary profit,
   (iv) a charitable body,
   (v) a public cemetery or crematorium,
   (vi) a public garden, public recreation ground or public reserve,
   (vii) a fire brigade, ambulance or mines rescue station,
   (viii) a private health facility (within the meaning of the Private Health Facilities Act 2007) not carried on for pecuniary profit,
   (ix) an authorised hospital within the meaning of the Mental Health Act 1990 not carried on for pecuniary profit,

(h) land owned by, or in trust for, any club or body of persons, and used primarily and principally for the purposes of any game or sport and not used for the pecuniary profit of the members of that club or body,

(i) land owned by, or in trust for, any club or body of persons, formed for promoting or controlling horse-racing, trotting-racing or greyhound-racing and used primarily and principally for the holding of meetings for horse-racing, trotting-racing or greyhound-racing,

(j) land used and occupied for the purpose of holding agricultural shows, or shows of a like nature and owned by, or held in trust for, a society which is established for the purpose of holding such shows and is not carried on for the pecuniary profit of its members and applies its revenues substantially towards the promotion or holding of such shows,

(k) land owned by a friendly society,

(l) association property that is vested in an association under the Community Land Development Act 1989 and is used primarily and principally:
  (i) as an open access way or private access way within a community scheme, precinct scheme or neighbourhood scheme under the Community Land Development Act 1989, or
  (ii) for the recreation of participants in such a scheme and their invitees, but is not used for a commercial purpose,

(m) land owned by a State owned corporation (within the meaning of the State Owned Corporations Act 1989) specified in the regulations to the extent, and from the date (whether that date is before or after the commencement of the regulations), prescribed by the regulations in respect of the corporation,

(n) land owned by any gas or electricity supply authority specified in the regulations (being an energy services corporation within the meaning of the Energy Services Corporations Act 1995, a distribution network service provider that holds an authorisation or licence to operate an electricity distribution system under the Electricity Supply Act 1995 or a gas distributor that holds an authorisation under the Gas Supply Act 1996) to the extent, and from the date (whether that date is before or after the commencement of the regulations), prescribed by the regulations in respect of such authority,

(o) land owned by the Returned Sailors, Soldiers and Airmen's Imperial League of Australia (New South Wales Branch) and being the site of Anzac House,

(p) land that is the subject of a biobanking agreement under Part 7A of the Threatened Species Conservation Act 1995 (see Act for p1, p2, & p3).

(q) land used solely as a police station,

(r) land that is exempt from taxation under the principal place of residence exemption, as provided for by Schedule 1A,

(r1) with respect to taxation leviable or payable in respect of the year commencing on 1 January 1987 or any succeeding year, land approved for multiple occupancy, and occupied, in accordance with an environmental planning instrument within the meaning of the Environmental Planning and Assessment Act 1979,

(t) with respect to taxation leviable or payable in respect of the year commencing on 1 January 1975 or any succeeding year, land owned by a co-operative under the Co-operatives Act 1992 that has as its objects any of the objects listed in section 7 of the Co-operation Act 1923.

(u) land that is used solely for the provision of an approved education and care service (within the meaning of the Children (Education and Care Services) National Law (NSW)), but only if:

(i) the service is provided by an approved provider under that Law and
(ii) the land is the place where children are educated or cared for by the service,

(v) land that is used solely for the provision of an approved education and care service (within the meaning of the Children (Education and Care Services) Supplementary Provisions Act 2011), but only if: (see Act for (i), (ii) & (iii)). Refer to section 10 for provisions under Parts 2, 3, 4 & 5).

Primary production land

Primary production land is defined under section 10AA of the Land Tax Management Act 1956 which further defines the uses of land which constitute primary production and the parameters which dictate for 'rural land'. An application for exemption from land tax for land used for primary production are made to the Office of State Revenue of the prescribed form, which comprises four parts:

1. Property details
2. Business or industry for which land is being used
3. Business details (only if land is zoned non-rural)
4. Farm improvements

Section 10AA – exemption for land used for primary production

(1) Land that is rural land is exempt from taxation if it is land used for primary production.
(2) Land that is not rural land is exempt from taxation if it is land used for primary production and that use of the land:
   (a) has a significant and substantial commercial purpose or character; and
   (b) is engaged in for the purpose of profit on a continuous or repetitive basis (whether or not a profit is actually made).
(3) For the purposes of this section, "land used for primary production" means land the dominant use of which is for:
   (a) cultivation, for the purpose of selling the cultivated produce; or
   (b) the maintenance of animals (including birds), whether wild or domesticated, for the purpose of selling them or their natural increase or bodily produce; or
   (c) commercial fishing (including preparation for that fishing and the storage or preparation of fish or fishing gear) or the commercial farming of fish, molluscs, crustaceans or other aquatic animals; or
   (d) the keeping of bees, for the purpose of selling their honey; or

(e) a commercial plant nursery, but not one at which the principal cultivation is the maintenance of plants pending their sale to the general public; or

(f) the propagation for sale of mushrooms, orchids or flowers.

(4) For the purposes of this section, land is "rural land" if:

(a) the land is zoned rural, rural residential, non-urban or large lot residential under a planning instrument; or

(b) the land has another zoning under a planning instrument, and the zone is a type of rural zone under the standard instrument prescribed under section 33A (1) of the Environmental Planning and Assessment Act 1979; or

(c) the land is not within a zone under a planning instrument but the Chief Commissioner is satisfied the land is rural land.

The following case defines what constitutes land used for primary production purposes in determining the application of this exemption. The dominant purpose test has now evolved and is used by state tax authorities in assessing the status of this exemption, in which the court/tribunal has defined the tolerances of the use and purpose in the following cases.

Maraya Holdings Pty Ltd v Chief Commissioner of State Revenue [2013] NSWSC 23

In these proceedings, the primary production use of the Property a beef cattle operation, is conducted by an operator on behalf of Maraya Holdings Pty Ltd ("Maraya"), on the Properties owned by the Plaintiffs and on additional properties owned by entities not related to the Plaintiffs.

The Properties are zoned non-rural and are small in area, being up to six hectares. The taxable value of the Properties was $26,558,700.00.

As dominant use of the land was conceded, the question before the Court was whether the cattle operation satisfied the commerciality test in section 10AA(2) of the Land Tax Management Act 1956 (LTMA). The evidence of the Plaintiffs was that in the relevant land tax years the herd was up to 40 head of cattle; in 2011 and 2012 it was up to 55 head of cattle. The operation was to graze and fatten cattle for market.

The parties agreed to briefing joint experts: Bill Hoffman as beef cattle expert and Mark Bryant as financial expert. The Plaintiffs did not accept the reports of either joint expert in their entirety. The parties engaged their own experts on drought issues.

Mr Hoffman opined that the cattle business was unlikely to produce a positive financial return over the foreseeable future, and that it cost more to produce a kilogram of beef than was received when it was sold. Mr Hoffman also expressed the view that the kilograms of beef produced per hectare were a valuable indicator of productivity. Maraya's beef operation produced on average 42.6 kilograms of beef per hectare, while Mr Hoffman's peer network of beef producers produced on average 228 kilograms of beef per hectare.

Mr Bryant identified two approaches to measuring financial return: the marginal benefit approach and the shared costs allocation approach. In his view, on neither bases did Maraya have a positive financial return in the six years in aggregate, and produced a large negative return in each tax year.

Justice Gzell concluded that the Plaintiffs did not meet the test that the primary production use of the land had a significant and substantial commercial purpose or character under section 10AA(2)(a) of the LTMA, having regard to the following factors in particular:

- 40 or even 55 cattle grazing on the Properties with or without the other properties does not constitute a serious or intense primary production use when the grazing areas are taken into account;
- accepting Mr Hoffman's expert evidence, the output of beef per hectare was very low;
- little attempt was made to improve the pastures;
- the operator only spent 1.5 to 2 hours per week and his son 1 hour per week on the cattle operation, and those times are consistent with a part-time operation;
- even with the omission of labour costs and holding costs, the cattle operation produced very small amounts of profit with respect to land valued at $26.5 million.
- Mr Bryant's expert evidence that the cattle operation was not and would not be commercially viable was accepted; Mr Bryant concluded that Maraya would have been financially better off had it not engaged in cattle trading, except for the possible land tax saving.

Additional concessions and allowances

A number of concessions and allowances exist which impact the assessment of both state land tax and local government rating in NSW.

Unutilised value allowance – Land Tax Management Act 1956 – division 3

Section 62I – purpose and interpretation of division

(1) This Division applies for the purposes of section 9A (concession for unutilised land value).

(2) Expressions used in this Division have the same meanings as in Division 2 of Part 8 of Chapter 15 of the Local Government Act *1993*, except to the extent that such a meaning would be inconsistent with the meaning given by this Act.

Section 62J – land that is eligible to have unutilised value ascertained

(1) Land is eligible to have an unutilised value allowance ascertained for its land value as at 1 July in a year if it satisfies the description in either of the following paragraphs as at midnight on 30 June in that year:
  (a) a parcel of land used or occupied solely as the site of a single dwelling-house and which is, under an environmental planning instrument, zoned or otherwise designated for use for the purposes of industry, commerce or the erection of residential flat buildings;
  (b) a parcel of land (which may comprise one or more lots or portions in a current plan within the meaning of the Conveyancing Act 1919) used or occupied solely as the site of a single dwelling-house and which is, under an environmental planning instrument, zoned or otherwise designated, so as to permit its subdivision for residential purposes;
  (c) a parcel of rural land (which may comprise one or more lots or portions in a current plan within the meaning of the Conveyancing Act 1919) which is zoned or otherwise designated under an environmental planning instrument so as to permit its use otherwise than as rural land, or its subdivision into two or more lots or portions, one or more of which has an area of less than 40 hectares.

Section 62K – unutilised value allowance to be ascertained on application of owner

(1) The owner of the land may apply to the Chief Commissioner for an unutilised value allowance to be ascertained for the land value of the land. The application must be in the form required by the Chief Commissioner and be accompanied by such supporting information as the Chief Commissioner may request.

(1A) If satisfied that the land to which such an application relates satisfies the description in any of the paragraphs of section 62J (1), the Chief Commissioner must refer the application to the Valuer-General for determination of an unutilised value allowance.

(2) The Valuer-General must then ascertain the allowance if the land is eligible to have that allowance ascertained.

(3) An allowance ascertained by the Valuer-General under this Division is to be entered by the Valuer-General in the Register in respect of the land value to which it relates and is to be shown in any assessment to which it is applicable.

Section 62L – how unutilised value allowance is ascertained

(1) The unutilised value allowance for a land value is the amount calculated by deducting from the land value of the land the value that the land would have if it could be used only as the site of a single dwelling-house.

(2) However, no account is to be taken of any portion of the land which is in excess of that which is reasonably necessary to be occupied or used in conjunction with the single dwelling-house.

Section 62M – unutilised value allowance to be re-ascertained in certain cases (refer to Act)

Section 62N – unutilised value allowance can be objected to (refer to Act)

Valuation of Land Act 1916

A number of additional allowances and concessions are available under the Valuation of Land Act 1916, of which a summary of these follows.

Sections 14X and 14BBA – apportionment factors for mixed use land

Under Division 5A, Sections 14BBA to 14BBE of the Valuation of Land Act 1916, an owner of the land may apply for an allowance for land that has a mixed use purpose; for example, a shop and residence. The apportionment factor is expressed as a percentage being the rental value part of the land that is occupied or used for non-residential purposes and the proportion that bears to the rental value of the mixed use land as a whole. Whilst not explicitly covered in the Act, the allowance is only provided where the residence component of the property is fully self-contained (has its own kitchen and bathroom) and does not utilise any part of the shop and, secondly, has separate access to the shop. An objection may also be lodged against the apportionment factor made.

Section 14L – allowances for profitable expenditure to or on land

Where expenditure has been incurred in relation to land, an application may be made to the Valuer-General for an allowance for that expenditure under section 14 of the Valuation of Land Act. Items for which an allowance for expenditure may be made include *Land Improvements*, on or to the land as defined under Section 4 of the Valuation of Land Act 1916, and may also include any visible and effective improvements which, although not on the land, provide the supply of water to the land or, for the purposes of draining the land, protect it from inundation or other provisions for the beneficial use of the land.

An allowance for profitable expenditure lasts for 15 years or expires if the property is sold within 15 years or if the expenditure was not incurred by an occupier or lessee of the land. The allowance must not exceed the cost of the improvements as at the date to which the allowance is being determined. An allowance may be contended to the Valuer-General with relevant supporting evidence of relevant expenditure. Alternatively, the Valuer-General will determine the allowance, if applicable, based on any information provided by the applicant. Provisions exist for objection to the amount of allowance provided by the Valuer-General.

Sections 14S–14W

A *subdivider*, as defined under section 14S, is to apply for an allowance prior to the registration of the plan of subdivision on the basis of owning all the land comprising lots in the plan. The allowance is the difference between the total of the land values of the lots had they been sold separately and the total of the land values of the lots had they been sold to one person. The allowance amounts to a discount for sale in one line.

New South Wales – council rating

Section 554 of the Local Government Act 1993 provides that all land in an area is rateable unless it is exempt from rating. Section 555 sets out what land is exempt from all rates. A concession for the postponement of rates is available under *Part 8 – Concessions, Division 2 – Other concessions.* The key sections of the Local Government Act 1993 set out the operation of this concession as follows:

Section 585 — who may apply for postponement of rates?

The rateable land described in any of the following paragraphs may apply to the council for a postponement of rates payable for the land in the current or following rating year (or in both years):

(a) a parcel of land on which there is a single dwelling-house used or occupied as such and which is zoned or otherwise designated for use under an environmental planning instrument for the purposes of industry, commerce or the erection of residential flat buildings, not being land referred to in paragraph (b) or (c);
(b) a parcel of land (which may comprise one or more lots or portions in a current plan) on which there is a single dwelling-house used or occupied as such and which is zoned or otherwise designated under an environmental planning instrument so as to permit its subdivision for residential purposes, not being land referred to in paragraph (c);
(c) a parcel of rural land (which may comprise one or more lots or portions in a current plan) which is zoned or otherwise designated under an environmental planning instrument so as to permit its use otherwise than as rural land, or its subdivision into two or more lots or portions, one or more of which has an area of less than 40 hectares.

Section 592 – interest on postponed rates

Interest accrues on parts of rates postponed under this Division as if the rates were overdue rates. For this purpose, the due dates for payment are taken to be the respective dates on which the parts of the payable rates became due.

Section 595 – rates to be written off after five years

(1) If five years have elapsed since the commencement of a rating year for which part of the rates levied on land have been postponed under this Division, the part postponed and any interest accrued on that part must be written off by the council.
(2) Nothing in this section affects the right of the council to recover rates and interest, even though they have been written off under this section, if it subsequently appears to the council that they should not have been written off.

Section 596 – change of circumstances

A rateable person for land for which an application has been made under this Division, but not determined, or for which a determination or redetermination of the attributable part of the land value is in force, must inform the council (within one month) if land used or occupied solely as a site for a single dwelling-house, or as rural land, ceases to be so used or occupied.

Section 601 – hardship resulting from certain valuation changes

(1) A ratepayer who, as a consequence of the making and levying of a rate on a valuation having a later base date than any valuation previously used by a council for the making and levying of a rate, suffers substantial hardship, may apply to the council for relief under this section.
(2) The council has discretion to waive, reduce or defer the payment of the whole or any part of the increase in the amount of the rate payable by the ratepayer in such circumstances for such period and subject to such conditions as it thinks fit.
(3) An applicant who is dissatisfied with a council's decision under this section may request the council to review its decision and the council, at its discretion, may do so.

Queensland

Land tax threshold

Queensland has two multi-tiered threshold structures for the assessment of land tax. For individuals, the tax-free threshold for 2015 is $599,999, which excludes land owned and used as the principal place of residence. For companies, trustees and absentee owners, the tax-free threshold is $349,999 for the 2015 land tax year. The tax-free thresholds are not indexed or adjusted annually by reference to movement of site values across the state of Queensland as per New South Wales. As at 2014, the tax-free threshold for individuals is higher than the tax-free threshold in New South Wales. The tax-free threshold and rates for land tax purposes in Queensland are set out under

Schedule 1 and 2 of the Land Tax Act 2010 and are re-determined at the discretion of the government.

Land tax exemptions

Several exemptions from land tax, in addition to the principal place of residence, are set out under *Part 6 –Exempt land, Division 2 – Charitable land,* and *Division 3 – Other exemptions,* under the Land Tax Act 2010. The main exemptions under these provisions are set out in the following sections of this Act:

Section 46 – meaning of exempt purpose

In this division –

exempt purpose means each of the following –

(a) activities of a religious nature;
(b) a public benevolent purpose;
(c) an educational purpose;
(d) conducting a kindergarten;
(e) the care of sick, aged, infirm, afflicted or incorrigible people;
(f) the relief of poverty;
(g) the care of children by –
  (i) being responsible for them on a full-time basis; and
  (ii) providing them with all necessary food, clothing and shelter; and
  (iii) providing for their general wellbeing and protection;
(h) another charitable purpose or promotion of the public good;
  (i) providing a residence to a minister or members of a religious order who is or are engaged in an object or pursuit of a kind mentioned in any of paragraphs (a) to (h).

Section 47 – exemption for land owned by or for charitable institution (refer to this section)

Section 51 – aged care facilities

(1) Land on which an aged care facility is located is exempt land.
(2) In this section, *aged care facility* means a facility at which residential care is provided by an approved provider within the meaning of the Aged Care Act 1997 (Cwlth), schedule 1.

Section 52 – government land

(1) Land owned by the Commonwealth or the State is exempt land.
(2) Land owned by a local government or public authority is exempt land unless the entity is subject to State taxation under an Act of the Commonwealth or a State.

Section 53 – land used for primary production (covered in detail later in the next section)

Section 54 – moveable dwelling parks

(1) This section applies to land that is used predominantly as a moveable dwelling park if more than 50 per cent of the sites in the moveable dwelling park are occupied, or solely available for occupation, for residential purposes for periods longer than six weeks at a time.
(2) The land is exempt land.
(3) In this section –
- caravan – see the Residential Tenancies and Rooming Accommodation Act 2008, section 7.
- manufactured home – see the Manufactured Homes (Residential Parks) Act 2003, section 10.
- moveable dwelling park means a place where caravans or manufactured homes are situated for occupation on payment of consideration.
- site, for a moveable dwelling park, means a site in the moveable dwelling park where a caravan or manufactured home is or is intended to be situated.

Section 55 – port authority land

Section 56 – recreational and public land

(1) This section applies to land that is –
    (a) owned by or held in trust for a person or society; and
    (b) used or occupied by the person or society solely as the site of one or more of the following –
        (i) a public library, institute or museum;
        (ii) a showground;
        (iii) a public cemetery or public burial ground;
        (iv) a public garden, public recreation ground or public reserve;
        (v) a public road;
        (vi) a fire brigade station.
(2) Also, this section applies to land that is solely the site of a building owned or held in trust for and occupied by a society, club or association not carried on for monetary profit.

Section 57 – retirement villages

(1) Land used for premises or facilities for residents of a retirement village is exempt land.
(2) In this section, retirement village – see the Retirement Villages Act 1999, section 5.

Section 58 – other exempt land

The following land is exempt land –

(a) land owned by or held in trust for a company registered under the Corporations Act that is a friendly society for the purposes of another law of the State or the Commonwealth;
(b) land owned by or held in trust for a trade union, if the land is not used to carry on a business for profit;
(c) land held by the trustees of the estates of the late James O'Neil Mayne and Mary Emilia Mayne.

Primary production land

Section 53 – land used for primary production

(1) This section applies to land or a part of land that is used solely for the business of agriculture, pasturage or dairy farming.
(2) The land or part of the land is exempt land if it is owned by any of the following –
    (a) an individual, other than a trustee or absentee;
    (b) a trustee of a trust, if all beneficiaries of the trust are persons mentioned in paragraph (a), (c) or (d);
    (c) a relevant proprietary company;
    (d) a charitable institution.
(3) For this section, if part of the land is exempt land, the commissioner must apportion the taxable value of the land between use for a purpose mentioned in subsection (1) and use for any other purpose.
(4) This section does not apply to land owned by the manager of a time-sharing scheme.
(5) In this section –
    - beneficiary includes a beneficiary in the first instance and a beneficiary through a series of trusts.
    - exempt foreign company – see the Corporations Act, section 9.
    - proprietary company – see the Corporations Act, section 9.
    - relevant proprietary company means a proprietary company –

        (a) that is not an exempt foreign company; and
        (b) in which no share or interest is held, whether directly or through interposed companies or trusts, by a body corporate other than a proprietary company that is not an exempt foreign company.

The exemption under section 53 of the Act is nondescript in defining the lands highest and best use. In the first instance it makes no reference to the zoning of land or planning regulations governing use of land, which hence places emphasis on the use elected by the owner of the land. There is no dominant purpose test criteria or use test. Further, there is no distinction in the granting of the exemption between land zoned rural and land zoned urban rural, in which the latter, the land may be ripe for subdivision. In its current form, the section may encourage land banking by using land for an alternate rural use. Several years of land banking would be achieved before a profit may be realised from primary production. Based on industry standards of the various rural uses of land, a DSE equivalent is needed within the definition of a rural land use for the gross production from the land on a per hectare basis.

Additional concessions and allowances

Further to the exemptions under the Land Tax Act 2010, and number of allowances and valuation directives, apply under the Land Valuation Act 2010, of which a summary follows:

Land Valuation Act 2010

PART 2 – DIVISION 5

The three allowances under this Act relate to the deduction for site improvement costs under *Sections 38–44, Exclusive use as a single dwelling house or for farming; Sections 45–48, Discounting for subdividing for land not yet developed Sections 49–51.*

PART 3 – DIVISION 3

Sections 56–59 address the basis on which one valuation for adjoining lots will apply.

Council rate exemptions

The Local Government Act 1993 sets out the provisions on which land is exempt from local government rating in Queensland under Section 93.

Section 93 – land on which rates are levied

(1) Rates may be levied on rateable land.
(2) Rateable land is any land or building unit in the local government area that is not exempted from rates.
(3) The following land is exempted from rates –
  (a) unallocated State land within the meaning of the Land Act;
  (b) land that is occupied by the State or a government entity, unless –
    (i) the government entity is a GOC or its subsidiary (within the meaning of the Government Owned Corporations Act 1993) and the government entity is not exempt from paying rates; or
    (ii) the land is leased to the State or a government entity by someone who is not the State or a government entity;

(c) land in a state forest or timber reserve, other than land occupied under –
  (i) an occupation permit or stock grazing permit under the Forestry Act; or
  (ii) a lease under the Land Act;
(d) Aboriginal land under the Aboriginal Land Act 1991, or Torres Strait Islander land under the Torres Strait Islander Land Act 1991, other than a part of the land that is used for commercial or residential purposes;
(e) the following land under the Transport Infrastructure Act –
  (i) strategic port land that is occupied by a port authority, the State or a government entity;
  (ii) strategic port land that is occupied by a wholly owned subsidiary of a port authority and is used in connection with the Cairns International Airport or Mackay Airport;
  (iii) existing or new rail corridor land;
  (iv) commercial corridor land that is not subject to a lease;
(f) airport land, within the meaning of the Airport Assets (Restructuring and Disposal) Act 2008, that is used for a runway, taxiway, apron, road, vacant land, buffer zone or grass verge;
(g) land that is owned or held by a local government, unless the land is leased by the local government to someone other than another local government;
(h) land that is –
  (i) primarily used for showgrounds or horseracing; and
  (ii) exempted from rating by resolution of a local government;
(i) land that is exempted from rating, by resolution of a local government, for charitable purposes;
(j) land that is exempted from rating under –
  (i) another Act; or (ii) a regulation, for religious, charitable, educational or other public purposes.

(4) The land mentioned in subsection (3)(f) stops being exempted land when either of the following events first happens –
(a) a development permit or compliance permit under the Planning Act comes into force for the land for a use that is not mentioned in subsection (3)(f);
(b) development within the meaning of the Planning Act (other than reconfiguring a lot) starts for a use that is not mentioned in subsection (3)(f).

Victoria

Land tax thresholds

Victoria has two multi-tiered threshold structures for assessment of its land tax. For individuals, the tax-free threshold for 2015 is $250,000, which excludes land owned and used as the principal place of residence. For trusts, the tax-free threshold is $25,000 for the 2015 land tax year. The tax-free thresholds are not indexed or adjusted annually by reference to movement of site values across the state. The land tax thresholds and rates for the assessment of land tax in Victoria are set out in Schedule 1 of the Land Tax Act 2005 and these are determined at the discretion of government.

Land tax exemptions

Part 4 of the Land Tax Act 2005 sets out the exemptions and concessions for land tax in Victoria. Part 4 comprises 8 Divisions, of which a summary of the exemptions and concessions follows:

Division 1 – principal place of residence (sections 52–63; see chapter 5)

Division 2 – primary production land (sections 64–70)

Section 64 sets out a comprehensive description of the geographic boundary of Melbourne city in delineating for land used for primary production purposes, both within and outside its metropolitan boundary. It further provides a detailed dictionary of definitions for the operation of the primary production definition. Within the dictionary are the uses defining primary production which follow:

greater Melbourne means the aggregate area consisting of –

(a) the area within the municipal district of each Council listed in Part 1 of Schedule 2; and
(b) the area within an urban growth boundary specified in a planning scheme that is in force in the municipal district of each Council listed in Part 2 of Schedule 2;

primary production means –

(a) cultivation for the purpose of selling the produce of cultivation (whether in a natural, processed or converted state); or
(b) the maintenance of animals or poultry for the purpose of selling them, their natural increase or bodily produce; or
(c) the keeping of bees for the purpose of selling their honey; or
(d) commercial fishing, including the preparation for commercial fishing or the storage or preservation of fish or fishing gear; or section 64
(e) the cultivation or propagation for sale of plants seedlings, mushrooms or orchids;

Section 65 – exemption of primary production land outside greater Melbourne

(1) Land outside greater Melbourne that is used primarily for primary production is exempt land.
(2) If a part of any land outside greater Melbourne is used primarily for primary production, then that part is exempt land even if an activity other than primary production is carried on on any other part of the land.

Section 66 – exemption of primary production land in greater Melbourne but not in an urban zone

Land is exempt land if the Commissioner determines that the land comprises one parcel that is –

(a) wholly or partly in greater Melbourne; and
(b) none of which is within an urban zone; and
(c) is used primarily for primary production.

Section 67 – exemption of primary production land in an urban zone in greater Melbourne

(1) Land is exempt land if the Commissioner determines that –
 (a) the land comprises one parcel that is –
  (i) wholly or partly in greater Melbourne; and
  (ii) wholly or partly in an urban zone; and
  (iii) used solely or primarily for the business of primary production; and
 (b) the owner of the land is a person specified in subsection (2).
(2) The owner of the land must be –
 (a) a natural person who is normally engaged in a substantially full-time capacity in the business of primary production of the type carried on on the land; or
 (b) a proprietary company (not acting in the capacity of trustee of a trust) –
  (i) in which all the shares are beneficially owned by natural persons; and
  (ii) the principal business of which is primary production of the type carried on on the land; or
 (c) a trustee of a trust (other than a discretionary trust) of which –
  (i) the principal business is primary production of the type carried on on the land; and
  (ii) each beneficiary is a natural person who is entitled under the trust deed to an annual distribution of the trust income; and
  (iii) at least one of the beneficiaries, or a relative of at least one of the beneficiaries, is normally engaged in a substantially full-time capacity in the business of primary production of the type carried on on the land;

The remaining parts of section 67 deal with the holding entity and specifically trusts and beneficiaries.

Section 68 – exemption of land being prepared for use for primary production

(1) Land is exempt land for a tax year if the Commissioner is satisfied that –
 (a) the land is being prepared for use primarily for primary production; and
 (b) the land will become exempt land under section 65, 66 or 67 within 12 months after the day on which the preparation referred to in paragraph (a) commenced.

(2) The Commissioner may extend the period referred to in subsection (1) (b) by a further period of 12 months.

Section 69 – application for exemption under section 66, 67 or 68

To obtain an exemption from land tax under section 66, 67 or 68, the owner of the land must –

(a) apply to the Commissioner for an exemption under the section; and
(b) give the Commissioner any information the Commissioner requests for the purpose of enabling the Commissioner to determine whether the land is exempt under the section.

Section 70 – parcels of land

(1) For the purposes of section 66, a part of a parcel of land is to be regarded as a separate parcel of land if that part is occupied separately from or is obviously adapted to being occupied separately from other land in the parcel.
(2) For the purposes of section 67, a part of a parcel of land is to be regarded as a separate parcel of land if –
 (a) that part is occupied separately from or is obviously adapted to being occupied separately from other land in the parcel; and
 (b) the owner of the parcel of land is the owner of the land within the meaning of section 67(2).

Division 3 – sporting, recreational and cultural land (sections 71–73)

Section 71 – land leased for sporting, recreational or cultural activities by members of the public

(1) Land vested in a person or body is exempt land if the Commissioner determines that –
 (a) it is leased for outdoor sporting, outdoor recreational, outdoor cultural or similar outdoor activities and is available for use for one or more of those activities by members of the public; and

(b) the proceeds from the leasing are applied exclusively by the person or body for charitable purposes.

(2) To obtain an exemption from land tax under this section, the owner of the land must –

(a) apply to the Commissioner for the exemption; and

(b) give the Commissioner any information the Commissioner requests for the purpose of enabling the Commissioner to determine whether the land is exempt under this section.

Division 4 – charities and health services (sections 74–74A)

Section 74 – charitable institutions and purposes

(1) Land is exempt land if the Commissioner determines that –

(a) it is used by a charitable institution exclusively for charitable purposes; or

(b) it is –

(i) owned by a charitable institution; and

(ii) vacant; and

(iii) declared by its owner to be held for future use for charitable purposes.

(2) If the Commissioner is satisfied that only a part of land is used by a charitable institution exclusively for charitable purposes –

(a) land tax is assessable on the remaining part of the land, unless another exemption applies to that part; and

(b) section 22 applies, if necessary, for that purpose.

(3) To obtain an exemption from land tax under this section, the owner of the land must –

(a) apply to the Commissioner for the exemption; and

(a) give the Commissioner any information the Commissioner requests for the purpose of enabling the Commissioner to determine whether the land is exempt under this section.

Division 5 – accommodation (sections 75–78A)

(1) Land is exempt land if the Commissioner determines that the land is used and occupied –

(a) as a rooming house (within the meaning of the Residential Tenancies Act 1997) that is registered under Part 6 of the Public Health and Wellbeing Act 2008; and

(b) primarily for low cost accommodation by people with low incomes, in accordance with guidelines issued by the Commissioner for the purposes of this section.

(2) The guidelines may include provisions with respect to the following –

(a) the circumstances in which accommodation is taken to be low cost accommodation;

(b) the types and location of premises in which low cost accommodation may be provided;

(c) the number and types of persons for whom the accommodation must be provided;
(d) the circumstances in which, and the arrangements under which, the accommodation is provided;
(e) maximum tariffs for the accommodation;
(f) periods within which tariffs may not be increased;
(g) the circumstances in which the applicant is required to give an undertaking to pass on the benefit of the exemption from taxation to the persons for whom the accommodation is provided in the form of lower tariffs.

(3) A guideline –
(a) may be of general or limited application;
(b) may differ according to differences in time, place or circumstances.

(4) To obtain an exemption from land tax under this section, the owner of the land must –
(a) apply to the Commissioner for the exemption; and
(b) give the Commissioner any information the Commissioner requests for the purpose of enabling the Commissioner to determine whether the land is exempt under this section.

(5) If the Commissioner is satisfied that only a part of land is used and occupied as a rooming house or primarily for low cost accommodation by people with low incomes –
(a) land tax is assessable on the remaining part of the land, unless another exemption applies to that part; and
(b) section 22 applies, if necessary, for that purpose.

(6) Without limiting the other ways in which land ceases to be exempt under this section, land ceases to be exempt if a person breaches an undertaking given as referred to in subsection (2)(g).

Division 6 – public government and municipal land

Section 79 – Crown land

(1) Land is exempt land if it is –
(a) the property of the Crown in right of Victoria; or
(b) vested in a Victorian Minister.

(2) Subsection (1) does not apply to land that is held by –
(a) a person who is entitled to the land under a lease of Crown land, unless the lease is a retail premises lease within the meaning of the Retail Leases Act 2003; or
(b) a licensee of vested land under Part 3A of the Victorian Plantations Corporation Act 1993; or
(c) a person who is entitled to the land under a licence of Crown land under which the person has a right, absolute or conditional, of acquiring the fee simple.

Division 7 – general exemptions (sections 83–86)

Division 8 – exemptions on transition easements (sections 87 and 88)

Local government rating exemptions

The Local Government Act 1989 sets out what land is rateable and what land is exempt under section 154 as follows:

Section 154 – what land is rateable?

(1) Except as provided in this section, all land is rateable.
(2) The following land is not rateable land –
 (a) land which is unoccupied and is the property of the Crown or is vested in a Minister, a Council, a public statutory body or trustees appointed under an Act to hold that land in trust for public or municipal purposes;
 (b) any part of land, if that part –
  (i) is vested in or owned by the Crown, a Minister, a Council, a public statutory body or trustees appointed under an Act to hold that land in trust for public or municipal purposes; and
  (ii) is used exclusively for public or municipal purposes;
 (c) any part of land, if that part is used exclusively for charitable purposes;
 (d) land which is vested in or held in trust for any religious body and used exclusively –
  (i) as a residence of a practising Minister of religion; or
  (ii) for the education and training of persons to be Ministers of religion; or
  (iii) for both the purposes in subparagraphs (i) and (ii);
 (e) land which is used exclusively for mining purposes;
 (f) land held in trust and used exclusively –
  (i) as a club for or a memorial to persons who performed service or duty within the meaning of section 3(1) of the Veterans Act 2005; or
  (ii) as a sub-branch of the Returned Services League of Australia; or
  (iii) by the Air Force Association (Victoria Division); or
  (iv) by the Australian Legion of Ex-Servicemen and Women (Victorian Branch).
(3A) For the purposes of subsections (2)(a) and (2)(b), any part of the land is not used exclusively for public or municipal purposes if –
 (a) it is used for banking or insurance; or
 (b) a house or flat on the land –
  (i) is used as a residence; and (ii) is exclusively occupied by persons including a person who must live there to carry out certain duties of employment; or

(c) it is used by the Metropolitan Fire Brigades Board.

(3A) For the purposes of subsection (2)(b), any part of land does not cease to be used exclusively for public purposes only because it is leased –

(a) to a rail freight operator within the meaning of the Transport (Compliance and Miscellaneous) Act 1983; or

(b) to a passenger transport company within the meaning of that Act.

South Australia

Land tax rates and thresholds

Section 8A of the Land Tax Act 1936 sets out the land tax scales and tax-free threshold for 2010/11 and the land tax years beyond.

Subsection (3), subject to this section, states that for the 2011/2012 financial year and for each subsequent financial year ("year x"), each of the thresholds will be adjusted to take into account increases in the site value of land according to the following formula:

$$\text{Threshold}_{\text{"year x"}} = \text{Threshold}_1 \times \text{Index Value}_{\text{"year x"}}$$ where –

"Threshold" $_{\text{"year x"}}$	represents each of the thresholds for the relevant financial year ("year x")
"Threshold" $_{\text{"1"}}$	represents each of the relevant thresholds set out in subsection (2) for the 2010/2011 financial year
"Index value" $_{\text{"year x"}}$	= Index value year x_{-1} x (1 + avg percentage change in site values year $_x$) where "Index value" $_{\text{"year x"}}$ is the index value for the relevant financial year ("year x") and the average percentage change in site values for that financial year is determined under subsection (4), and with the index value for the 2010/2011 financial year being 1.

Land tax exemptions

The imposition and exemptions for State land tax in South Australia are set out under Section 4 of the Land Tax Act 1936. Primary production land is included as an exemption within this Section.

Section 4 — imposition of land tax

(1) Taxes are imposed on all land in the State, with the following exceptions:

(a) land of the Crown that is not subject to –

(i) a perpetual lease; or

(ii) an agreement for sale or right of purchase;

(b) park land, public roads, public cemeteries and other public reserves;

(c) land used solely for religious purposes, solely for the purposes of a hospital subsidised by the Government of the State or by any library or other institution administered by the Libraries Board of South Australia;

(d) land that is –

(i) owned by an association whose objects are or include the supplying to necessitous or helpless persons of living accommodation, food, clothing, medical treatment, nursing, prematernity or maternity care or other help, either without cost to such persons or in return for payments or services the amount or value of which is, in the Commissioner's opinion, substantially less than the value of the accommodation, food, clothing, treatment, nursing, care or help supplied; and

(ii) solely or mainly used for all or any such purposes;

(e) land that is –

(i) owned by an association which receives an annual grant or subsidy from money voted by Parliament; and

(ii) in the Commissioner's opinion, solely or mainly used for the purposes for which the grant or subsidy is made;

(f) land that is let to or occupied by an association of the kind mentioned in paragraph (d) or (e), and that is used solely or mainly for purposes mentioned in those paragraphs, and for which the association pays either no rent or other sum or a rent or other sum that, in the Commissioner's opinion, is a nominal one;

(g) land that –

(i) is owned by an association whose object is, or whose objects include, the conservation of native fauna and flora; and

(ii) is, in the opinion of the Commissioner, used without profit to the association or any other person, solely or mainly as a reserve for the purpose of conserving native fauna or flora;

(h) land that is owned or occupied without payment by any person or association carrying on an educational institution otherwise than for pecuniary profit, and that is occupied and used solely or mainly for the purposes of such an institution (but this exemption does not extend to land or buildings held as an investment and not being the site or grounds of the institution);

(i) land that is owned by –

(i) a municipal or district council; or

(ii) a controlling authority established under Part 19 of the Local Government Act 1934 ; or

(iii) the Renmark Irrigation Trust;

(j) land that is owned by an association established for a charitable, educational, benevolent, religious or philanthropic purpose

(whether or not the purpose is charitable within the meaning of any rule of law) and is declared by the Commissioner to be exempt from land tax on the ground –
(i) that the land is or is intended to be used wholly or mainly for that purpose; or
(ii) that the whole of the net income (if any) from the land is or will be used in furtherance of that purpose;

(k) land that is owned by –
(i) an association that holds the land wholly or mainly for the purpose of playing cricket, football, tennis, golf, bowling or other athletic sports or exercises; or
(ii) an association that holds the land wholly or mainly for the purpose of horse racing, trotting, dog racing, motor racing or other similar contests; or
(iii) an association of former members of the armed forces or of dependents of former members of the armed forces that holds the land for the social or recreational purposes of its members; or
(iv) an association of employers or employees, registered under a law of the Commonwealth or of the State, relating to industrial conciliation and arbitration that occupies the land for the purposes of the association; or
(v) an association that holds the land wholly or mainly for the recreation of the local community; or
(vi) an association that holds the land for the purpose of agricultural shows, and exhibitions of a similar nature; or
(vii) an association that holds the land for the purpose of preserving buildings or objects of historical value on the land; or
(viii) a prescribed association or an association of a prescribed kind that is declared by the Commissioner to be exempt from land tax on the ground that the whole of the net income (if any) from the land is used in furtherance of the objects of the association and not for securing a pecuniary profit for the association or any of its members;
(l) land used for primary production other than such land that is situated within a defined rural area;
(m) land that is owned by a prescribed body and used for the benefit of the Aboriginal people;
(n) land that is wholly exempt from land tax under section 5.

Subsections (2 to 5) provide further qualifications.

Local government rating

Exemptions of local government rates in South Australia are addressed under Section 147 of the Local Government Act 1999.

Section 147 — rateability of land

(1) All land within the area of a council is rateable, except for land within a specific exemption (see especially subsection (2)).
(2) The following is not rateable:
  (a) unalienated Crown land;
  (b) land used or held by the Crown or an instrumentality of the Crown for a public purpose (including an educational purpose), except any such land –
    (i) that is held or occupied by the Crown or instrumentality under a lease or licence; or
    (ii) that constitutes domestic premises;
  (c) land (not including domestic or residential premises) occupied by a university established by statute;
  (d) land that is exempt from rates or taxes by virtue of the Recreation Grounds Rates and Taxes Exemption Act 1981;
  (e) land within the area of the District Council of Coober Pedy that is subject to a mining lease under the Mining Act 1971 or a precious stones tenement under the Opal Mining Act 1995;
  (f) land occupied or held by the council, except any such land held from a council under a lease or licence;
  (g) land occupied by a subsidiary where the land is situated in the area of the council that established the subsidiary or a constituent council (as the case may be);
  (ga) land occupied or held by an emergency services organisation under the Fire and Emergency Services Act 2005;
  (h) land that is exempt from council rates under or by virtue of another Act.
(3) If land is divided by a strata plan under the *Strata Titles Act 1988* –
  (a) rates will be assessed against the units and not against the common property; but
  (b) the equitable interest in the common property that attaches to each unit will be regarded, for the purpose of valuation, as part of the unit.
(4) If land is divided by a primary, secondary or tertiary plan of community division under the *Community Titles Act 1996* –
  (a) in the case of the division of land by a primary plan – rates will be assessed against the primary lots that are not divided by a secondary plan and against the development lot or lots (if any);
  (b) in the case of the division of land by a secondary plan – rates will be assessed against the secondary lots that are not divided by a tertiary plan and against the development lot or lots (if any);
  (c) in the case of the division of land by a tertiary plan – rates will be assessed against the tertiary lots and a development lot or lots (if any).

(5) If land is divided by a primary, secondary or tertiary plan of community division under the Community Titles Act 1996 –

Tasmania

Land tax thresholds and rates

The land tax rates and thresholds are set out under *Schedule 1 – Rate of land tax* of the Land Tax Rating Act 2000. The review of the rates and thresholds are determined at the discretion of the government.

Land tax exemptions

State land tax exemptions are set out under the Land Tax Act 2000, *Division 2 – Exempt land,* of which a summary of these follows:

Section 17 – exempt Crown and public lands

(1) Land tax is not payable in respect of the following:
  (a) Crown land;
  (b) public roads and public cemeteries that are not the property of any joint stock or public company;
  (c) public recreation grounds and reserves held by a State Government body;
  (d) land on which is built a public library or public museum;
  (e) . . . .
  (f) land owned by any association or society used solely by it for holding public exhibitions and not for profit or gain.
(2) For the purpose of subsection (1) –
  (a) land owned by or vested in a Government Business Enterprise is not –
    (i) Crown land; or
    (ii) land the property of and occupied by or on behalf of the Crown; or
    (iii) land vested in trust for public purposes.

Section 18 – exempt trust land

Land tax is not payable in respect of the following:

(a) land used for purposes related to a medical establishment or convalescent home and owned by, in trust for or vested in any person or body having the management or control of the medical establishment or convalescent home;
(b) land owned by, in trust for or vested in a religious denomination or religious society and used solely –

(i) for religious, charitable or educational purposes; or
(ii) for the support of aged or infirm clergy or ministers of the religious denomination or religious society or their spouses, widows, widowers or dependent children; or
(iii) as a place of worship for members of the religious denomination, religious society or a religious order; or
(iv) as a place of residence for clergy or ministers of the religious denomination or religious society or for members of a religious order;

(c) land owned by, in trust for or vested in a religious denomination or religious society, the proceeds of which are applied for a purpose specified in paragraph (b);
(d) land owned by, in trust for or vested in a charitable institution and that is –
  (i) exempt from the payment of income tax under the Income Tax Assessment Act 1997 of the Commonwealth; and
  (ii) used solely for charitable purposes;
(e) land vested in trust for public purposes;
(f) land used solely for non-profit educational purposes and owned by, in trust for or vested in a person or body having the ownership, management or control of an educational institution;
(g) land owned by, in trust for or vested in an association of ex-servicemen or of dependents of ex-servicemen and used for the purposes of the association;
(h) land owned by, in trust for or vested in a community service organisation if –
  (i) the organisation is exempt from payment of income tax under the Income Tax Assessment Act 1997 of the Commonwealth; and
  (ii) the land is not primarily used to raise income for the organisation.

Section 19 – other exempt land

Land tax is not payable in respect of the following:

(a) principal residence land or primary production land owned 50 per cent or more by a person in receipt of –
  (i) a current Pensioner Concession Card issued under a relevant Act of the Commonwealth; or
  (ii) a card that is prescribed to be equivalent to the card referred to in subparagraph (i);
(b) principal residence land or primary production land, 50 per cent or more of which is owned by a person who is –
  (i) in receipt of a special rate pension under the Veterans' Entitlements Act 1986 of the Commonwealth; and
  (ii) totally and permanently incapacitated;

(ba) land used for the purposes of a retirement village;
(bb) land used for purposes which are ancillary to the purposes of a retirement village;
(c) Aboriginal land within the meaning of the Aboriginal Lands Act 1995 used principally for Aboriginal cultural purposes;
(d) ....
(e) land in respect of which land tax was not levied pursuant to section 10(1)(q)(ix) of the Land and Income Taxation Act 1910.

Section 19A – partially exempt land: conservation covenants

Section 19B – partially exempt land: public parks and gardens

Section 19C – partially exempt land: flood levees

Local government rates

Local government rating exemptions are set out under section 87, Division 1 of Part 9 of the Local Government Act 1993 as follows:

Section 87 – exemption from rates

(1) All land is rateable except the following, which are exempt from general and separate rates, averaged area rates, and any rate collected under section 88 or 97:
  (a) land owned and occupied exclusively by the Commonwealth;
  (b) land held or owned by the Crown that –
    (i) is a national park, within the meaning of the Nature Conservation Act 2002; or
    (ii) is a conservation area, within the meaning of the Nature Conservation Act 2002; or
    (iii) is a nature recreation area, within the meaning of the Nature Conservation Act 2002; or
    (iv) is a nature reserve, within the meaning of the Nature Conservation Act 2002; or
    (v) is a regional reserve, within the meaning of the Nature Conservation Act 2002; or
    (vi) is a State reserve, within the meaning of the Nature Conservation Act 2002; or
    (vii) is a game reserve, within the meaning of the Nature Conservation Act 2002; or
    (viii) ....
    (ix) is a public reserve, within the meaning of the Crown Lands Act 1976; or
    (x) is a public park used for recreational purposes and for which free public access is normally provided; or

(xi) is a road, within the meaning of the Roads and Jetties Act 1935; or
(xii) is a way, within the meaning of the Local Government (Highways) Act 1982; or
(xiii) is a marine facility, within the meaning of the Marine and Safety Authority Act 1997; or
(xiv) supports a running line and siding within the meaning of the Rail Safety National Law (Tasmania) Act 2012;

(c) land owned by the Hydro-Electric Corporation or land owned by a subsidiary, within the meaning of the Government Business Enterprises Act 1995, of the Hydro-Electric Corporation on which assets or operations relating to electricity infrastructure, within the meaning of the Hydro-Electric Corporation Act 1995, other than wind-power developments, are located;
(d) land or part of land owned and occupied exclusively for charitable purposes;
(da) Aboriginal land, within the meaning of the Aboriginal Lands Act 1995, which is used principally for Aboriginal cultural purposes;
(e) land or part of land owned and occupied exclusively by a council.

(2) The owner of any land referred to in subsection (1) may agree to pay general or separate rates or an averaged area rate.
(3) Land occupied by a joint authority or single authority to which Part 3A applies is not exempt from rates or averaged area rates.

Western Australia

State land tax in Western Australia is administered under the provisions of the Land Tax Assessment Act 2002. Part 3 of the Act addresses exemptions concessions and rebates. A summary of the provisions of Part 3 now follow.

Part 3 – Exemptions, concessions and rebates

Division 2 – private residential property (see Chapter 5)

Division 3 – rural business land

Section 29 – land used solely or principally for rural business, exemption for:

(1) Land (except land in a non-rural zone) is exempt for an assessment year if, at midnight on 30 June in the previous financial year, it is or was used solely or principally on a commercial basis to produce income to the user from the sale of produce or stock in the course of carrying out one or more of the following kinds of rural business –

(a) an agricultural business, silvicultural business or reafforestation business;
(b) a grazing business, horse-breeding business, horticultural business, viticultural business, apicultural business, pig-raising business or poultry-farming business.

(2) However, land used as holding paddocks for stock is not exempt unless it is used in the course of carrying out a rural business.

(3) Land in a non-rural zone that is used by the owner of the land for a rural business or rural businesses is exempt from land tax for an assessment year if more than one third of the owner's total net income for the previous financial year was derived from the owner's carrying out a business or businesses of that kind in the State.

(4) However, even if subsection (3) does not apply to land in a non-rural zone used for a silvicultural business, reafforestation business or both, the land is exempt for an assessment year if –
(a) it is at least 100 hectares in area; and
(b) at midnight on 30 June in the previous financial year, it is fully stocked for the purposes of the business or businesses.

Section 30 – other rural business land, concession for

If land of a kind referred to in section 29(3) or (4) is not exempt only because less than one third of the owner's total net income for a financial year was derived from carrying out a rural business or rural businesses of that kind in the State, or because the land is less than 100 hectares in area, then the land tax on the land is payable at 50 per cent of the rate imposed for the assessment year by the Land Tax Act 2002.

Division 4 – Crown land and other land used for public purposes – Sections 31–39

31. Land owned by Crown, public authority etc., exemption for
32. Land owned by religious bodies, exemption for
33. Land owned by educational institutions, exemption for
34. Land used for public or religious hospitals, exemption for
35. Mining tenements, exemption for
36. Land used for various public purposes, exemption for
37. Land owned by public charitable or benevolent institutions, exemption for
38. Land owned by non-profit associations, exemption or concession for
38A. Land used as aged care facility, exemption for
39. Land used for retirement villages, exemption for

Division 4A – land used for non-permanent residences (sections 39A and 39B)

39A. Land to which s. 39B applies
39B. Dwelling park land, concessions for

Division 5 – other exemptions and concessions (sections 40–43A)

40. Land owned by veteran's surviving partner or mother, exemption for
41. Land under conservation covenant, exemption for
42. Land vacated for sale by mortgagee, one-year exemption for
43A. Newly subdivided land, concession for

Local government rates

Local government rating exemptions in Western Australia, as addressed under section 6.26 of the Local Government Act 1995, as follows:

Section 6.26 – rateable

(1) Except as provided in this section all land within a district is rateable land.
(2) The following land is not rateable land –
  (a) land which is the property of the Crown and –
    (i) is being used or held for a public purpose; or
    (ii) is unoccupied, except –
      (I) where any person is, under paragraph (e) of the definition of owner in section 1.4, the owner of the land other than by reason of that person being the holder of a prospecting licence held under the Mining Act 1978 in respect of land, the area of which does not exceed 10 ha, or a miscellaneous licence held under that Act; or
      (II) where, and to the extent and manner in which, a person mentioned in paragraph (f) of the definition of owner in section 1.4 occupies or makes use of the land; and
  (b) land in the district of a local government while it is owned by the local government and is used for the purposes of that local government other than for purposes of a trading undertaking (as that term is defined in and for the purpose of section 3.59) of the local government; and
  (c) land in a district while it is owned by a regional local government and is used for the purposes of that regional local government other than for the purposes of a trading undertaking (as that term is defined in and for the purpose of section 3.59) of the regional local government; and

(d) land used or held exclusively by a religious body as a place of public worship or in relation to that worship, a place of residence of a minister of religion, a convent, nunnery or monastery, or occupied exclusively by a religious brotherhood or sisterhood; and
(e) land used exclusively by a religious body as a school for the religious instruction of children; and
(f) land used exclusively as a non-government school within the meaning of the School Education Act 1999; and
(g) land used exclusively for charitable purposes; and
(h) land vested in trustees for agricultural or horticultural show purposes; and
(i) land owned by Co-operative Bulk Handling Limited or leased from the Crown or a statutory authority (within the meaning of that term in the Financial Management Act 2006) by that co-operative and used solely for the storage of grain where that co-operative has agreed in writing to make a contribution to the local government; and
(j) land which is exempt from rates under any other written law; and
(k) land which is declared by the Minister to be exempt from rates.

Subsections (3 to 6) provide further qualification to (1) and (2).

Northern Territory

The Local Government Act is currently under review in Northern Territory. At present, the rating of land by local government is covered under the provisions of the Local Government Act 2008. Land exempt from local government rating is set out under Section 144 of the Act as follows.

Section 144 – exempt land

(1) The following land is exempt from rates:
  (a) Crown land occupied by the Territory for a public purpose (other than the provision of public housing);
  (b) land of the council, other than such land leased for a purpose that does not give rise to an exemption under some other provision of this section;
  (c) a public place consisting of:
    (i) a park, garden or reserve; or
    (ii) a playground or sports ground; or
    (iii) a cemetery; or
    (iv) a road;
  (d) land belonging to a religious body consisting of:
    (i) a church or other place of public worship; or
    (ii) a place of residence for a minister of religion associated with a church or other place of public worship; or

(iii) a place of residence for the official head in the Territory of the religious body; or
(iv) an institution for religious teaching or training;

(e) a public hospital;
(f) land used for a non-commercial purpose by a public benevolent institution or a public charity;
(g) a kindergarten, Government school as defined in section 4(1) of the Education *Act*, non-Government school registered under the Education Act, or a university or other tertiary educational institution;
(h) land recognised by the council as a youth centre;
(i) a public library or public museum;
(j) the common property:
(i) in a units plan or building development plan registered under the Real Property (Unit Titles) Act; or
(ii) of a scheme formed under the Unit Title Schemes Act;
(k) land owned by a Land Trust or an Aboriginal community living area association except:
(i) land designated in the regulations as rateable; or
(ii) land subject to a lease or a licence conferring a right of occupancy; or
(iii) land used for a commercial purpose;

(2) land exempted from rates under another Act.
(3) If land is used for two or more different purposes, and one or more, but not all, of the purposes are exempt, the land is not exempt from rates unless the non-exempt purpose is merely incidental to the exempt purpose.

Example

An allotment consists of a public museum containing a cafeteria. The existence of the cafeteria would not negative the exemption. However, if it were a restaurant attracting customers in its own right, it would do so.

(4) In deciding whether land is used for a commercial or non-commercial purpose, the fact that the user is a public benevolent institution or a public charity is irrelevant: the question is to be decided according to the nature of the use and not the nature of the user.

Land tax (not imposed in Northern Territory)

Australian Capital Territory

Land tax exempted from section 9 generally

(1) The following parcels of land are exempt from land tax imposed under section 9:
(a) a parcel of residential land owned by an individual, if the parcel is exempted under section 13 (Decision on compassionate application) in relation to the parcel; Note An exemption under s13 is for 1 year or less.

(b) a parcel of rural land;
(c) a parcel of land owned by the housing commissioner under the Housing Assistance Act 2007;
(d) a parcel of land owned or leased by an entity declared under the Duties Act 1999 section 73A (transfers etc. to entities for community housing);
(e) a parcel of land leased for a retirement village;
(f) a parcel of land leased for a nursing home;
(g) a parcel of land leased for a nursing home and a retirement village;
(h) a parcel of land leased by a religious institution or order to provide residential accommodation to a member of the institution or order and allow the member to perform his or her duties as a member of the institution or order;
(i) a parcel of land, other than a parcel of residential land leased to a corporation or trustee, being used for a purpose prescribed under the regulations.

Section 11 – land exempted from land tax

(1) The following parcels of land are exempted from land tax:
 (a) a parcel of land held under a development lease by a corporation;
 (b) a parcel of residential land owned by a not-for-profit housing corporation.
(2) A parcel of land owned by a corporation carrying on business as a builder or land developer is exempt from land tax in relation to the parcel for two years, beginning on the first day of the first quarter after the corporation becomes the owner of the parcel if –
 (a) the parcel is used by the corporation only to construct new residential premises; and
 (b) the new residential premises are to be sold by the corporation when finished.

Subsections (2) and (3) set out definitions which qualify (1) and (2).

Rate exemptions

Rates Act 2004 – section 41 exemption from rates

(1) The Minister may exempt the owner of a parcel of land from payment of rates, owing for any period in relation to the parcel of land, or from payment of a stated part of the rates.
(2) An exemption is a notifiable instrument. *Note*: A notifiable instrument must be notified under the Legislation Act.
(3) The Minister may make guidelines for the exercise of a function under this section.
(4) Guidelines are a disallowable instrument.

Section 42 – remission of rates

(1) The Minister may remit the rates or a part of the rates payable for a parcel of land if the Minister is satisfied that it is fair and reasonable in the circumstances.
(2) The Minister may make guidelines for the exercise of a function under this section.
(3) Guidelines are a disallowable instrument.

7 Objections, appeals and enforcement

Introduction

In contrast to other taxes where the tax base is readily definable by reference to income, consumption, capital gains or turnover, land tax has an additional layer of complexity in that the base of the tax is determined by independent valuation as the first step in the assessment process. This is further compounded by the fact that, unlike other taxes, in which the taxpayer has a perceived level of control or input through the lodgement of an income tax, GST or other form of return, no such taxpayer input exists with land tax. Once ownership of land is declared, land taxes are solely assessed by the government or their contractor without any reference to or input from the taxpayer.

In the absence of taxpayer input, perceived control, the lack of predictability of assessments and potential for fluctuations in value impact the tax payable, which has little relevance to the taxpayer's ability to pay and heightens speculation about the valuation process. To this end, the objection process serves as an important taxpayer outlet and in some cases constitutes taxpayer participation and input in the land tax assessment process. This can be compounded by a lack of information relating not only to the process used to determine land value, but also the evidence relied upon, which is an important part of the information to be provided to the taxpayer.

A 2008 review of NSW state taxes highlighted the weaknesses in the taxation of land under the principles of transparency and simplicity, as discussed in Chapter 2. Contributing to poor performance against the criteria of transparency and simplicity are concerns that taxpayers have poor information when a property exceeds the threshold and subsequently becomes liable for the land tax. Whilst the three-year averaging of values has added a further level of complexity to this tax, the key issue is making available to taxpayers the information used to value land. Here, the sales data and related information used when valuing land has been identified as key in providing greater transparency and understanding for taxpayers as to how their land value and tax was determined (NSW Ombudsman 2005 and Walton 1999).

Among factors impacting simplicity and transparency are how land taxes are imposed, monitored and enforced. The objection and appeal process extends beyond the valuation process to include the application of the tax in the first instance, the determination and assessment of an allowance and whether the property is liable or exempt from this tax. As a key and increasingly important source of revenue for state and local government, several disparate issues have evolved in the approach to the way challenges are run against the imposition of recurrent land taxes.

Some of the novel approaches to challenging this tax have resulted in devastating cash flow consequences for governments and have exposed flaws in the mechanics used to assess this tax. There is little doubt that, as revenue from recurrent land taxes expands, wholesale changes to assessment procedures, concessions and allowances, and the valuation process will be required. The objection and appeal processes are the present breakwaters for dealing with challenges against the tax; however, one of the key areas for reform will be the way objections and appeals are administered going forward. This in turn is impacted by the way land tax policy is designed and applied in the first instance.

This chapter examines the grounds of objection and appeal processes, timeframes and the longevity of matters which have taken several years to pass through the administrative hurdles and subsequent court processes. These challenges have resulted in the retrospective refunds of tax revenue and bring to the fore the newest principle of good tax design, *revenue robustness*. This chapter defines the basis and content of a valid objection against the value of land and addresses some of the emerging challenges confronting tax and valuation administrators.

The chapter is organised on a state-by-state basis, examining each of the provisions commencing with the Valuation of Land/Land Valuation statutes which govern the first step, the valuation process. This is followed by the sections of the respective Local Government Acts which set out the objection and appeal provisions governing the impost of local government rates. The relevant provisions governing objections and appeals/review of state land tax under the Land Tax/Land Tax Management and Tax Administration Acts is the final tax reviewed.

New South Wales land values – objections and appeals

Personal views on land tax need to be set aside when undertaking an objective assessment of land values assigned to property. An objective assessment is even more crucial for valuers or other professionals who may be advising a client in relation to land values and related matters. In assigning land values to property, the Valuer-General's land value is based on sales evidence, with adjustments made for differences between the sales evidence and the property to which a land value is assigned.

The premise that too much land tax is being paid, that the increase in land tax or a land value over the previous year is too high, or that it is higher than the general market are not sound bases for determining the correctness of a land value. It may well be that the Valuer-General's land value moved 15 per cent from one year to the next for a particular property, when in fact the market moved only 10 per cent. The Valuer-General's land value may have been low in the previous year or for a number of years and has been adjusted in line with the market in that year. Size of increase is not alone a basis of objection.

A similarly flawed method of determining the correctness of a land value is to compare one land value to another land value. This does not establish that either land value is correct, or is either too high or too low. It may be that both the properties' land values are conservative or high. Comparing one valuation to another is not a method of valuation. The only sound basis for determining the accuracy of a property's land value is through the use of sales evidence.

A Valuation Objection Form may be downloaded from the NSW Department of Lands website. This form provides a process for an objector to follow in correctly identifying the property on which the objection may be made and the grounds of objection. In some cases, objections are lodged in letter format to the Valuer-General. In these cases, it is important that the property is correctly identified, along with the grounds of objection and supporting evidence.

Land tax is applicable to ownership at midnight on 31 December of each year in New South Wales. Land tax assessments are usually sent out to land tax liable parties between January and March each year. The impact of receiving a land tax assessment is far greater than the announcement of any changes to land tax. A good example was the removal of the threshold in the 2005 land tax year. It is not until people can hold their land tax liability in their hand that they fully grasp and understand the gravity of changes and the impact these changes have on their circumstances.

Upon receipt of a land tax assessment, parties liable to pay land tax have a specific period of time to respond should they wish to challenge a land value. Section 35 of the Valuation of Land Act 1916 stipulates the time period as 60 days from the date of service of the notice. If a party receives a land tax assessment with the last date for objection printed on their notice being less than 60 days from the date of receipt, it is important to respond to that assessment by the last date permitted on the assessment notice. Land tax payers will need to seek advice in relation to the validity of objections lodged after the last permitted date, particularly where it can be shown that 60 days was not, in fact, provided from the date of service of the notice. Section 35A of the Valuation of Land Act makes provision for out of time objections to land values to be considered by the Valuer-General.

As discussed in Chapter 3, land values are assessed by the Valuer-General annually and provided to Office of State Revenue for the assessment of land tax. If an objection to a land tax assessment relates to the land value, the objection to the land value should be lodged with Valuer-General. Objections to the grounds set out below are made to the Valuer-General.

Grounds of objection to land values

The grounds for objecting to the Valuer-General's land value are set out under section 33 and 34 of the Valuation of Land Act 1916.

In relation to land values, the grounds upon which an objection may be made under the Act are:

1) that the values assigned are too high or too low;
2) that the area, dimensions or description of the land are not correctly stated;
3) that the interests held by various persons in the land have not been correctly apportioned;
4) that the apportionment of the valuations is not correct;
5) that lands which should be included in one valuation have been valued separately;
6) that lands which should be valued separately have been included in one valuation; and
7) that the person named in the notice is not the lessee or owner of the land.

In relation to stratum, the only grounds upon which objection may be taken under the Act are:

1) that the values assigned are too high or too low;
2) that the situation, description or dimensions of the stratum are not correctly stated;
3) that strata which should be included in one valuation have been valued separately;
4) that strata which should be valued separately have been included in one valuation; and
5) that the person named in the notice is not the lessee, occupier or owner of the stratum.

An objection to a strata unit cannot be made by an individual unit owner. Objections must be lodged by the secretary of the strata scheme or the owners' corporation. Notices of valuation are not issued to individual unit owners, but to the secretary of the owners' corporation.

The most common ground of objection is to a land value being too high. Prior to considering this ground, we will consider objections on the ground of a land value being too low.

Land value too low

It is unimaginable that an owner would object to a land value being too low. Objections to land values being too low are usually lodged by owners, lessors, or head lessors of land who lease land from which the rent for that land is determined as a percentage of the land value as determined by the Valuer-General. This is not common; however, parts of Sydney are held in leasehold, including the Rocks area, Sydney Cove and railway stations where retail space is leased on overpasses and walkways. Land is sometimes leased where the owner has sought a return from the land itself. In these cases, the land value is the basis upon which a rental may be determined.

In the above circumstances, dual objections may sometimes be lodged against a land value. An objection may be lodged by the lessor that the land value is too low, while a second objection is lodged by the lessee on the basis that the land value is too high. In some cases, a third party, namely council, may also lodge an objection in a General Valuation year where the land is being re-valued for the assessment of council rates. Council's objection would usually result on the basis of the land value being too low.

Land value too high

By far, the most objections lodged with the Valuer-General are against land values being too high. The primary issue of concern is the impact of the land value ascribed to land, being the basis on which land tax is levied. Objections against land values being too high vary from statements in letter format to letters of objection with accompanying valuation rationale and supporting evidence attached to the Valuation Objection Form.

All objections to land values are to be considered by an officer of the Valuer-General. The degree of success in having a land value considered for a reduction is to first determine that the land value is, in fact, too high or incorrect. This can only be done through the use of sales evidence. Property indexes and the use of statistics are only as good as the correctness of the underlying values that these tools and techniques are applied to. The most fundamental method of measuring the correctness of a land value is by benchmarking it against sufficient relevant sales evidence. This is the basis upon which objections lodged against land values are assessed.

Objections

The three key elements to an objection against a land value are the correct identification of the land being objected against, the grounds of the objection and the basis and evidence supporting an objector's contended land value.

In addition to the above three elements of an objection, an examination of the objection process reveals two important steps.

Step 1 – Refuting the land value assigned by the Valuer-General and the grounds upon which the land value is refuted.

Step 2 – The objector's contended land value and the basis and supporting evidence of the contended land value are put forward.

In many cases an objector addresses the first step of the process, but fails to address the second step, providing any compelling evidence to support a contended land value. In part, the argument is that the sales evidence used by the Valuer-General is not provided to land tax payers. The reality is that, in order to obtain the sales evidence, the land tax payer needs in many cases to proceed from being a land tax payer to becoming an objector, and in some cases a litigant, in order to obtain sales information.

When preparing an objection to support a contended land value, the format and detail of the information supporting the objection is important. The essentials, as briefly covered earlier, are a description of the subject property being objected against, sales evidence to support the contended land value and a detailed comparison of the sales to the subject property with adjustments where necessary. The following is an example of a format that a valuer may use when preparing a statement of evidence in the event that the objection proceeds to the Land and Environment Court.

Statement Format

- Address of the subject property
- Legal description of the subject property
- The date at which the land value is being challenged. Under the current system this is known as the base date and is always 1 July in the year the land value is being objected against.
- Contended land value
- Description of the land (size, shape, dimensions and access)
- Location of the property
- Description of the subject land (very important, particularly if there is an attribute that is not known to the Valuer-General that may impact on value, e.g., slope, shape, adjoining development, changes to flight paths or road patterns)
- Description of the improvements
- Town planning
- Grounds of objection
- Valuation approach/method
- Summary of sales evidence and comparability of each sale against the subject property being objected against

Annexed to the report should be:

- Subject property
  - Deposited plan
  - Certificate of title

  - Section 149 certificate as at the date of valuation (if applicable)
  - Contamination report (if applicable)
  - Heritage details (if applicable)
- Sales evidence (each sale)
  - Photograph
  - Title details and encumbrances
  - Deposited plan
  - Zoning details
  - Description of improvements
  - Description of land
  - Sale date (exchange and settlement)
  - Sale price
  - Discussion with either selling agent or purchaser
  - Commentary on how the sale compares to the subject property being objected against and adjustments made between these for inherent and external features.

It is important for sales evidence provided in an objection to be discussed and compared with the subject property. Sales evidence may be inferior, superior or directly comparable to the subject property. Using a suitably qualified valuer to prepare and argue sales evidence would greatly increase the chance of successfully challenging a land value *only after the land value has been determined to be too high based on sales evidence.*

Objection and response process

1) An objection to a land value is lodged in writing, with the grounds of the objection specified, and is signed by or on behalf of the objector. Section 33, Valuation of Land Act.
2) An objection to a land value must be made within the time specified under Section 35 of the Valuation of Land Act. At present, this is 60 days from date of service of notice.
3) An officer of the Valuer-General will view the objection and delegate a party who will consider and determine the objection. In accordance with Section 35B of the Valuation of Land Act, *the person who considers the objection must be a different person from, and not subordinate to, the person who made the decision against which the objection is lodged.* The person who is delegated this task is a valuer.
4) As a matter of process, the delegate of the Valuer-General considering the objection will briefly view the property that objections have been lodged against.
5) There are three possible outcomes arising out of the inspection and consideration of the objection:
   - The objection is disallowed.
   - The objection is successful at the land value contended by the objecting party.

- The objection is reduced, but not by the full amount contended in the objection.

The Valuer-General will notify the parties of the outcome of their objection in writing. In the event of the objector being dissatisfied with the outcome of their objection by the Valuer-General, they may refer the matter onto the Land and Environment Court for hearing. There is a cost in listing the matter in the court and, under section 38 of the Valuation of Land Act, an objector has 60 days after receiving a notice of outcome from the Valuer-General regarding their objection to list the matter in the court.

Objection without basis or evidence

An objection against a land value assessed without sales evidence or basis does not shed any further information or evidence of how or why the land value is considered to be too high or too low. If no additional evidence is provided, the officer considering the objection will have no further evidence than was originally used at the time the land value was assessed by the Valuer-General; therefore, in most instances the reviewing valuer may reach the same conclusion as the original valuation recommendation.

It is at this point that a percentage of objectors accept the response from the Valuer-General, being a letter of reply indicating that their objection has been considered but disallowed. The letter may go on to say that if the parties are dissatisfied with the decision, they may refer the matter on to the Land and Environment Court and challenge the land value there. It is at this stage that most objections end.

Objection with sales evidence

Objections with sales evidence and a basis for a land value contended by the objector are likely to draw more attention than objections without evidence. In years past, sales evidence used by the Valuer-General's Office and ranked by the Land and Environment Court in matters before it was selective, with greater emphasis placed on vacant land sales and/or near vacant land sales. As discussed earlier, with the outcome of the High Court ruling on sales evidence, the type of sales evidence admissible in determining land value has expanded, particularly where only a few vacant land sales exist in the locality of the property being assessed.

Sales evidence comparable to the subject property may need to be adjusted for improvements, location and time. Whilst most property owners have an idea about values, property valuers are trained to determine the value of property to a greater degree of accuracy. It may be to the objector's advantage to engage the services of a valuer to assist in determining the accuracy of a land value if in doubt as to the accuracy of a site or land value. In the event that sales evidence is produced by an objector which has been analysed and shown to support a reduction in the land value, a reduction in the land value either in part or to the objector's contended land value may be achieved.

Procedural fairness

Once an objection has been lodged against a land value, it is important to realise that the objector has commenced the first step in engaging the Valuer-General in a process that may lead to a challenge in the Land and Environment Court. The Valuer-General receives thousands, and sometimes tens of thousands, of objections to land values across NSW each year. At the objection stage, the Valuer-General does not know which objections are likely to proceed to challenges in the Land and Environment Court and treats each objection with caution. Most objectors will object to the Valuer-General and seek sales evidence supporting the land value assigned to their property. The Valuer-General is not obliged to provide sales evidence; however, on the Land and Property Information web site, reference is made to Department of Housing sales figures by locality. This primarily assists with showing the movements in trends by locality.

There are four points in time during which sales evidence may be requested from the Valuer-General:

1. Before receiving land tax assessment notices, which state the Valuer-General's land value, usually in November and December;
2. Between receiving assessment notices and objecting to a land value;
3. After objecting to a land value and before lodging the matter in the Land and Environment Court;
4. After lodging the matter in the Land and Environment Court, but before the hearing.

In the latter two cases, it may be more difficult to communicate and obtain information freely and it is more likely objectors will receive a letter stating their objection has been recorded and is being considered. As mentioned, at no stage is the Valuer-General obliged to provide sales information to objectors. Many objectors feel that this is an unfair practice and feel that the process lacks transparency, as they are not provided with all the information required to make an informed decision as to whether to contest a land value.

As highlighted in the previous four points above, the timing of request for information is an important part of the inquiry process. The provision of sales evidence is perhaps the greatest criticism of the valuation process that underpins the land tax system. This was highlighted as an item to be addressed under recommendation 6.24 of the NSW Ombudsman report issued in October 2005.

Land and environment court

When an objector is dissatisfied with the outcome of an objection by the Valuer-General, the objection must be referred to the Land and Environment Court within 60 days after the objection has been dealt with by the Valuer-General. Depending on the format of the objection lodged with the Valuer-General, the document lodged with the court may also be used as part of the statement of evidence to be tendered to the court. Parties attending the Land and Environment Court should consider seeking legal advice if they have not already done so at the objection stage of their action.

A person who is entitled to appear before the Court may appear in person, be represented by a barrister, solicitor or (except in proceedings in Class 5, 6 or 7 of the Court's jurisdiction) by an agent authorised by the person in writing (Section 63 Land and Environment Court Act 1979). Expert valuers who prepare objections for owners to the Valuer-General must be prepared to attend court if the matter is referred to court by the owner. It is important for valuers undertaking land value objection work, should the objection be disallowed and the matter proceeds to court, to attend and provide expert evidence and assistance to the court.

Council rates – objections and appeals

Objections and appeals to local government rates in NSW are made under the provisions of the Local Government Act 1993. While the grounds of appeal concerning rates are limited to those specified in section 574, opportunity is given at different points in the rate-making process for objections, submissions (including submissions by way of objection), and applications to be made to a council concerning rates. The opportunities to object includes the following points, which are supported by the relevant sections which follow:

- public notice of the draft operational plan;
- application for change of category for purposes of ordinary rate; and
- deferral and reduction of rates.

Rateable land

Section 554 of the Local Government Act provides that all land in an area is rateable unless it is exempt from rating. Section 555 sets out what land is exempt from all rates.

Section 526 – appeal against declaration of category

(1) A rateable person who is dissatisfied with:
   (a) the date on which a declaration is specified, under section 521, to take effect; or

(b) a declaration of a council, under section 525, may appeal to the Land and Environment Court.

(2) An appeal must be made within 30 days after the declaration is made.

(3) The Court, on an appeal, may declare the date on which a declaration is to take effect or the category for the land, or both, as the case requires.

Section 574 – appeal on question of whether land is rateable or subject to a charge

(1) A person who has an estate in land or who is the holder of a licence or permit for land under the Crown Lands Act 1989, in respect of which rates and charges notice is served, may appeal to the Land and Environment Court:

(a) in the case of a rate-against the levying of the rate on the ground that the land or part of it is not rateable, is not rateable to a particular ordinary rate or is not rateable to a particular special rate; or

(b) in the case of a charge-against the levying of the charge on the ground that the land is not subject to any charge (excluding a charge limited under section 503 (2)) or is not subject to the particular charge.

(2) An appeal may not be made under this section on the ground that land has been wrongly categorised under Part 3.

(3) An appeal must be made within 30 days after service of the rates and charges notice.

(4) If the Land and Environment Court determines that only a part of the land is rateable, it is required to determine the value of that part.

State land tax – objections and reviews

Objections and reviews for state land tax are addressed under the Taxation Administration Act 1996, which covers interest and penalty tax, objections and reviews, as well as hardship provisions and the enforcement of the state taxation in New South Wales. This Act makes provision for the assessment and reassessment of taxation under Part 3. Part 5 addresses interest and penalty tax a précis of the key provisions follow.

Part 5 – interest and penalty tax

The equivalence of interest on outstanding tax is the lost opportunity cost of money held by the taxpayer over the time that the tax should have been paid. Under section 22, the interest is determined on a mix of a market component rate and the premium component.

Provisions distinguishing circumstances under which penalty tax payable may be reduced apply under sections 28–30 for disclosure made before and during an investigation. Further, an increase penalty tax provision exists for concealment under section 30.

Section 21 – interest in respect of tax defaults

(1) If a tax default occurs, the taxpayer is liable to pay interest on the amount of tax unpaid, calculated on a daily basis from the end of the last day for payment until the day it is paid at the interest rate from time to time applying under this Division.
(2) Interest is payable under this section in respect of a tax default that consists of a failure to pay penalty tax under Division 2, but is not payable in respect of any failure to pay interest under this Division.

Additional provisions apply under sections 22–25.

The equivalence of interest on outstanding tax is the lost opportunity cost of money held by the taxpayer over the time that the tax should have been paid. Under section 22, the interest is determined on a mix of a market component rate and the premium component.

Section 26 – penalty tax in respect of certain tax defaults

(1) If a tax default occurs, the taxpayer is liable to pay penalty tax in addition to the amount of tax unpaid.
(2) Penalty tax imposed under this Division is in addition to interest.
(3) Penalty tax is not payable in respect of a tax default that consists of a failure to pay:
    (a) interest under Division 1; or
    (b) penalty tax previously imposed under this Division.

Additional provisions apply under sections 27–33.

Part 7 – collection of tax

Division 1 – general

Section 43 – tax payable to the Chief Commissioner

Tax that is payable is payable to the Chief Commissioner.

Section 44 – recovery of tax as a debt

If the whole or part of the tax payable by a taxpayer has not been paid to the Chief Commissioner as required, the Chief Commissioner may recover the amount unpaid as a debt to the Chief Commissioner.

Section 45 – joint and several liability

(1) If two or more persons are jointly or severally liable to pay an amount under a taxation law, the Chief Commissioner may recover the whole of the amount from them, or any of them, or any one of them.

Section 46 – collection of tax from third parties

(1) The Chief Commissioner may require any of the following persons, instead of the taxpayer, to pay tax that is payable but remains unpaid:
  (a) a person by whom any money is due, is accruing or may become due to the taxpayer;
  (b) a person who holds or may subsequently hold money for or on account of the taxpayer;
  (c) a person who holds or may subsequently hold money on account of some other person for payment to the taxpayer; and
  (d) a person having authority from some other person to pay money to the taxpayer.

Subsections 2–9 address the service of notice and payment of the tax.

Part 10 – objections and reviews

Section 86 – objections

(1) A taxpayer who is dissatisfied with:
  (a) an assessment that is shown in a notice of assessment served on the taxpayer; or
  (b) any other decision (within the meaning of the Administrative Decisions Review Act 1997) of the Chief Commissioner under a taxation law may lodge a written objection with the Chief Commissioner.

Subsection (2) sets out the provisions under which a taxpayer may not lodge such an objection (refer to the Act for full detail).

Section 87 – grounds for objection

(1) The grounds for the objection must be stated fully and in detail, and must be in writing.
(2) The grounds for the objection, in the case of a reassessment, are limited to the extent of the reassessment.

Section 88 – onus of proof and evidence on objection

On an objection, the objector has the onus of proving the objector's case.

Section 89 – time for lodging objection

(1) An objection must be lodged with the Chief Commissioner not later than 60 days after the date of service of the notice of the assessment or

the date on which the decision referred to in section 86 (1)(b) is served on the taxpayer, except as provided by section 90.

(2) An objection is taken to have been lodged with the Chief Commissioner when it is served on the Chief Commissioner.

Section 94 – recovery of tax pending objection

The fact that an objection is pending does, not in the meantime, affect the assessment or decision to which the objection relates and tax may be recovered as if no objection were pending.

Section 96 – review by Civil and Administrative Tribunal

(1) A taxpayer may apply to the Civil and Administrative Tribunal for an administrative review under the Administrative Decisions Review Act *1997* of a decision of the Chief Commissioner that has been the subject of an objection under Division 1 if:
   (a) the taxpayer is dissatisfied with the Chief Commissioner's determination of the taxpayer's objection; or
   (b) 90 days (not including any period of suspension under section 92) have passed since the taxpayer's objection was served on the Chief Commissioner and the Chief Commissioner has not determined the objection.

(2) However, a taxpayer cannot apply to the Civil and Administrative Tribunal under this section for an administrative review in respect of a decision of a kind prescribed by the regulations as an exempt decision for the purposes of this section.

(3) A taxpayer who has applied to the Supreme Court under section 97 for a review of a decision cannot apply to the Civil and Administrative Tribunal under this section in respect of the same decision. However, the taxpayer may do so if the earlier application is withdrawn with the approval of the Supreme Court for the purpose of enabling the Civil and Administrative Tribunal to deal with the matter.

Subsections 4–6 address matters relating to the Administrative Decisions Review Act 1997 and Civil Administrative Tribunal Act 2013 (refer to Act for details).

Section 97 – review by Supreme Court

(1) A taxpayer may apply to the Supreme Court for a review of a decision of the Chief Commissioner that has been the subject of an objection under Division 1 if:
   (a) the taxpayer is dissatisfied with the Chief Commissioner's determination of the taxpayer's objection; or

   (b) 90 days (not including any period of suspension under section 92) have passed since the taxpayer's objection was served on the Chief Commissioner and the Chief Commissioner has not determined the objection.
(2) A taxpayer who has applied to the Civil and Administrative Tribunal under section 96 for an administrative review in respect of a decision cannot apply to the Supreme Court under this section in respect of the same decision. However, the taxpayer may do so if the earlier application is withdrawn with the approval of the Civil and Administrative Tribunal for the purposes of enabling the Supreme Court to deal with the matter.

Under section 97, a review by the Supreme Court is taken to be an appeal for the purposes of the Supreme Court Act *1970* and the regulations and rules made under that Act, except as otherwise provided by that Act or those regulations or rules.

Section 99 – time for making application for review

(1) An application for review following a determination by the Chief Commissioner of an objection must be made not later than 60 days after the date of issue of the notice of the Chief Commissioner's determination of the objection. The court or tribunal to which the application is to be made may allow a person to apply for a review after that 60-day period.
(2) An application for review following a failure of the Chief Commissioner to determine an objection within the relevant 90-day period may be made at any time after the end of that period.

Section 100 – provisions relating to applications for review

(1) An application for review following a failure of the Chief Commissioner to determine an objection cannot be made unless the applicant has given written notice of the proposed application to the Chief Commissioner not less than 14 days before it is made.
(2) The applicant's and respondent's cases on an application for review are not limited to the grounds of the objection.

Further, under this section the onus of proof remains on the taxpayer to prove their case. This provision further applies to appeals from a decision of the Civil and Administrative Tribunal in an application for review to an Appeal Panel of the Tribunal.

Section 106A – constitution of Hardship Review Board

(1) There is to be a Hardship Review Board consisting of:
   (a) the Chief Commissioner; and
   (b) the Auditor-General; and
   (c) the Secretary of the Treasury.

Section 106B – waiver of tax

(1) The Hardship Review Board may, if authorised by a taxation law to do so, waive the payment of tax, either wholly or in part, if it is satisfied that:
  (a) the person liable to pay it is in such circumstances that the exaction of the full amount of tax would result in serious hardship for the person or the person's dependants; or
  (b) the person liable to pay it has died and that person's dependants are in such circumstances that the exaction of the full amount of tax would result in serious hardship for them.

(2) The Chief Commissioner may exercise the functions of the Hardship Review Board under this section if the amount of the unpaid tax is less than $2,000 in any particular case for any financial year. Section 106C Deferral and the writing-off of tax the Hardship Review Board may direct the Chief Commissioner:
  (a) to extend the time for payment of tax under section 47; or
  (b) to write off tax under section 110.

Section 106CA – refunds of tax

The Chief Commissioner is to make such refund of tax already paid as is necessary to give effect to a decision of the Hardship Review Board under this Division.

Victoria

Site, capital improved and net annual values – objections and appeals

The objection and appeal process, including grounds of objections to values, are addressed under Divisions 3 and 4, Part III of the Valuation of Land Act 1960. The key provisions of the process follow.

Section 16 – who may object

(1) A person aggrieved by a valuation of any land made or caused to be made by a valuation authority may lodge a written objection to that valuation, on any one or more of the grounds set out in section 17, with the authority that gave the notice of valuation.

(2) An objection must –
  (a) contain the prescribed information (if any); and
  (b) give particulars of the bases of valuation to which objection is made; and
  (c) state the grounds on which the objection is based.

(3) An objection must not be disallowed merely because of a failure to comply with subsection (2).

(3A) In addition to the requirements of subsection (2), if –
 (a) a ground for the objection is that the value assigned is too high or too low; and
 (b) the value assigned is not less than the prescribed amount – the objection must state the amount that the objector contends is the correct value.

(3B) An amount stated in an objection in accordance with subsection (3A) is not binding on the objector.

(4) A person who is given a notice of valuation is deemed to be a person aggrieved by the valuation whether or not the valuation is used for the purposes of a rate or tax levied by the authority that gave the notice.

(5) A person is deemed to be a person aggrieved by a valuation of land if –
 (a) the person is liable for or required to pay any rate or tax in respect of land; and
 (b) the person has not been given a notice of valuation.

Further subsections (6A–C) set out who is an aggrieved person.

Section 17 – provides the grounds for objection

The grounds for an objection are –

(a) that the value assigned is too high or too low;
(b) that the interests held by various persons in the land have not been correctly apportioned;
(c) that the apportionment of the valuation is not correct;
(d) that lands that should have been included in one valuation have been valued separately;
(e) that lands that should have been valued separately have been included in one valuation;
(f) that the person named in the notice of valuation, assessment notice or other document is not liable to be so named;
(g) that the area, dimensions or description of the land including the AVPCC allocated to the land are not correctly stated in the notice of valuation, assessment notice or other document.

Section 18 – time for lodging objection

An objection must be lodged –

(a) within 2 months after the notice of valuation is given; or
(b) in the case of a person referred to in section 16(5) –
 (i) if a notice of assessment of the rate or tax was served on the person by a rating authority, within 2 months after the notice was served; or

(ii) if a notice of assessment of the rate or tax was not served on the person by a rating authority and the person is the occupier of the land, within 4 months after the date of issue specified on the notice; or

(c) in the case of a person referred to in section 16(6A) – within 2 months after receiving the notice of assessment of land tax.

Section 20 – exchange of information on certain objections

(1) This section applies to an objection if –
  (a) a ground for the objection is that the value assigned is too high or too low; and
  (b) the value assigned is not less than the prescribed amount.
(2) Within one month after the objection is lodged with the authority, the valuer to whom the objection has been referred must give the objector the prescribed information concerning the valuation that is the subject of the objection.
(3) Within one month after receiving the prescribed information under subsection (2), the objector may lodge a written submission concerning the valuation with the valuation authority.

Section 21 – determination of objection

(1) An objection must be determined in accordance with section 20 and this section.
(2) If an objection has been lodged with an authority that is not the valuation authority that caused the valuation to be made, the authority must refer the objection to valuation authority.
(2A) A valuation authority with whom a valuation has been lodged or to whom an objection has been referred under subsection (2) must refer the objection to the valuer who made the valuation.
(2B) A valuer to whom an objection has been referred must provide the objector with a reasonable opportunity to discuss his or her objection with the valuer.
(3) Within 4 months after receiving the objection, the valuer must –
  (a) if he or she considers that no adjustment in the valuation is justified – give the objector written notice of that decision; or
  (b) if he or she considers that an adjustment in the valuation is justified –
    (i) recommend accordingly to the valuer-general; and
    (ii) give the objector and the valuation authority a copy of the recommendation.
(4) The valuer-general, after consultation if practical with the valuer, must determine the objection as follows –
  (a) the valuer-general may disallow the recommended adjustment in whole or part if, in his or her opinion, the adjusted valuation is not correct; or

(b) in any other case, the valuer-general must confirm the recommended adjustment.

(5) Within two months after receiving the recommendation, the valuer-general must give written notice of his or her decision to the objector, the valuer, the valuation authority and any other rating authority that uses or proposes to use the valuation.

(6) Subject to any review or appeal under Division 4, the decision of the valuer-general must be given effect by the rating authority and every other rating authority using that valuation.

(7) If section 20 applies, subsection (2B) does not require the valuer to provide a reasonable opportunity for the objector to discuss the matter with him or her unless the objector lodges a submission under section 20(3).

Section 22 – application to VCAT for review

(1) An objector who is dissatisfied with the decision of a valuer or the valuer-general on the objection may apply to VCAT for review of the decision.

(2) If the valuer for a valuation authority has not given an objector notice of a decision on the objection or a copy of a recommendation under section 21(3)(b)(ii) within 4 months after the objection was lodged with the valuation authority, the valuer is deemed to have made a decision that no adjustment in the valuation is justified.

(3) If the valuer-general has not given an objector notice of a decision under section 21(5) within 2 months after a copy of a recommendation was given to the objector under section 21(3)(b)(ii), the valuer-general is deemed to have made a decision disallowing the recommended adjustment.

(4) An application under this section must be made –

(a) in the case of an application in respect of a deemed decision referred to in subsection (2) – within 9 months after the date on which the objection was lodged with the valuation authority;

(b) in the case of an application in respect of a deemed decision referred to in subsection (3) – at any time after the end of the 2 month period referred to in that subsection;

(c) in any other case – within 30 days after the date notice of the decision is given to the objector.

(5) An applicant under this section must serve a copy of the application on the valuation authority.

(6) The valuation authority must, within 1 month after being served with a copy of the application, forward to the principal registrar of VCAT the notice of objection, copies of any notices given under section 21 in connection with the objection and any information given or submissions lodged under section 20 in connection with the objection.

(7) The principal registrar of VCAT must notify the valuer-general of an application under this section.

(8) Despite subsection (2), the valuer for a valuation authority may give an objector notice of a decision on the objection or a copy of a recommendation under section 21(3)(b)(ii) more than 4 months after the objection was lodged with the valuation authority.

(9) Despite subsection (3), the valuer-general may give an objector notice of a decision under section 21(5) more than 2 months after a copy of a recommendation was given to the objector under section 21(3)(b)(ii).

Section 23 – appeal to Supreme Court

(1) The President of VCAT, on his or her own initiative or on the application of a party, may refer a matter that is the subject of an application under section 22 to the Supreme Court to be treated as an appeal to the Supreme Court if the President is satisfied that the matter raises questions of unusual difficulty or of general importance.

(2) The principal registrar of VCAT must notify the valuer-general of a referral to the Supreme Court under subsection (1).

(3) In addition to subsection (1), a matter that is or could be the subject of an application under section 22 may be treated as an appeal to the Supreme Court if, on the application of any party, the Court is satisfied that the matter raises questions of unusual difficulty or of general importance.

(4) For the purposes of subsection (3), a "party" includes a person who would be a party if the matter were the subject of an application under section 22.

(5) The prothonotary must notify the valuer-general of an application to the Supreme Court under subsection (3).

(6) Nothing in this section limits the application of section 77 of the Victorian Civil and Administrative Tribunal Act 1998.

Section 24 – grounds of review or appeal

(1) On a review or appeal the objector's case is limited to –
   (a) the grounds of the objection; and
   (b) any other grounds set out in the application for review or appeal –
   unless VCAT or the Court (as the case requires) otherwise orders.

(2) If a ground for the objection or application is that the value assigned is too high or too low, the application for review or appeal (as the case requires) must state the amount that the objector contends is the correct value.

Section 25 – powers on review or appeal

(1) On a review or appeal, VCAT or the Court (as the case requires) may –
   (a) by order, confirm, increase, reduce or otherwise amend any valuation; and
   (b) make any other order it thinks fit.

An appeal to the Court of Appeal from an order of the Court under this section lies only on a question of law.

Council rates – reviews, appeals and ministerial discretion

Reviews and appeals to local government rating in Victoria are addressed under the Local Government Act 1989. These extend to include appeals by ratepayers, reviews by the Minister for Local Government and the Victorian Administrative Appeal Tribunal (VCAT) and relevant Court. The key provisions governing rating reviews and appeals follows.

Section 183 – review of differential rating by VCAT

(1) An owner or occupier of land whose interests are affected by a decision of the Council to classify or to not classify that land as land of a particular type or class for differential rating purposes may apply to VCAT for review of the decision.
(2) An application for review must be made within 60 days after the day on which the owner or occupier receives the first notice under section 158(3) following the decision.
(3) The provisions of Part III of the Valuation of Land Act 1960 apply to a review under this section with any necessary modifications.

Section 184 – appeal to County Court

(1) A person who is aggrieved –
    (a) by a rate or charge imposed by a Council under this or any other Act; or
    (b) by anything included or excluded from such a rate or charge – may appeal to the County Court for a review of the rate or charge.

This section excludes appeals against the value or valuation process used to assess the rate under the Valuation of Land Act 1960. The person has 60 days to appeal after receiving the rate notice and the grounds of appeal are limited to: a) the land was not rateable; b) the rate or charge was calculated incorrectly; or c) the person levied was not liable to be rated.

Section 185 – application to VCAT

(1) A person may apply to VCAT for review of a decision of a Council imposing a special rate or special charge on that person.
(2) The person must apply –
    (a) within 30 days after the date of issue of a notice to the person of the special rate or special charge; and
    (b) on the ground that –
        (i) the works and projects or the period of maintenance for the purposes of which the special rate or special charge was

imposed are not or will not provide a special benefit to that person; or

(ii) the basis of distribution of the rate or charge amongst those persons who are liable to pay it is unreasonable; or

(iii) if the planning scheme for the area contains any relevant policies or specific objectives, the works and projects proposed for the construction of a road or for the drainage of any land are inconsistent with those policies or objectives; or

(iv) if the planning scheme for the area does not contain any relevant policies or specific objectives, the works and projects proposed for the construction of a road or for the drainage of any land are unnecessary, unreasonable, excessive, insufficient, unsuitable or costly, having regard to the locality or environment and to the probable use of the road or drainage of the land.

Section 185B – minister may give directions concerning rates and charges

(1) The Minister may, by Order published in the Government Gazette, direct a Council specified in the Order that the Council's general income in respect of a financial year –

(a) is not to exceed the Council's general income in respect of a specified previous financial year; or

(b) is not to exceed a specified percentage of the Council's general income in respect of a specified previous financial year.

(2) The Minister may specify a percentage of more than, or less than, 100% under subsection (1)(b).

(4) An Order does not have effect in respect of a financial year unless it is published in the Government Gazette at least 1 month before the start of that financial year.

(5) Subsection (4) also applies to any Order that amends a previous Order made under this section unless –

(a) the amendment is to correct a typographical error or a mathematical miscalculation (or anything stemming from a mathematical miscalculation) or any error that is apparent on the face of the previous Order; or

(b) the amendment has the sole effect of removing or reducing a restriction placed, or to be placed, on one or more Councils by an Order made under this section.

(6) If a provision in an Order is inconsistent with a provision in a previous Order, the provision of the later Order prevails.

(7) For the purpose of giving effect to a direction –

(a) an Order may specify how changes in the number of rateable properties in a municipal district between 2 relevant periods are to be taken into account;

(b) if the boundaries of a municipal district have changed, or the Council responsible for a municipal district has been restructured or reconstituted, an Order may specify what the general income of the Council is to be taken as having been –
  (i) in the financial year in which the change or restructuring occurred;
  (ii) in any subsequent financial year in respect of which the Council did not make a declaration under section 158 or was deemed to have made a declaration;

(c) if a Council did not make a declaration under section 158 during a financial year, an Order may specify what the general income of the Council is to be taken as having been in that financial year.

Section 185C – councils must comply with Minister's direction

(1) A Council must comply with any direction under section 185B that applies to it.
(2) If a Council fails to comply with such a direction –
  (a) the failure does not affect the validity of any rates or charges levied in respect of the financial year in respect of which the failure occurred; but
  (b) any rate or charge that contributes to general income and that is declared, with respect to the following financial year is invalid for all purposes unless –
    (i) before it is declared, the Council gives the Minister all the information the Minister requires regarding it and the proposal to declare it; and
    (ii) the Minister approves its making in writing; and
    (iii) it complies with any conditions specified by the Minister in granting that approval.

State land tax

Objections, appeals and reviews of land tax are covered by Part 10, Divisions 1, 2 and 3 of the Taxation Administration Act 1997. The key provisions of the objection and appeals process follow.

Section 96 – objection

(1) A taxpayer may lodge a written objection with the Commissioner if the taxpayer is dissatisfied with –
  (a) an assessment other than a compromise assessment; or (ca) a valuation made for or on behalf of the Commissioner under section 21(1)(b) of the Land Tax Act 2005 by the Valuer-General or a valuer nominated by the Valuer-General that is used by the Commissioner in an assessment of land tax;

Section 97 – grounds for objection

(1) The grounds for the objection must be stated fully and in detail, and must be in writing.
(2) The grounds for the objection, in the case of a reassessment, may only relate to tax liabilities specified in the reassessment to the extent that they are additional to or greater than those under the previous assessment.
(3) A taxpayer cannot object to an assessment of land tax on any ground relating to the value of the land if –
    (a) the assessment is based on a valuation made by a valuation authority under the Valuation of Land Act 1960; and
    (b) that valuation was not made for or on behalf of the Commissioner under section 21(1)(b) of the Land Tax Act 2005.
(4) Nothing in subsection (3) limits or affects the right of any person to object to a valuation in accordance with Part III of the Valuation of Land Act 1960.
(5) Deals with objections under the *Planning and Environment Act 1987.*

Section 98 – onus of proof on objection

On an objection, the objector has the onus of proving the objector's case.

Section 99 – time for lodging objection

(1) An objection must be lodged with the Commissioner within 60 days after the date of service of the notice of the assessment or decision on the taxpayer, except as provided by section 100.
(2) An objection is taken to have been lodged with the Commissioner when it is received by the Commissioner.

Under section 100, the Commissioner may permit an objection that is outside the 60-day period subject to the provisions in that section.

Section 100A – objections concerning the value of property

(1) If an objection concerns the value of any property, the Commissioner may refer the matter to the Valuer-General or another competent valuer for valuation of the property.
(2) The objector must pay the cost of a valuation under subsection (1) if, at the time the objection was lodged –
    (a) the objector had not provided any information to the Commissioner as to the value of the property; or
    (b) the objector –
        (i) had provided information to the Commissioner as to the value of the property; and

(ii) the valuation under subsection (1) exceeds the value provided by the objector by 15% or more; and

(iii) the valuation, as determined on the objection, or on appeal or review, exceeds the value provided by the objector by 15% or more.

(3) Subsection (1) does not apply to an objection to an assessment of land tax or an objection referred to in section 96(1)(ca), however, the Commissioner must consult the Valuer-General before determining an objection referred to in section 96(1)(ca).

Section 101 – determination of objection

(1) The Commissioner must consider an objection and either allow the objection in whole or in part or disallow the objection.

(2) The Commissioner may determine an objection that is subject to a right of review or appeal at any time before the hearing of the review or appeal proceedings commences.

The Commissioner has power to suspend a determination of an objection under section 102, subject to the provisions of that section. Further, a pending objection review or appeal does not mitigate the payment of the tax. The tax may be recovered as if no objection or appeal were pending under section 104, and interest is payable under section 105 following an unsuccessful objection.

Section 106 – right of review or appeal

(1) If –

(a) a taxpayer is dissatisfied with the Commissioner's determination of the taxpayer's objection; or

(b) 90 days (not including any period of suspension under section 102) have passed since a taxpayer's objection was received by the Commissioner and the Commissioner has not determined the objection –

the taxpayer, in writing, may request the Commissioner to refer the matter to the Tribunal or to treat the objection as an appeal and cause it to be set down for hearing at the next sittings of the Supreme Court.

Subsection (2) sets out further detail relating to a taxpayer's request in relation to subsection (1) and section 107.

Under section 107, the Commissioner may request further and better particulars. Further provisions set out the particulars for setting down the objections for hearing.

Section 108 – must tax be paid before a review or appeal is determined?

(1) A taxpayer may make a request under section 106(1) for a review or appeal, whether or not the tax to which it relates has been paid.
(2) If a taxpayer makes a request before the tax has been paid, the Commissioner may apply to the Supreme Court for an order that the tax, or a specified part of it, be paid before the review or appeal proceed.
(3) On an application under subsection (2), the Supreme Court may order that the review or appeal not proceed until the tax, or a specified part of it, is paid.
(4) The Supreme Court may make an order under subsection (3) only if satisfied that it is reasonably likely that, unless the order is made, any tax payable by the taxpayer in accordance with the determination of the review or appeal would not be paid within the period within which it is payable.

Section 109 – grounds of review or appeal

On a review or an appeal –

(a) the taxpayer's case is limited to the grounds of the objection; and
(b) the Commissioner's case is limited to the grounds on which the objection was disallowed – unless the Tribunal or Court otherwise orders.

Further, Section 110 specifies that the taxpayer has the onus of proving the taxpayers case.

Queensland

Site values – appeals and objections

Objections and appeals to Site Values are made under the relevant provisions of the Land Valuation Act 2010. Queensland has made quantum progress in objections and reviews to the valuation of land for rating and taxing purposes. The objective is to facilitate the resolution of the objection if warranted.

Section 105 sets out the right to object and allows objections for different purposes (rating and taxing); however, it has an innovative provision which only allows a second objection where a difference exists between the two valuations. This, in essence, prohibits a second objection to a previous objection that may have been failed. Section 109 sets out the objection period, referred to as the usual objection period, which is 60 days.

Late objections may be considered by the Valuer-General under Section 111(3) as follows:

The Valuer-General must accept the objection if satisfied it was not made within the usual objection period because of –

(a) the owner's mental or physical incapacity; or
(b) an extreme circumstance; or
(c) an extraordinary emergency; or
(d) another reason the valuer-general considers satisfactory in the circumstances. *Note – See also chapter 5 (Internal and external reviews).*

Section 112 sets out what constitutes a properly made object as follows:

(1) An objection is properly made only if –
  (a) it is in the approved form; and
  (b) it relates to only 1 valuation, unless section 107 applies; and
  (c) either –
    (i) it is signed by the objector; or
    (ii) it is signed by an agent or representative of the objector, for the objector, and is accompanied by the objector's written consent to the objection; and
  (d) it complies with all of the requirements under section 113; and
  (e) it is accompanied by the fee prescribed under a regulation.
(2) The approved form must state –
  (a) the matters that the valuer-general considers appropriate about the objection process, including, for example, about objections that are not properly made; and
  (b) that the valuer-general cannot consider or decide an objection that is not properly made.

Further provisions of this section set out details of non-compliant objections.

Section 113 sets out the required content of objections as follows:

(1) An objection must state all of the following –
  (a) the objector's address for service of any notices concerning the objection; *Note – See also section 153(2) (Address for service for objections).*
  (b) information that identifies the land, including, for example, its property identification number, real property description or property address shown on the valuation notice;
  (c) if the valuation is more than the relevant amount – the valuation sought;
  (d) the grounds of objection to the valuation (each an objection ground for the valuation);
  (e) the information the objector seeks to rely on to establish each objection ground;

(f) if an objection ground concerns the comparability of the sale of any other land –
  (i) details of the sale; and
  (ii) the reasons why the objector contends the sale is comparable to the valuation of the objector's land; and
  (iii) the basis of comparison between the objector's land and the land the subject of the sale.

(2) Also, if an objection ground –
  (a) is a deduction application or about a decision on a deduction application concerning the land; or
  (b) concerns a claim for a higher site improvement deduction than that stated in the valuation notice; the objection must –
  (c) state the amount the objector claims; and
  (d) state the matters mentioned in section 41(2)(a); and
  (e) be accompanied by the documents mentioned in section 41(2)(b).

Part 3 of Chapter 3 of the Land Valuation Act sets out the provisions for conferences to which the objectives of conferencing under Section 120 follow:

(1) This part provides for the holding of conferences about properly made objections (an objection conference).
(2) An objection conference's purposes are to –
  (a) encourage the settlement of disputes about the objection by facilitating and helping the conduct of negotiations between the parties; and
  (b) promote between the parties an open exchange of information relevant to any dispute; and
  (c) give the parties information, relevant to the dispute, about the operation of this Act; and
  (d) help in the settlement of the dispute in any other way.

Section 121 – conditions for holding conference

An objection conference can not be held for an objection if the objection –

(a) is not properly made; or
(b) has been decided by the valuer-general.

Section 122 – conference by agreement

If a valuation is not more than the following, the valuer-general and the objector may agree to participate in an objection conference –

(a) if an amount of more than $5m has been prescribed under a regulation – the prescribed amount;
(b) otherwise – $5m.

Section 123 – when conference is required

(1) If a valuation is more than the amount mentioned in subsection (2), the valuer-general must –
    (a) offer, to the objector, to participate in an objection conference; and
    (b) if the objector accepts the offer – participate in the objection conference.
(2) For subsection (1) the amount is –
    (a) if an amount of more than $5m has been prescribed under a regulation – the prescribed amount; or
    (b) otherwise – $5m.
(3) The offer may be verbal or by notice.

Council rates – objections and appeals

Provisions covering objections and appeals to local government rates in Queensland are addressed under the Local Government Act 1993, Sections 984 to 991. These provisions allow objections and appeals primarily to the categorisation of land for rating purposes.

Section 984 – owner's objection to categorisation

(1) An owner of rateable land categorised under division 1 may object to the categorisation of the land on the sole ground that, having regard to the criteria decided by the local government by which rateable land is categorised, the land should have been included, as at the date of issue of the relevant rate notice, in another rating category.
(2) The objection must be made by giving notice of the objection to the local government.
(3) The notice of the objection must –
    (a) be given within 30 days after the date of issue of the rate notice or any further period the local government allows; and
    (b) be in the form approved by the local government; and
    (c) nominate the rating category in which the owner claims the land should have been included; and
    (d) specify the facts and circumstances on which the claim is based.

Section 985 – decision on owner's objection

(1) If the owner of rateable land objects to the categorisation of the land, a person authorised by the local government for the purpose must –

(a) consider the categorisation of the land; and
(b) consider the facts and circumstances on which the claim is based.

(2) The person may –
(a) allow the objection; or
(b) disallow the objection; or
(c) decide that the land should be included in another rating category.

(3) The person must decide the objection and give written notice of the decision to the owner within 60 days after the end of the period within which the objection had to be made.

(4) The notice must include the reasons for the decision.

Section 987 – right of appeal against decision

If the owner of rateable land is aggrieved by –

(a) the decision on an objection to the categorisation of the land; or
(b) the failure by the local government to allow a further period to give a notice of objection; the owner may appeal to the Land Court against the decision or failure.

Section 988 – where and how to start appeal

(1) The appeal must be started by filing a notice of appeal in the Land Court registry.

(2) The notice of appeal must –
(a) be filed within 42 days after the owner received notice of the decision or failure; and
(b) be in a form approved by the Land Court.

(3) The owner must give a copy of the notice of appeal to the local government within seven days after the notice of appeal is filed in the Land Court registry.

(4) Failure to comply with subsection (3) does not affect the making of the appeal or the jurisdiction of the Land Court to decide the appeal, but costs of any adjournment caused by the failure may be awarded against the owner of the land.

Section 989 – constitution and procedure of land court

(1) When exercising jurisdiction in an appeal under this division, the Land Court –
(a) is constituted by one member; and
(b) is not bound by rules of evidence.

(2) The appeal must be conducted as directed by the Land Court with a view to its prompt disposal.

Section 990 – decision on appeal by Land Court

(1) In deciding an appeal against a decision on an objection to the categorisation of land, the Land Court may –
  (a) set aside the decision and decide that the land should be included in a different rating category; or
  (b) disallow the appeal.
(2) In deciding an appeal against a failure to allow a further period to give a notice of objection, the Land Court may –
  (a) allow a further period to give the notice; or
  (b) disallow the appeal.
(3) If the Land Court sets aside the decision on the objection, the land is taken to be included in the category decided by the Land Court for the period for which the relevant rate notice is issued.

State land tax – appeals and objections

The right to object to an assessment is made under section 63 of the Taxation Administration Act 2001. The grounds of objection under section 64 provide that an objection may be made on any grounds. Section 65 sets out that an objection must:

(a) be in writing; and
(b) state in detail the grounds on which the objection is made; and
(c) be accompanied by copies of all material relevant to decide the objection; and
(d) be lodged within 60 days after the assessment notice for the assessment to which the objection relates is given to the taxpayer.

Further, the Commissioner may extend the 60-day period.

Like all of the other states, the onus of proof is on the objector under section 66 of the Act. The Commissioner must allow the objection fully or in part or disallow the objection under Section 67. Under section 68:

(1) The Commissioner must give written notice to the objector of the Commissioner's decision on the objection.
(2) If the objection is allowed in part or disallowed, the notice must state the following –
  (a) the decision;
  (b) the reasons for the decision;
  (c) that the taxpayer has a right to –
    (i) appeal to the Supreme Court; or
    (ii) apply, as provided under the QCAT Act, to QCAT for a review of the Commissioner's decision; and

(d) how, and the period within which, the taxpayer may appeal or apply for the review.

A taxpayer following the result of an objection may appeal the decision of the Commissioner under section 69 as follows:

(1) This section applies to a taxpayer if –
  (a) the taxpayer is dissatisfied with the Commissioner's decision on the taxpayer's objection; and
  (b) the taxpayer has paid the whole of the amount of the tax and late payment interest payable under the assessment to which the decision relates.
(2) The taxpayer may, within 60 days after notice is given to the taxpayer of the Commissioner's decision on the objection –
  (a) appeal to the Supreme Court; or
  (b) apply, as provided under the QCAT Act, to QCAT for a review of the Commissioner's decision.
(3) QCAT may not, under the QCAT Act, section 61(1)(a), extend the period under subsection (2) within which the taxpayer may apply to QCAT for the review.

Under section 69A of the Act, the Commissioner may undertake a reassessment after the appeal or review commences.

Lodging an appeal to the Supreme Court is set out under Section 70 as follows:

(1) An appeal to the Supreme Court is started by giving written notice of the appeal to the Commissioner within seven days after the notice of appeal is filed.
(2) The notice of appeal must be filed within 60 days after notice is given to the taxpayer of the Commissioner's decision on the objection.
(3) The Supreme Court must not extend the time for filing the notice.
(4) The notice of appeal must state fully the grounds of the appeal and the facts relied on.
(5) The grounds of an appeal to the Supreme Court are limited to the grounds of objection unless the court otherwise orders.

South Australia

Annual, capital and site values – objections and appeals

Provision for objections and appeals against values are made under the Valuation of Land Act 1971 Part 4 – Objections, appeals and reviews, Divisions 1 to 3. Section 23 of the Act requires the Valuer-General to serve notice of valuation under this Act on the owner or occupier or both, as the Valuer-General considers appropriate.

Section 24 – Objection to valuation sets out the time frame and ground of object to values as follows:

(1) Subject to this section, a person who is dissatisfied with a valuation of land in force under this Act may, by notice in writing served personally or by post on the Valuer-General, object to the valuation.

(1a) After notice of a valuation (whenever made) is first served after the commencement of this subsection on the owner or occupier of the land, an objection to the valuation may only be made by the owner or occupier so served within 60 days after the date of service of the notice.

(1b) However, if the owner or occupier is served with a further notice of the valuation, the person so served will have a further right to object to the valuation provided that –

   (a) the further notice is the first notice of the valuation served on the person under the Act under which the notice is served; and

   (b) the objection is made within 60 days after the date of service of that further notice.

(1c) A person may not make an objection to a valuation if the Valuer-General has previously considered an objection by that person to the valuation.

(1d) refer to Act.

(1e) Despite any other provision of this section, the Valuer-General may, for reasonable cause shown by a person entitled to make an objection to a valuation, extend the period within which the objection may be made.

(2) A notice of objection under subsection (1) must contain a full and detailed statement of the grounds on which the objection is based.

Section 25 – Valuer-General to consider and decide upon objection

(1) The Valuer-General must, as soon as practicable, consider any objection made under this Act and may either allow or disallow the objection.

Section 25B – review by valuer

(1) A person who is dissatisfied with the decision of the Valuer-General upon an objection under this Part may, within 21 days of the day on which he or she receives notice of the decision, apply for a review of the valuation.

(2) An application under this section –

   (a) must be made in the prescribed manner and form; and

   (b) must be lodged at the office of the Valuer-General or served by post on the Valuer-General; and

   (c) must be accompanied by the prescribed fee.

(3) No application for review of a valuation may be made under this section if the objection to the valuation involves a question of law.

(4) Where due application for review of a valuation is made under this section, a land valuer (in this section referred to as "the valuer") must be selected in accordance with the regulations from the appropriate panel of land valuers to conduct the review.
(5) Subject to this section, the valuer must, in conducting a review under this section, take into account –
  (a) the matters set out in the application for review; and
  (b) any representations of the applicant and the Valuer-General made under subsection (7); and
  (c) any other matter that the valuer considers relevant to the review of the valuation.
(6) The matters to be considered upon a review under this section must be confined to questions of fact and must not involve questions of law.
(7) The valuer must afford the applicant and the Valuer-General a reasonable opportunity to make representations to the valuer on the subject matter of the review.
(8) Representations may be made under subsection (7) personally, by a land valuer acting on behalf of the applicant or the Valuer-General, or by any other representative.
(9) Subject to subsection (10), the valuer must, upon the determination of the review, confirm, increase or decrease the valuation.
(10) A valuer must not make any alteration to a valuation under subsection (9), which has the effect of increasing or decreasing the valuation by a proportion of one-tenth or less.

Section 25C – right of appeal

(1) A person who is dissatisfied with –
  (a) the decision of the Valuer-General upon an objection under Division 1; or
  (b) the decision of a land valuer, upon a review under Division 2, may, in accordance with the appropriate rules of the Supreme Court, appeal to the Land and Valuation Court against the decision.
(2) The right of appeal conferred by subsection (1)(b) may be exercised by the Valuer-General.
(3) Upon an appeal under this section, the Land and Valuation Court –
  (a) may confirm, increase or decrease the valuation to which the appeal relates; and
  (b) may make such orders in relation to incidental or ancillary matters (including costs) as it thinks just.

Local government rates – objection and appeals

Under *Section 156 – basis of differential rates*, a ratepayer may object to the land use assigned to their land under subsection (9) and further appeal the matter to the Land and Valuation Court under subsection (12). Objections to valuations made by council under Section 169 are dealt with under in the following section (objections and appeals to land values).

Section 186 sets out that the recovery of rates is not affected by and objection review or appeal.

Section 256 – rights of review

(1) An order must include a statement setting out the rights of a person to appeal against the order under this Act.
(2) A statement is sufficient for the purposes of subsection (1) if it includes the information specified by the regulations.
(3) A person to whom an order is directed may, within 14 days after service of the order, appeal against the order to the District Court.

State land tax – objections and appeals

Objections and appeals to land tax in South Australia are dealt with under the Taxation Administration Act 1996 Part 10 – Divisions 1 and 2. The right to object to an assessment is made under Section 82, while the right to appeal is made under Section 92 of the Act.

Section 83 requires the grounds of an objection to be stated fully and in detail in the notice of objection, and a right to object to a reassessment is set out under Section 84. The objector has the onus of proving their case under Section 85 and the time frame for lodging an objection is 60 days from the date of service on the taxpayer, as specified under Section 86. The Minister has discretion to extend the 60-day period to object in line with the provisions of Section 87. The Minister may refer the objection onto the Crown Solicitor for advice or seek further information from the taxpayer among other criteria in the determination of the objection, as set out under Section 88. The Minister must reply in writing with the decision and reasons for the decision under Section 89. Sections 90 and 91 deal with interest on refunds and recovery of tax pending objection.

Section 93 addresses appeals and requires that tax is paid before an appeal can be lodged; however, the Minister has power to make exceptions in these case. The Minister's decision is non-reviewable.

Section 94 – time for appeal

(1) An appeal must be made by a person not later than 60 days after the date of service on the person of notice of the Minister's determination of the person's objection.
(2) However, if –
    (a) 90 days (not including any period of suspension under section 88) have passed since the person's objection was lodged with the Minister; and

(b) the Minister has not determined the objection and served notice of the determination on the person, the person may appeal at any time provided that the Commissioner is given not less than 14 days written notice of the person's intention to make the appeal.

Section 95 permits the Supreme Court an extension of up to 12 months after date of service of the result of an objection by the Minister.

Section 96 – grounds of appeal

(1) The appellant's and respondent's cases on an appeal are not limited to the grounds of the objection or the reasons for the determination of the objection or the facts on which the determination was made.
(2) However, if the objection was to a reassessment, any limitation of the matters to which the objection could relate under Division 1 applies also to the appeal.

The appellant has the onus of proving their case under Section 97.

Tasmania

Land values, capital values and assessed annual values – objections and appeals

Objections and appeals to values used to assess state land tax and local government rates are made under the Valuation of Land Act 2001, Parts 5 to 8. The key provisions governing objections and appeals to values are as follows.

A notice of valuation is issued by the Valuer-General under Section 27 of the Act and must be given to the owner of any land, at his or her last known address, notice of every valuation of that land made under section 11, 18, 20 or 21, unless the Valuer-General has made suitable arrangements with the relevant rating authority for notices of valuations to be given to the owners of land. Further, the notice is to be in writing and via the approved form, and is to contain the words, "The notified valuations are determined under the Valuation of Land Act 2001 and for no other purpose".

Objections to valuations are made under Section 28 of the Act by an owner of land who is dissatisfied with either:

(a) a valuation of that land made under section 11, 18, 20 or 21; or
(b) the provision of a certificate under section 44 –

may, within 60 days after receipt of a notice under section 27 or the provision of that certificate, post to or lodge with the Valuer-General an objection, in an approved form, against the relevant valuation stating fully and in detail the grounds on which he or she relies and stating any changes to the values specified in that valuation or certificate which he or she considers should be made.

(2) A rating authority may, within 60 days after it is notified of a valuation made under this Act, by notice in an approved form posted to or lodged with the Valuer-General, object to the valuation and, where a rating authority makes any such objection, it must cause a copy of the objection to be served on every person who is liable for payment of any rates or charges payable to the rating authority in respect of that land.

The grounds on which objections may be made are listed under Section 29 of the Act as follows:

Without limiting section 28(1), (2) or (3), an objection under this Part may be made on any one or more of the following grounds, but on no other ground:

(a) that the land value, capital value or assessed annual value assigned to any land is too high or too low;
(b) that the interests of the several persons having an interest in any land have not been correctly apportioned;
(c) that the apportionment of any valuation is not correct;
(d) that lands which should be included in the one valuation have been valued separately;
(e) that lands which should be valued separately have been included in the one valuation;
(f) that the person named in any notice under section 27 is not an owner of the land to which the notice relates;
(g) that the area, dimensions or particulars of any land are not correctly described.

Section 30 – consideration of objections by Valuer-General

(1) On receipt of an objection under this Part, the Valuer-General must, with all reasonable despatch, consider the objection and make any necessary amendments to any or all of the values shown in the valuation roll as he or she thinks fit.
(2) On the determination of any such objection, the Valuer-General must give to the person by whom the objection was made notice, in writing, of his or her decision on the objection.
(3) Where the Valuer-General amends the valuation, the Valuer-General must, in accordance with section 24, enter the amended valuation on the valuation roll and give notice of the amended valuation in the manner provided in section 27(2) to the person making the objection who may have the matter referred to the court as provided by subsection (4), but that person is not entitled to further object to the amended valuation.

(4) Within 30 days after receipt of a notice under subsection (2), the person by whom the objection was made may, by notice in writing served on the Valuer-General, require him or her to refer the objection –
   (a) whatever the amount of the valuation, to the court; or
   (b) if the valuation objected to exceeds an amount prescribed by the regulations for capital value, land value or assessed annual value, to the Supreme Court.

(5) Within 30 days after receipt of a notice under subsection (2), the person by whom the objection was made may –
   (a) with the agreement of the Valuer-General, refer the objection to the Supreme Court; or
   (b) in the case of a municipal rating valuation, with the agreement of the Valuer-General, refer the objection to an arbitrator.

(6) The Valuer-General and the person making the objection may agree that the objection is to be referred to arbitration on such terms and conditions as may be agreed including, without limitation –
   (a) the appointment of an arbitrator; and
   (b) whether the decision of the arbitrator is to be final; and
   (c) the representation of the parties; and
   (d) the costs of the arbitration; and
   (e) security for costs of the arbitrator.

Part 6 – Section 31 – arbitrators for fresh valuations

(1) The Minister may appoint a panel of persons to act as arbitrators for the purposes of fresh valuations.
(2) A person appointed under subsection (1) is to be a person who, in the opinion of the Minister, has sufficient experience in the valuation of land to act as an arbitrator.
(3) Subject to this section, an arbitrator holds office subject to such terms and conditions as are specified in the instrument of appointment.
(4) An arbitrator may resign office by notice in writing addressed to the Minister.
(5) An arbitrator holds office for such term, not exceeding 4 years, as may be specified in the instrument of appointment.

Part 7 – land valuation court

Section 34 – constitution of court

(1) The Land Valuation Court, as continued under the Land Valuation Act 1971, is further continued for the purposes of this Act, notwithstanding the repeal of that Act effected by section 66 of this Act.
(2) The court is to be constituted by a magistrate.
(3) The court continues to be a court of record and to have an official seal, which is to be judicially noticed.

Part 8 – powers of Supreme Court

Section 40 – appeals to Supreme Court

(1) There is a right of appeal to the Supreme Court, which is to be heard by way of rehearing, from any decision of the Land Valuation Court on the hearing of an objection under this Act.
(2) Except as provided by subsection (1), the decision of the Land Valuation Court on the hearing of an objection is final.

Council rates – reviews, objections and appeals

Objections, reviews and appeals to council rates are addressed under the Local Government Act 1993.

Local Government Act 1993

PART 9 OF THE ACT – DIVISION 5 – SEPARATE RATES AND CHARGES UNDER SECTION 100 A COUNCIL MAY LEVY A SEPARATE RATE OR CHARGE

(1) A council may, by absolute majority, make a separate rate or separate charge in respect of land, or a class of land, within a part of its municipal area.
(2) A separate rate or separate charge may be made –
    (a) in addition to any other rates or charges; and
    (b) in respect of a financial year or part of a financial year; and
    (c) for the purpose of planning, carrying out, making available, maintaining or improving any thing that in the council's opinion is, or is intended to be, of particular benefit to –
        (i) the affected land; or
        (ii) the owners or occupiers of that land.

(subsections operate in addition)

Section 105A – review of separate rate or charge

(1) A review of a separate rate or separate charge is to include an assessment of the particular benefit of the separate rate or separate charge to –
    (a) the affected land; or
    (b) the owners or occupiers of that land.
(2) Before undertaking a review, the council is to –
    (a) notify the ratepayers of the affected land of its intention to conduct the review; and
    (b) publish notification of that intention in a daily newspaper circulating in its municipal area.
(3) Notification published under subsection (2)(b) is to include an invitation to the ratepayers, owners and occupiers of the affected land to

make written submissions in respect of the review within 30 days after publication of the notification.

(4) In deciding, following a review under this section, whether or not to make or continue a separate rate or separate charge, a council must take into account –
  (a) any submissions made under subsection (3); and
  (b) the outcomes of the review.

Division 9 – liability and payment of rates

SECTION 123 – OBJECTIONS TO RATES NOTICE

(1) A person may object to a rates notice on the ground that –
  (a) the land specified in the rates notice is exempt from the payment of those rates; or
  (b) the amount of those rates is not correctly calculated having regard to the relevant factors; or
  (c) the basis on which those rates are calculated does not apply; or
  (d) he or she is not liable for the payment of the rates specified in the rates notice; or
  (e) he or she is not liable to pay those rates for the period specified in the rates notice.
(2) An objection is to be –
  (a) made in writing within 28 days after receipt of the rates notice; and
  (b) lodged with the general manager.
(3) The general manager may –
  (a) amend the rates notice as the general manager considers appropriate; or
  (b) refuse to amend the rates notice.
(4) A person may appeal to the Magistrates Court (Administrative Appeals Division) for a review if the general manager –
  (a) fails to amend the rates notice within 30 days after lodging the objection; or
  (b) refuses to amend the rates notice.

Local Government (Rates and Charges Remissions) Act 1991

In the case where the Commissioner contends that a ratepayer is or was not entitled to a remission of their rates, the ratepayer may object to that contention or action under Section 4C.

Section 4C – objections

(1) A person may object to a decision of the Commissioner under section 4B.
(2) An objection is to –
  (a) be in writing; and

(b) be lodged with the Commissioner within 60 days after the objector receives a written notice under section 4B or as otherwise allowed by the Commissioner under section 4D.

State land tax – reviews and appeals

Objections, reviews and appeals are addressed under Part 10 – Divisions 1–4 of the Taxation Administration Act 1997. A taxpayer may object to an assessment other than a compromise assessment, as per Section 80, or against a decision made by the Commissioner other than a non-reviewable decision. A decision of the Commissioner does not include a refusal by the Commissioner under subsection (1A). Subsection (2) sets out that an objection is to be in writing, stating in full the grounds of objection and be lodged with the Commissioner and must be lodged within the prescribed time period defined in the Act under Sections 82 (within 60 days of notice of service) and 83, which deal with out of time objections.

Subsection (3) of 80 states that the grounds for objection to a reassessment are limited to the reassessment. Further, subsection (4) states that a court or administrative review body, including the Magistrates Court (Administrative Appeals Division), does not have jurisdiction or power to consider any question concerning an assessment or decision of the Commissioner except as provided by this Part. Section 81 highlights that the onus of proof is on the objector to prove their case. The Commissioner may allow the objection in full or in part or disallow the objection under Section 84.

Section 85 provides under subsection (1) that the Commissioner, by written notice served on the objector, may suspend the determination of an objection for any period during which the objector, or another person having information relevant to the objection, fails to provide information relevant to the objection that the Commissioner has requested under a taxation law. Section 86 states under (1) that the Commissioner is to give notice to the objector of the determination of the objection, and under (2) that the notice of the determination is to state the grounds for disallowing an objection or for allowing an objection in part only. Section 87 allows the recovery of the tax pending an objection, review or appeal as follows: The fact that an objection, review or appeal is pending does not affect the assessment or decision to which the objection, review or appeal relates and tax may be recovered as if no objection, review or appeal were pending.

Division 2 of Part 10 provides the right of reviews and appeals. Section 89 states:

(1) A taxpayer may apply to the Magistrates Court (Administrative Appeals Division) for a review of or appeal to the Supreme Court against –
   (a) the Commissioner's determination of the taxpayer's objection; or
   (b) the Commissioner's failure to determine the taxpayer's objection.
(2) An application for review or an appeal –

(a) is to be made within 60 days after the date of service on the taxpayer of the notice of the Commissioner's determination of the objection for a review or appeal under subsection (1)(a); or
(b) is not to be made before 90 days after the objection was lodged for a review or appeal under subsection (1)(b).

(3) A taxpayer is not to apply for a review or make an appeal unless –
(a) the whole of the amount of the tax to which the review or appeal relates has been paid; or
(b) if that amount has not been paid, the Commissioner allows the taxpayer to apply for a review or make the appeal.

(4) A taxpayer must give written notice to the Commissioner of the intention to apply for a review or make an appeal under subsection (1)(b) at least 14 days before the application for a review or appeal is made.

Section 90 addresses the grounds of review or appeal as follows:

Unless the Magistrates Court (Administrative Appeals Division) or Supreme Court otherwise orders, on a review or an appeal –

(a) the taxpayer's case is limited to the grounds of the objection; and
(b) the Commissioner's case is limited to the grounds on which the objection was disallowed.

Section 91 places the onus of proving their case on the taxpayer. Section 92 addresses orders of court, to which the Supreme Court may make any order it thinks fit and by order, confirm, reduce, increase or vary the assessment or decision. Further, the costs of the appeal are in the discretion of the Supreme Court. Section 93 gives effect to a decision on review or appeal as follows:

(1) Within 60 days after the decision on review or appeal becomes final, the Commissioner must take any action that is necessary to give effect to the decision.
(2) If an appeal to a court, from a decision of the Supreme Court on an appeal or a decision of the Magistrates Court (Administrative Appeals Division) on a review, is not instituted within 30 days after the day on which the decision is made, the decision is taken to be final at the end of that period.

Division 3 deals with refunds following successful objection, review or appeal. Section 94 addresses refund of amount, and Section 95 addresses payment of interest.

Western Australia

Gross rental values and unimproved values – objections and reviews

The objection and review provisions governing the valuation of land for rating and taxing purposes are under Part IV of the Valuation of Land Act 1978. The grounds of objection and time frame for lodging objections are addressed under Section 32 of the Act. A dissatisfied objector may refer the outcome of their objection to the State Administrative Review Tribunal (SAT) for further review under Section 33 of the Act. Following any referral of a valuation onto the SAT, the Valuer-General is obliged to notify all rating and taxing authorities of the objection and review.

Section 32 – objections to valuation

(1) Any person liable to pay any rate or tax assessed in respect of land, who is dissatisfied with a valuation of such land made under Part III, may serve upon the Valuer-General or any rating or taxing authority a written objection to the valuation –
  (a) in the case of land that is the subject of a general valuation, within 60 days after the date on which the making of the valuation was notified in the Government Gazette, under section 21 or section 22; and
  (b) in any case where the valuation is the basis of the assessment by a rating or taxing authority of any rate or tax, within 60 days after the issue of such an assessment.

(1a) In subsection (1), *person liable to pay any rate or tax assessed in respect of land* includes the authorised representative of such a person.

(2) An objection to a valuation of land shall –
  (a) describe the relevant land so as to identify it; and
  (b) identify the valuation objected to; and
  (c) set out fully and in detail the grounds of objection and the reasons in support of those grounds of objection.

(3) An objection to a valuation of land may be made on the ground that the valuation is not fair or is unjust, inequitable or incorrect, whether by itself or in comparison with other valuations in force under this Act.

(4) A person may not make more than one objection to the one valuation during any period of 12 months.

(5) Where an objection to a valuation is served on a rating or taxing authority, that authority shall, as soon as is practicable, refer the objection to the Valuer-General and advise him of the date on which the objection was served on that authority.

(6) The Valuer-General may, for reasonable cause shown by a person entitled to make an objection, extend the time for service of the objection

for such period as the Valuer-General considers reasonable in the circumstances and whether or not the time for service of the objection has already expired.

(7) The Valuer-General shall, with all reasonable dispatch, consider any objection and may either disallow it or allow it, wholly or in part.

(8) The Valuer-General shall promptly serve upon the person by whom the objection was made written notice of his decision on the objection and a brief statement of his reasons for that decision.

(9) Where the Valuer-General decides to allow an objection, wholly or in part, he shall also advise the person by whom the objection was made of any consequent amendment of valuation; and where the Valuer-General decides to disallow an objection, wholly or in part, he shall also advise that person of the time within which and the manner in which a review of the valuation may be sought.

Section 33 – State Administrative Tribunal review of valuation, after objection

(1) Any person who is dissatisfied with the decision of the Valuer-General on an objection by that person may, within 60 days (or such further period as the Valuer-General, before or after the expiry of that time, for reasonable cause shown by the person, allows) after service of notice of the decision of the Valuer-General, serve on the Valuer-General a notice requiring that the Valuer-General refer the valuation to the State Administrative Tribunal for a review.

(2) Upon receipt of such notice the Valuer-General shall promptly refer the valuation to the State Administrative Tribunal for a review.

Section 34 – Valuer-General to advise rating and taxing authorities of objections and review

The Valuer-General shall promptly advise every rating or taxing authority obliged to adopt or use or which has adopted any valuation –

(a) of receipt by him of an objection to the valuation; and

(b) of any allowance by him of an extension of time for service of an objection to the valuation; and

(c) of his decision on an objection to the valuation and the reasons therefor; and

(d) of any amendment of the valuation consequent upon his allowance, wholly or in part, of an objection to the valuation; and

(e) of receipt by him of a notice requiring him to refer the valuation to the State Administrative Tribunal for a review.

Section 36A – new matters raised on SAT review

(1) Upon a review by the State Administrative Tribunal on a referral under section 33 or 35, the State Administrative Tribunal may consider –
 (a) grounds in addition to those stated in the notice of objection; and
 (b) reasons in addition to any reasons previously given for the Valuer-General's decision that is under review.

(2) The State Administrative Tribunal is to ensure, by adjournment or otherwise, that each party and any other person entitled to be heard has a reasonable opportunity of properly considering and responding to any new ground or reason that the State Administrative Tribunal proposes to consider in accordance with subsection (1).

Local government rates – objections and reviews

The objection and review of local government rates is addressed under Local Government Act 1995 Part 6, Division 6, Subdivision 7.

Section 6.76 – grounds of objection

(1) A person may, in accordance with this section, object to the rate record of a local government on the ground –
 (a) that there is an error in the rate record –
  (i) with respect to the identity of the owner or occupier of any land; or
  (ii) on the basis that the land or part of the land is not rateable land; or
 (b) if the local government imposes a differential general rate, that the characteristics of the land recorded in the rate record as the basis for imposing that rate should be deleted and other characteristics substituted.

(2) An objection under subsection (1) is to –
 (a) be made to the local government in writing within 42 days of the service of a rate notice under section 6.41; and
 (b) identify the relevant land; and
 (c) set out fully and in detail the grounds of objection.

(3) An objection under subsection (1) may be made by the person named in the rate record as the owner of land or by the agent or attorney of that person.

(4) The local government may, on application by a person proposing to make an objection, extend the time for making the objection for such period as it thinks fit.

(5) The local government is to promptly consider any objection and may either disallow it or allow it, wholly or in part.

(6) After making a decision on the objection, the local government is to promptly serve upon the person by whom the objection was made written notice of its decision on the objection and a statement of its reason for that decision.

Where a person is not satisfied with the decision of a local government to an objection, they may appeal to the State Administrative Tribunal within the prescribed timeframe as follows:

Section 6.77 – review of decision of local government on objection

Any person who is dissatisfied with the decision of a local government on an objection by that person under section 6.76 may, within 42 days (or such further period as the State Administrative Tribunal, for reasonable cause shown by the person, allows) after service of notice of the decision, apply to the State Administrative Tribunal for a review of the decision.

Section 6.78 – review of decision to refuse to extend time for objection

A person who is dissatisfied with a decision of the local government to refuse to extend the time for making an objection against the rate record may apply to the State Administrative Tribunal for a review of the decision.

Section 6.79 – new matters raised on review

Upon a review by the State Administrative Tribunal under section 6.77 or 6.78, the State Administrative Tribunal may consider, (a) grounds in addition to those stated in the notice of objection; and (b) reasons in addition to any reasons previously given for the local government's decision that is under review.

Section 6.81 – objection not to affect liability to pay rates or service charges

The making of an objection under this Subdivision does not affect the liability to pay any rate or service charge imposed under this Act pending determination of the objection.

Section 6.82 – general review of imposition of rate or service charge

(1) Where there is a question of general interest as to whether a rate or service charge was imposed in accordance with this Act, the local government or any person may refer the question to the State Administrative Tribunal to have it resolved.

(2) Subsection (1) does not enable a person to have a question relating to that person's own individual case resolved under this section if it could be, or could have been, resolved under section 6.76.
(3) The State Administrative Tribunal, dealing with a matter referred to it under this section, may make an order quashing a rate or service charge which, in its opinion, has been improperly made or imposed.

State land tax – objections and appeals

The right to object to a land tax assessment is provided for under Section 34 of the Taxation Administration Act 2003. Under Section 35, an objection must be in writing; it must set out fully and in detail the grounds on which the taxpayer objects to the assessment or decision; and it must be lodged in accordance with section 115.

Section 36 states that an objection to an official assessment must be lodged within 60 days after –

(a) the assessment notice is issued; or
(b) if the assessment is indicated by endorsement in accordance with section 23(2)(b) – the date on which the document was endorsed; or
(c) if a taxpayer has requested a statement of grounds in accordance with section 25(2)(a) within 30 days of the issue of the assessment – the date on which the Commissioner serves a statement of the grounds; or
(d) if the assessment is an interim assessment – the date on which the three-year period referred to in section 34(2)(ca) ends.

Section 37 of the Act sets out the considerations to be made by the Commissioner in assessing an objection as follows:

(a) the grounds set out in the objection and any other relevant written material submitted by the taxpayer; and
(b) if the objection is against an interim assessment – any other information relevant to considering the objection that was obtained by the Commissioner before the assessment notice for the interim assessment was issued; and
(c) if the objection is not against an interim assessment – any other information relevant to considering the objection, whether obtained by the Commissioner before or after the objection was lodged.

Section 37 further provides that the onus of proof is on the taxpayer to prove their case.

Section 38 addresses the time limit for the determination of an objection by the Commissioner being 90 days. If the Commissioner fails to determine an objection within 120 days of the day that the objection was lodged with the Commissioner, the taxpayer may, by written notice to the Commissioner, require the Commissioner to apply to the State Administrative Tribunal for directions as to any or all of the matters referred to in this section, including but not limited to –

(a) the length of the decision period;
(b) the time for a taxpayer to comply with a request for information;
(c) the information to be provided by the taxpayer;
(d) the time for the Commissioner to seek advice and assistance from an external agency.

Section 39 sets out the provisions for reassessment on determination of an objection, where the Commissioner allows an objection in whole or in part. Section 40 allows a right of review by the State Administrative Tribunal as follows:

(1) A person dissatisfied with the Commissioner's decision on an objection or on an application for an extension of time for lodging an objection may apply to the State Administrative Tribunal for a review of the decision.
(2) A person ceases to be entitled to apply to the State Administrative Tribunal for a review of a decision on an objection against an interim assessment if the assessment following the interim assessment is made before the person makes an application under subsection (1) for a review of the decision.

Section 42 provides that an application for review to the State Administrative Tribunal must be made within 60 days after notice of decision is served on the taxpayer. Section 43 sets out the constitution of the State Administrative Tribunal among other related provisions.

Section 43A – appeal from decision of State Administrative Tribunal

(1) An appeal from a decision of the State Administrative Tribunal can be brought on a question of law, of fact or mixed law and fact, without having first obtained leave to appeal.
(2) The appeal has to be instituted in accordance with the rules of the Supreme Court and within the period of 28 days after –
    (a) the day on which the Tribunal's decision is made; or
    (b) if the Tribunal gives oral reasons for the decision and the appellant then requests it to give written reasons under section 78 of the State Administrative Tribunal Act 2004, the day on which the written reasons are given to the appellant.

Northern Territory

Objections and appeals to values

Objections and the review of values used to assess local government rates are addressed under the provisions of the Valuation of Land Act – Part V.

Section 18 of the Act provides 30 days for objections to values, which, by comparison with the states, is half the time period permitted around Australia. Grounds of objections to values (unimproved, improved or annual) are set out under section 19 as follows:

(a) that the values determined are too high or too low;
(b) that the description of the land is not correctly stated;
(c) that parcels of land that should be included in one valuation have been separately valued; or
(d) that parcels of land that should be separately valued have been included in the one valuation.

The Valuer-General shall disallow, allow or allow in part the objection under Section 20. Section 20A allows a person to object to the decision of the Valuer-General made under Section 20. Sections 20B–M Part VA of the Act set out the provisions for the establishment of a valuation board of review. Appeals from the Valuation Board of Review are made to the Land and Valuation Review Tribunal under Section 25 as follows:

Section 25 – application for review

(1) An objector who is dissatisfied with the decision of a Board on his objection may, within 30 days after the date of posting of the notice of the decision, by writing, request the Board to refer the decision to the Tribunal for review.
(2) Upon receipt of a request under subsection (1), the Board shall forthwith refer the decision to the Tribunal.
(3) Upon such a reference, the objector is limited to the grounds stated in the objection.
(4) If the valuation has been varied by the Board after considering the objection, the valuation as varied shall be the valuation to be dealt with on the reference.

Section 31 – decision of Tribunal not to be challenged

A decision of the Tribunal shall not be challenged, appealed against, reviewed, quashed or called into question. It shall not be subject to prohibition, mandamus, certiorari or injunction in the Supreme Court on any account whatever.

Local government rates – appeals

Land tax is not imposed in Northern Territory. Local council rates are the only recurrent tax imposed on land and these are assessed under the Local Government Act, which commenced 1 July 2008. Rates and charges are administered under Chapter 11 of the Act.

Section 154 – correction of record

(1) A person may apply to the council for the correction of an entry in the assessment record.
(2) The application may be made on any one or more of the following grounds:
   (a) the entry wrongly classifies an allotment that is not rateable as rateable land;
   (b) the entry should, but does not, classify an allotment as urban farm land;
   (c) the entry wrongly records the use of an allotment;
   (d) the entry contains some other relevant misclassification or misdescription of an allotment;
   (e) the entry wrongly records ownership or occupation of an allotment;
   (f) the entry wrongly designates the applicant as principal ratepayer for an allotment;
   (g) the entry contains some other relevant error.
(3) The application:
   (a) must be in writing; and
   (b) must state the applicant's interest in the allotment to which the application relates; and
   (c) must state the nature of the amendment that should, in the applicant's opinion, be made.
(4) If the application is uncontroversial, the CEO may decide the application on behalf of the council but, if it raises matters of possible controversy, the application is to be dealt with by the council or a council committee.
(5) The CEO must notify the applicant, in writing, of the decision on the application as soon as practicable.
(6) If the council or council committee decides to reject the application in whole or part, the decision is reviewable.

Australian Capital Territory (ACT)

Objection appeals and redetermination of rates

Objections, reviews and redetermination of rates and land tax are addressed under the Rates Act 2004 and the Taxation Administration Act 1999, of which the main provisions of each follow.

Rates Act 2004

Part 2 – unimproved value of rateable land

Section 11 – Redetermination error

(1) This section applies if an error was made in relation to a determination of the unimproved value of a parcel of land as at a particular date.
(2) The Commissioner may redetermine the unimproved value of the parcel as at that date.
(3) The redetermination applies to the parcel for the period –
    (a) beginning on 1 July in the calendar year in which the relevant date when the redetermination is made falls; and
    (b) ending on 30 June in the next calendar year.
(4) In this section: 'error', in relation to a determination, includes –
    (a) an error in making a valuation on which the determination is based; and
    (b) the duplication of an error in relation to an earlier determination.

Section 11A – redetermination – change of circumstances

(1) This section applies if a change of circumstances happens in relation to a parcel of land that affects the unimproved value of the land.
(2) The Commissioner may redetermine the unimproved value of the parcel as at a date if the unimproved value as at that date is used in calculating the average unimproved value of the land for the year in which the change of circumstances happens.
(3) The Commissioner may also redetermine the unimproved value of the parcel as at a later date if a determination of the unimproved value as at that date did not take the change of circumstances into account.
(4) A redetermination under subsection (2) applies to the parcel for the period –
    (a) beginning on the day the change of circumstances happened; and
    (b) ending on 30 June in the next calendar year.
(5) A redetermination under subsection (3) applies to the parcel for the period –
    (a) beginning on 1 July in the calendar year in which the relevant date when the redetermination is made falls; and
    (b) ending on 30 June in the next calendar year.

Section 12 – recording, notification and publication of determinations

(1) The Commissioner must record particulars of each determination of the unimproved value of a parcel of land.
(2) The Commissioner must give written notice of the amount determined as the unimproved value of a parcel to the owner.

(3) As soon as practicable after each annual redetermination under section 10 of the unimproved values of parcels of rateable land, the Commissioner must arrange for the unimproved values to be available to the public.

Section 71 – objections relating to valuations – general

(1) This section applies to an objection to an assessment if the objection relates to the valuation on which the assessment is based.
(2) The objection must be made within 60 days after the day the Commissioner gives notice under section 12(2) of the amount determined as the unimproved value of the parcel.

Section 72 – objections relating to valuations – unit owners

For a unit subdivision, if an objection to an assessment relates to the valuation on which the assessment is based, the Taxation Administration Act, section 100(1) (Objection) –

(a) applies to the owners corporation as if the assessment were served on the owners corporation; and
(b) does not apply to a unit owner.

Tax Administration Act 1999

Section 100 – objection

(1) A taxpayer may lodge a written objection with the Commissioner if the taxpayer is dissatisfied with –
    (a) an assessment, other than a compromise assessment, that is shown in a notice of assessment served on the taxpayer; or
    (b) a decision mentioned in schedule 1 or schedule 2; or
    (c) a decision under a tax law that is prescribed under the law for this section.

Section 101 – grounds for objection

(1) The grounds for the objection must be stated fully and in detail, and must be in writing.
(2) The grounds for the objection, for a reassessment, are limited to the extent of the reassessment.
(3) The burden of showing that an objection should be sustained lies with the taxpayer making the objection.

Section 102 – time for lodging objection

An objection must be lodged with the Commissioner not later than 60 days after the date that the notice of the assessment, or of the decision objected to, is served on the taxpayer, except as provided by section 103.

Section 104 – determination of objection

(1) The Commissioner must consider an objection and either allow the objection in whole or in part or disallow the objection.
(2) The Commissioner must take such steps as are necessary (for example, by delegating the functions given by this section) to ensure that the individual who considers the objection is not the individual who made the assessment or decision against which the objection was lodged.

Note – The Commissioner's decision in relation to an objection is a reviewable decision (see s 107A), and the Commissioner must give a reviewable decision notice to the taxpayer (see s 108).

Section 105 – recovery of tax pending objection or review

The fact that an objection or review is pending does not affect the assessment or decision to which the objection or review relates, and tax may be recovered as if no objection or review were pending.

Division 10.2 – Sections 107–109 address the notification and review of decisions

SECTION 109 – GIVING EFFECT TO ACAT DECISION

(1) Within 60 days after the day an ACAT decision becomes final, the Commissioner must take any action, including amending any relevant assessment that is necessary to give effect to the decision.
(2) For this section, an ACAT decision becomes final when a period of 30 days has passed after the day a relevant decision is made, and no appeal against the relevant decision has been begun within the 30-day period.

Summary

Part 2 demonstrates the need for reform across all facets of recurrent land tax in Australia. These reforms range from the manufacture of value on which the base of the tax is assessed to the tier of government imposing and collecting the tax and the outdated exemptions afforded to residential property. It will be further demonstrated in Chapter 12 that a fast-emerging and compelling case exists for the reform of the bases of value on which recurrent land taxes are assessed, particularly in the capital cities where most of this tax revenue is collected.

Chapter 5 reviewed the principal place of residence exemption status from state land tax within the statutory frameworks of each of the state's Land Tax Acts. It was highlighted that this exemption was relevant at the time the states recommenced taxing land during the 1950s to minimise competition with revenue collected from local government rates. However, the overall recurrent land tax revenue collected from the principal place of residence in Australia reflects poorly on sub-national government's tax effort.

Through the exemption status of the principal place of residence from state land tax and limiting increases in local government rating in some states, homeowners enjoy a low land tax environment. While the states shelter themselves behind the supply side argument as the contributor to higher housing prices, they are loathe to address tax reform of the principal place of residence, which is the second largest contributor to a lack of housing affordability in Australia. While much debate is focused on tax expended from negative gearing, this pales into insignificance compared with the tax expended on the principal place of residence.

Further, while much focus has centred on the generous tax concessions afforded to superannuation contributions, these pale into insignificance when compared with the amount of unstructured superannuation money assigned to the principal place of residence. This asset class escapes state land tax, capital gains tax, and is under-taxed through low local government rates around Australia. What must be avoided at all costs are narrow bolt-on tax reforms by the states through levies tacked onto local government rates, while an inefficient land tax system coexists. The fiscal reform of sub-national government commences with reforming the recurrent taxation of the principal place of residence.

Chapter 6 built on Chapter 5 through the review of a number of concessions and allowances that exist across the states. Following the principal place of residence, the land tax-free threshold is the second largest land tax expenditure, which is afforded to property investors and owners of property which does not qualify as the principal place of residence. Table 7.1 demonstrates the benefits afforded to investors using median house and unit prices as at 2014. It shows the number of residential properties an investor may own at the point at which the aggregate land or site value reaches the land tax threshold in each state for the 2015 land tax year. A brief summary of the inputs is located in the paragraph prior to this table.

The need for a land tax threshold applicable to residential property in the capital cities of Australia has long passed. Its initial role of promoting

Table 7.1 Investment property relative to land tax thresholds

City	*Housing type*	*Land : Improved value ratio*	*Sale price median 2014 March*	*Land value median*	*Tax-free threshold*	*No properties*
Sydney	House	40%	$787,728	$315,091	$432,000	1.37
Sydney	Unit	30%	$551,741	$165,522	$432,000	2.61
Melbourne	House	40%	$597,510	$239,004	$250,000	1.05
Melbourne	Unit	30%	$416,525	$124,958	$250,000	2.00
Brisbane	House	40%	$470,382	$188,153	$599,999	3.19
Brisbane	Unit	30%	$361,724	$108,517	$599,999	5.53
Adelaide	House	40%	$457,087	$182,835	$300,000	1.64
Adelaide	Unit	30%	$280,213	$84,064	$300,000	3.57
Perth	House	40%	$618,160	$247,264	$300,000	1.21
Perth	Unit	30%	$406,197	$121,859	$300,000	2.46
Hobart	House	40%	$326,653	$130,661	$24,999	0.19
Hobart	Unit	30%	$250,616	$75,185	$24,999	0.33

investment to increase the supply of rental housing stock ended over a decade ago. The states must contribute to the tax reform agenda of Australia, despite their narrow tax base; reforming taxes on residential housing is key to this reform. The efficient reform of existing tax revenues should not be viewed as a zero sum gain by the states. That is, offsetting stamp duty revenue with increases in state land tax is an oversimplistic perspective in addressing the tax reform needed.

Part 2 has demonstrated that land tax statutes comprise provisions which are largely designed to manage the significant carve-outs afforded to residential property through the principal place of residence exemption. It further highlights the impact of the tax-free thresholds and raft of exemptions which render this tax applicable to approximately 20 per cent of property in Australia. In addition to the governing statutes is the cost of managing and auditing these carve-outs and exemptions. In contrast, it is local government rating which is conceptually the more efficient and robust tax source, capturing up to 98 per cent of rateable property.

However, local government rating systems are not without issue in battling the use of rating differentials and the fast-emerging phenomenon resulting from low to high density housing, a matter further addressed in Chapter 12. Part 3 sets the stage for the direction of structural tax reform of sub-national government in Australia. The review of land tax in five countries, two of which are federated structures similar to Australia, paves the way for reform and demonstrates how land tax may contribute further to sub-national government tax revenue.

The inputs for Table 7.1 comprise the median price for units and houses during 2014 and the corresponding land tax-free threshold for the 2015 land tax year has been used. Land tax is assessed on values determined the year prior to land tax year in most states. The land to improved value ratio is that approximate relationship between the land value component of units and houses for each city. The methodology and rationale for the determination of this ratio is again discussed in Chapter 12 and underpins the necessity for the reform of the base on which recurrent land tax is assessed.

Part 3

Recurrent land taxation – international case studies

Introduction

Land tax is an international tax and is assessed on numerous bases, of which value is the dominant base used in the OECD economies of the world. A review of the history and evolution of the current operation of this tax in several countries demonstrates that it is a tax under continual change and evolution which is marked by ongoing challenges, reviews and reform. The question that often arises is, 'which is the best land tax system and why is in not adopted globally?' The answer to this question is that there is no one perfect or one-fit system for all, as the operation of land tax around the world is driven and determined by the circumstances and factors impacting the specific jurisdictions at any point in time.

The previous Part demonstrated that land tax operates on several different bases and across two tiers of government, with different legislative provisions applying across Australia. This is due to the specificity of matters that have arisen in various states which are at different phases in their economic development and political decentralisation and evolution. These factors are no different between countries and international jurisdiction which levy this tax. While this may be the case, there are similarities in the strengths, limitations and challenges confronting many jurisdictions. This part of the book is dedicated to the review of land tax in its various forms in several jurisdictions and provides two key objectives. The first objective is the examination of the current status and operation of this tax within these jurisdictions, while the second is the examination of recurring issues and matters common across some jurisdictions.

The jurisdictions examined were selected on the basis of their tax effort from recurrent land tax which was either greater or similar to Australia's and the base on which the tax is assessed. Over the period of 2007 to 2014, the author visited each of the countries and cities covered in this Part of the book with the exception of Ireland. These countries were visited as part of this research, which provided a rich source of factual local knowledge and understanding of the operation of land tax in these countries. This also enhanced the limitations of texts which rely on secondary sources undertaken through literature reviews. Detailed interviews were undertaken, along with external examinations of land and buildings and the review of key literature. The three primary criteria for the selection of these countries are an amalgam of the use of land and improved value as the base of the tax, fiscal decentralisation of the tax to local government and tax effort raised from this source.

Common among the factors shaping land tax policy across the world are the aging population and the transition from an industrial to service-based economy across the OECD, a transition also impacting Australia. The evolution of the world's economies from the broadacre production of food and the use of land for secondary production and heavy industry to the service sector, retailing and commerce production have resulted in changes to the urban formation of land uses. Intensification of the use of land for housing and office buildings has revolutionised the way land value for tax purposes is and will continue to be measured. An international review of land tax within and across international jurisdictions brings to the fore some of the challenges now confronting Australia in the administration of this tax. It provides a broad spectrum and contribution to the way this tax may be assessed and administered in the future.

The chapters which follow in Part 3 review this tax in the United States, Canada, New Zealand, Denmark and parts of the United Kingdom. The key factors examined are the base of the tax and its application to property, its administration by government, taxpayer safety nets and the review and appeal processes, with a brief focus of tax expenditure through concessions also included.

8 United States and Canada

Overview

This chapter is a review of the property tax in the states of New York and California and more specifically addresses its operation in the cities of New York City (NYC) and Los Angeles (LA). In Canada, the provinces reviewed are British Columbia and Ontario. The objective is to examine the basis of value on which the tax is assessed, how the base of the tax is determined, concessions and allowances, objections and appeals and the organisational governance of the tax. The property tax operates in each state of America and across the provinces of Canada and is governed by state law, while its administration is primarily undertaken by local government. In the United States and Canada, this tax is assessed on improved value and, in contrast to Australia, is referred to as the 'property tax', which is the label used in this Part.

United States

The United States has a constitution-based federal republic structure of government, comprising 50 states and one district which are spread across six time zones. Local governments generally include two tiers: counties (also known as boroughs in Alaska and parishes in Louisiana), municipalities or cities/towns. In some states, counties are divided into townships. Municipalities can be structured in many ways, as defined by state constitutions, and are called townships, villages, boroughs, cities or towns. Various kinds of districts also provide functions of local government outside county or municipal boundaries, such as school districts or fire protection districts.

Municipal governments – those defined as cities, towns, boroughs (except in Alaska), villages and townships – are generally organised around a population centre and, in most cases, correspond to the geographical designations used by the United States Census Bureau for reporting of housing and population statistics. Municipalities vary greatly in size, from the millions of residents of NYC and LA to the 287 people who live in Jenkins,

Minnesota. Municipalities generally take responsibility for parks and recreation services, police and fire departments, housing services, emergency medical services, municipal courts, transportation services (including public transportation) and public works (streets, sewers, snow removal, signage and so forth). Whereas the federal government and state governments share power in countless ways, a local government must be granted power by the state. In general, mayors, city councils and other governing bodies are directly elected by the people.

Property tax was introduced in North America soon after its colonisation in Plymouth, Massachusetts in 1620, where the tax was imposed on land, buildings and personal property. Between 1834 and 1896, uniformity in taxation became a major constitutional issue in the United States and 31 states adopted constitutional uniformity clauses. Its implementation has been used to fund the provision, maintenance and renewal of local and state infrastructure and local services. In later years in the United States, the property tax has gained momentum with its earmarking to school education, while in Canada it has been aligned with a variety of local government services.

At a national level in both the United States and Canada, property tax revenue has reached 3.5 per cent of GDP, being among the highest in the OECD. While serving as an important tax for sub-national government, it is a highly visible yet poorly understood tax by taxpayers and even to some who administer the tax. In both of these nations, the property tax is assessed on the broad definition of improved value, being the amalgam of both land and improvements on the land. This is where any similarity in the operation of this tax ends, both within and across these countries. The United States, and specifically NYC and LA, is first reviewed followed by Canada, and specifically the provinces of Ontario and British Columbia, where a comparison is made, followed by the strengths and challenges of the tax, which are addressed in the conclusion of this chapter.

New York City

New York City (NYC) as at 2014, comprises approximately 8.4 million people spread across five boroughs (The Bronx, Brooklyn, Queens, Staten Island and Manhattan) and raises in excess of US$18 billion in property tax per annum. A borough is a similar geographic unit to a ward, which is a geographic unit within local government areas in Australia. The property tax collected in NYC equated to over 80 per cent of the total combined state land tax and local government rate revenue collected across the whole of Australia over the same period. The property tax among other services is used to maintain two rail systems, two eight-lane tunnels and a third rail system below the Hudson River that connects Manhattan Island with Jersey City, New Jersey.

The property tax in this metropolis is a serious tax and well-established source of revenue used to service the fast-evolving services aligned with a sophisticated city. While no more popular than the equivalent tax in Australia, its purpose in NYC is not understood to be a tax dedicated to fixing roads or collecting rubbish, but extends to include policing, transportation and regeneration of parts of the city and elements of the school education system. While debate surrounds some of these programs, the tax is nonetheless viewed by taxpayers as the funding and fabric of their city. That said, the tax itself is one that has evolved and has been the subject of many attempts of reform at the operational and fiscal policy levels.

The rationale for reform becomes apparent in the following sections of this chapter, in the review and analysis of the split in revenues from the various classes of property and the mechanisms used to manage increases in revenue from property. It further defines the many forms used to define the term *value* and how this is codified within the complex approach to taxing property. To give context to the complexity of the application of this tax, the following inputs of the tax are provided, followed by examples of the mechanics used in assessing this tax across the various land uses in NYC.

Basis of value and the tax

In layman speak, the base of the property tax in NYC is improved value, which encompasses land and buildings. Both a land value and improved value are determined each year for each parcel. In the case of high-rise, multi-unit housing, one parcel may have a number of properties within it. For example, a high-rise unit development may comprise 100 flats; however, is one parcel, to which the allocation of value is distributed based on the distribution determined by the building cooperative or other structure, used for managing the entitlement of the building? The concept is similar to that used in Australia in its application to strata title.

While the tax is assessed on improved value, this determinant loses relevance as property is classified differently; secondly, as it is held for increasingly longer periods of time, the tax becomes further distorted at the individual lot level. Further, the application of the tax is impacted by the percentage of taxable value used, which changes for each of the various classes of property and, in some instances, also varies within the various subclasses of property. Table 8.1 sets out these taxable values and demonstrates their use in assessing the property tax.

Assessing the property tax in New York City

In demonstrating the operation of the property tax in NYC, the following scenarios are used. They show how the tax is initially determined and how a tax bill is then re-determined from one tax year to the next.

Components of the tax:

Tax levy:	Total amount of tax to be collected in a taxing jurisdiction.
Assessment:	The taxable value of the individual property.
Rate:	Rate applied to the tax assessment (taxable value).
Taxpayer's property tax bill:	The bill determined by applying the rate to the assessment.

On closer review, it is noted that different tax rates apply to different classes of property, depending on their classification (how they are grouped). Further, assessment ratios apply which regulate the size of the increase in the assessment from one year to the next. The classes and definitions of property within each of these classes follow.

Property class tax and definition structure:

Class 1:	One, two and three-family homes
Class 1A:	Residential condominiums of three stories or less which were always condominiums
Class 1B:	Certain vacant land zoned for residential use
Class 1C:	Residential condominiums with no more than three dwelling units which had been classified in tax Class 1 on a prior assessment roll
Class 1D:	One-family houses located on cooperatively owned land of bungalow colonies in existence before 1940
Class 2:	All other residential condominiums, rental apartments with 11 or more units, all cooperatives
Class 2A:	Four- to six-unit rental apartment houses
Class 2B & 2C:	Seven- to 10-unit rental apartment houses and condominiums and co-ops 2 to 10 units, primarily residential
Class 3:	Utility property (water, power, etc.)
Class 4:	All other property (commercial and manufacturing properties in the city, including office buildings)
Class 4A:	Certain railroad property

Increases for physical reasons, such as new construction or alteration of improvements, are not restricted by the assessment limitations described above. Cooperatives and condominiums in Class 2 are subject to special valuation rules set forth in the Real Property Tax Law section 581. Transition assessments for Class 3 were eliminated by state legislation which sunset on 31 December 1986. Of further note in the application to the above Maximum Permissible Assessments (MPAs) is that when a property is purchased, the rate applies to the assessment and there is no application of the MPA in the first year, which differs from an identical property either in the same

Maximum Permissible Assessment (MPA) – Revised as at July 2012

Tax class	*Target assessment ratio*	*Assessment limitation*	*1st year effective*
Class 1	6%	6% a year, 20% over any 5 yrs	1982/83
Class 1A	6%	6% a year, 20% over any 5 yrs	1986/87
Class 1B	6%	Not to exceed 112% of tax assmt	1989/90
Class 1C	6%	6% a year, 20% over any 5 yrs	1991/92
Class 1D	6%	6% a year, 20% over any 5 yrs	1990/91
Class 2	45%	None, but increases phased in over 5 yrs	1982/83
Class 2A	45%	8% a year, 30% over any 5 yrs	1984/85
Class 2B and 2C	45%	8% a year, 30% over any 5 yrs	1986/87
Class 3	45%	None	1982/83
Class 4	45%	None but increases phased in over 5 yrs	1982/83
Class 4A	45%	None	1992/93

street, in the case of a house – or in the same building, in the case of an apartment – that was purchased three years earlier. This point has become the subject of debate in NYC and highlights the differential between new property owners in contrast to existing long-term property owners.

The following scenario is used to demonstrate, firstly, the calculation used in the assessment of this tax, followed by the intricacies in the adjustment of an assessment from one year to the next. It is one of the more complex adjustment processes noted by the author across several jurisdictions.

Scenario

In applying the above property classes and MPAs, we first construct a hypothetical scenario which sets out the process demonstrating how the tax levy is determined, followed by the determination of tax bills using two of the above classes of property. The third point is to demonstrate how a property tax bill would alter from one year to the next, subject to changes in the tax levy, assessments and rates. While the above classes and MPAs are from NYC as at July 2012, a simple hypothetical scenario is used to highlight the operation of this tax.

In this hypothetical case in NYC, the school district, local and county government raised collectively $135 million in property tax in the 2014 tax year. The budget is determined based on revenue estimates needed by school districts, municipalities and counties to fund public services. Once budgets are determined, these authorities subtract the revenues from all other sources and the remaining shortfall revenue is raised from the property tax. This amount is referred to as the *tax levy* and represents the total amount of property tax to be collected. The tax levy is divided by the total assessed value within a jurisdiction to determine the tax rate. The tax rate is applied to each assessment to arrive at the individual property tax bill.

In summary, a number of factors may impact changes in property tax bills, the key changes of which are:

1) Local budget increases
2) Less tax revenue from other sources
3) Changes in the total value of property within an assessed district
4) Changes in the way the tax levy is divided between the different municipalities.

Determining the tax levy and an individual tax bill for two classes of residential property are set out in Table 8.1. In these examples, which are in the same location, Property 1 is a Class 1, two-family home with a market value of $2,000,000. The assessment of $120,000 is derived by applying 6 per cent as shown in the above MPA figures to the market value of $2,000,000. Property 2 is a Class 2 residential condominium with a market value of $1,000,000. This assessment of $450,000 is derived by applying 45 per cent as shown in the above MPAs. Both Property 1 and 2 were purchased in 2014.

Table 8.1 Determination of a property tax bill NYC (Class 1 and Class 2 property)

2014	*Total taxable assessments*	*Projected expenses*	*All other revenue sources*	*Property tax levy*
School district budget	$3,000,000,000	$80,000,000	$30,000,000	$50,000,000
Municipality budget	$2,000,000,000	$10,000,000	$6,000,000	$4,000,000
County budget	$30,000,000,000	$200,000,000	$120,000,000	$80,000,000
Property 1 – 2014		**Rate**	**Assessment**	**Tax bill**
School district budget	$50,000,000 / $300,000,000	= $16.67 / $1,000 of assessed value	$16.67 x ($120,000 / $1,000)	**$2,000.40**
Municipality budget	$4,000,000 / $2,000,000,000	= $2.00 / $1,000 of assessed value	$2.00 x ($120,000 / $1,000)	**$240**
County budget	$80,000,000 / $30,000,000,000	= $2.67 / $1,000 of assessed value	$2.67 x ($120,000 / $1,000)	**$320.40**
Total tax				**$2,560.80**

Table 8.1 (continued)

2014	*Total taxable assessments*	*Projected expenses*	*All other revenue sources*	*Property tax levy*
Property 2 – 2014		**Rate**	**Assessment**	**Tax bill**
School district budget	$50,000,000 / $300,000,000	= $16.67 / $1,000 of assessed value	$16.67 x ($450,000 / $1,000)	**$7,501.50**
Municipality budget	$4,000,000 / $2,000,000,000	= $2.00 / $1,000 of assessed value	$2.00 x ($450,000 / $1,000)	**$900**
County budget	$80,000,000 / $30,000,000,000	= $2.67 / $1,000 of assessed value	$2.67 x ($450,000 / $1,000)	**$1,201.50**
Total tax				**$9,603.00**

Concessions and allowances

In line with other national and international property tax jurisdictions, NYC has a number of concessions and allowances applicable to the levying of the tax, of which a summary follows:

- Clergy allowance
- Veterans allowance
- Aged persons concession
- Rent control concession

Los Angeles, California

California is the third largest state in land area and the largest state economy in the United States, also being one of the top 10 largest economies in the world, with a population of approximately 39 million people. Originally, California was part of Mexico until 1850, when it was declared a state of the United States and is now an important contributor to the US food bowl. Sacramento is the capital of California, with Los Angeles (LA), San Francisco, San Jose and San Diego among the major cities of this State. The property tax has been part of California's tax mix since it joined the United States and has been assessed on the improved value of property (land and buildings combined) since the tax was introduced.

Like the other US states, California imposes the property tax at the state level, which is totally administered and collected at the local level by either cities or the 58 counties. In LA, the tax is collected by the Office of the Tax Collector. The revenue collected from this tax is shared between state, counties, cities and school districts, and the split among these entities is determined by state law and the annual state budget. There are approximately

30 million assessments undertaken annually across the state of California. The City of Los Angeles is the largest city of California, with a population of approximately 3 million people. The Los Angeles County Assessor's Office is responsible for the assessments of 10 counties across the LA metropolitan area, which has a population of approximately 10 million people.

Proposition 13

During 1978, nearly two-thirds of California's voters passed Proposition 13, reducing property tax rates on homes, businesses and farms by about 57 per cent. In the environment prior to Proposition 13, the property tax rate throughout California averaged a little less than 3 per cent of market value. Additionally, there were no limits on increases for the tax rate or on individual ad valorem charges. (*Ad valorem* refers to taxes based on the assessed value of property.) Some properties were reassessed between 50 per cent to 100 per cent in just one year and their owners' property tax bills increased accordingly.

Rising home prices, leading to an increase in property taxes, coupled with legislative inaction, were trends that generally existed through much of the five years predating Proposition 13. When the California legislature adjourned in the fall of 1977 without passing any significant property tax reforms, even though 22 different reform plans were proposed, voters quickly signed the circulating initiative petitions for the Jarvis-Gann proposition (Jarvis-Gann became known as Proposition 13 because of its number on the 1978 ballot). Proponents argued that the proposition was both a property tax relief measure and a necessary constraint upon the size of government. The legislature reconvened and passed a potential reform (which necessitated a constitutional change) that would appear along with Proposition 13 on the June ballot, where Proposition 13 easily passed.

Although poorly written, the basic rules of Proposition 13 were relatively straightforward. The maximum property tax rate was set at 1 per cent of the value of the property. The value of the property was set at its 1975–76 level, but was allowed to increase by the rate of inflation, up to 2 per cent each year. Property could be revalued only upon a change of ownership. No new ad valorem property taxes could be imposed. Any special taxes (which were not defined) needed to be approved by two-thirds of the voters. Finally, the distribution of the property taxes that were collected was to be done 'according to law', and since no such law existed, one had to be created. Prior to the adoption of Proposition 13, local agencies established their own separate property tax rates and received the proceeds of the tax. For the first time in the state's history, the state was put in charge of allocating the proceeds of the locally levied property tax, with the rate and base defined by the state-wide initiative.

Proposition 13 tax reform

Under Proposition 13 tax reform, property tax value was rolled back and frozen at the 1976 assessed value level. Property tax increases on any given property were limited to no more than 2 per cent per year, as long as the property was not sold. Once sold, the property was reassessed at 1 per cent of the sale price, and the 2 per cent yearly cap became applicable to future years. This allowed property owners to finally be able to estimate the amount of future property taxes, and determine the maximum amount taxes could increase as long as he or she owned the property. Proposition 13 initiated sweeping changes to the California property tax system. Proposition 13 capped, with limited exceptions, ad valorem property tax rates at 1 per cent of full cash value at the time of acquisition. Prior to Proposition 13, local jurisdictions independently established their tax rates and the total property tax rate was the composite of the individual rates, with few limitations. Prior to Proposition 13, jurisdictions established their tax rates independently and property tax revenues depended solely on the rate levied and the assessed value of the land within the agency's boundaries. (California Constitution Article 13A)

Reassessment upon change of ownership

Prior to Proposition 13, if homes in a neighbourhood sold for higher prices, neighbouring properties might have been reassessed based on the newly increased area values. Under Proposition 13, the property is assessed for tax purposes only when it changes ownership. As long as the property is not sold, future increases in assessed value are limited to an annual inflation factor of no more than 2 per cent. Proposition 13 requires any increases in state taxes to be approved by a two-thirds vote of each house of the legislature. Further, the same provisions apply to local government for proposed increases in taxes for a designated or special purpose to be approved by two-thirds of voters.

Concessions and allowances

The property tax applies to all property in the state unless otherwise exempted by state statute. A homeowner exemption of $7,000 off the assessed value is available for owner-occupied housing and must be applied for by owners. Tax rates vary by property types and tax rate area. By state law, all property is assessed by a 1 per cent levy. An additional voter indebtedness is added, plus direct assessments. An average estimate is 1.25 per cent plus direct assessments.

The legislature may determine the manner in which a person of low or moderate income who is 62 years of age or older may postpone ad valorem property taxes on the dwelling owned and occupied by him or her as his or her principal place of residence. The legislature may also provide for the manner in which a disabled person may postpone payment of ad valorem property taxes on the dwelling owned and occupied by him or her as his or her principal place of residence.

In the case of age abatements, the legislature shall provide by for subventions to counties, cities and counties and cities and districts in an amount equal to the amount of revenue lost by each by reason of the postponement of taxes and for the reimbursement to the state of subventions from the payment of postponed taxes. Provision shall be made for the inclusion of reimbursement for the payment of interest on, and any costs to the state incurred in connection with, the subventions.

A summary of main exemption available in California follows:

- Homeowner exemption
- Veteran's and disabled veteran's exemption
- Institutional exemption
- Builder's exclusion
- Disaster relief
- Eminent domain – Proposition 3
- Decline in value – Proposition 8
- Parent to child exclusion – Proposition 58
- Grandparent to grandchild exclusion – Proposition 193

Basis of assessment

It was stated earlier that the property tax is assessed on improved value; however, more specifically, the statutory basis of value is the *assessed value.* The basis of value is set out under section 1 of Article XIII of the California Constitution, which establishes the foundation of the fair market value standard, which follows:

> Unless otherwise provided by this Constitution or the law of the United States, all property is taxable and shall be assessed at the same percentage of fair market value. When a value standard other than fair market value is prescribed by this Constitution or by statute authorised by this Constitution, the same percentage shall be applied to determine the assessed value. The value to which the percentage is applied, whether it be fair market value or not, shall be known for property tax purposes as the as the full value.

This requires that property be valued at its most productive value, also known as its *highest and best use.* Computer Assisted Mass Appraisal (CAMA) is used in some counties, including LA, for a few appraisal functions, with up to 38 variables used in multi-regression analysis.

Assessments and reassessments

Property tax appraisers in California are required by the California Constitution statutes and regulations to appraise property at market value. That means that property must be appraised on the highest and best use value. Within the state of California, it is the State Board of Equalization or the

County Assessor that values property for tax purposes. All are government employees either by the state or counties (local jurisdictions). Within Los Angeles County, all assessments are undertaken by government employee assessors.

Since the passage of Proposition 13 in 1978, property is assessed at its fair market value as of the date it is acquired. The purchase price generally becomes the taxable 'base value' as of that date. From that point forward, the taxable value of the property is limited to no more than a 2 per cent increase per year. For example, if the purchase of a property gets assessed at $500,000, the annual taxes would be based on $500,000 the first year, trended to $510,000 ($500,000 x 1.02) the second year and trended to $520,200 ($510,000 x 1.02) the third year, etc.

Additionally, new construction is a trigger for a new assessment, which may result from any of the following:

- new structures;
- area added to existing structures;
- new items added to an existing structure such as bathrooms, fireplaces or central heating/air conditioning;
- physical alterations resulting in a change in use;
- rehabilitation, renovation or modernisation that converts an improvement to the substantial equivalent of a new improvement; and
- land development (grading, engineered building pad, infrastructure).

Examples: new homes, room additions, patio covers, pools, spas and decks may be considered new construction.

The property owner is notified by the either an assessment change notice or a supplemental change notice. The notice informs the property owner of the old assessed value and the new assessed value. Additionally, on the notice are the contact information of the local assessor's office and the assessment appeals information.

Objections and reviews

There is no objection or appeal process against the property tax itself, with appeals and objections confined to assessments (values) used to determine the tax. There is an informal appeal process where taxpayers may contact the local assessor's office with a grievance, where the appraisers review the value on request. A formal appeal process exists with the Los Angeles County Appeals Board, where the taxpayer does not have to wait for the outcome of the informal appeal before proceeding to the Appeals Board; instead, these may be lodged concurrently.

Once an appeal is filed, the property owner/applicant will have a hearing heard by either a hearing officer or a three-member board. At the hearing, based on the evidence presented, the hearing officer or three-member board will make a determination of value for the appealed property. Appeals from the Appeals Board are made to either the superior court or beyond (state court or federal court) for procedural issues.

The grounds of objection to assessments/values are as follows:

- Decline in value (the market value at the lien date: January 1)
- Ownership issues (whether the property should have been reassessed)
- Base year value issues (the value of the property on the date of transfer ownership)
- New construction value (the value of the property on the date of completion)

Summary

While there has been much focus on Proposition 13 in California internationally, there are numerous different systems in place across the United States used in the operation and assessment of the property tax. It was demonstrated through the examples in NYC that there is room for reform, and that the assessment of the property tax is complex and not easily understood by many taxpayers. The two states reviewed in this chapter highlight the diversity that exists and further highlight that in countries as well-advanced as the United States there is no exact or correct approach for this tax. More importantly, it demonstrates that what exists in these, and no doubt in many other jurisdictions across the United States, is that systems are dictated by the local circumstances that exist at specific points in time.

In the late 1970s, the fear of escalating property values and tax bills led to the preemptive strike to introduce Proposition 13. This has impacted all property in California and perhaps inadvertently affected the evolution of a well-designed property tax system which may also contribute to more tax revenues across the state and particularly in the capital cities of California. Similarly in New York, the use of MPAs and Target Assessment Ratios for the various classes of property demonstrates the need for reform in better aligning tax policy with the principles of good tax design, particularly that of simplicity.

There is no doubt that the United States is one of the leading countries in the imposition of the property tax worldwide. What appears to be emerging from the review of the two states in this chapter is the realignment of the tax revenues it raises from this source. This brings to the fore the challenge of any tax reform, the debate on the political economy of where and how tax revenue is to be raised and how it might be recalibrated. Other options for the reform of this tax would better reflect the capacity to pay as taxpayer circumstances change over time, as does the mix of property uses in highly urbanised locations.

Canada

Canada is a constitutional monarchy, with a head of state represented by the Office of Governor General. Canada has a federated structure of government with the office of Prime Minister as its head of government. Sub-national government comprises 10 provinces and 3 territory governments, as well as a local government system. The population of Canada as at July 2014 was 35,540,000, with over 25 million of the population located within its metropolitan areas. Canada, like Australia, has a highly urbanised population with a similar structure of government, though they are at opposite ends of the fiscal decentralisation spectrum. Canada is one of the more fiscally decentralised federations, while Australia is the second least fiscally centralised federation in the OECD.

Like Australia, Canada imposes stamp duty on the purchase of property at a significantly lower rate and levies a capital gains tax on property, with the principal place of residence exempt from this tax. Capital gains tax applies only on the sale of investment properties (i.e., capital gains tax does not apply on the sale of a principle residence). It is applied as part of the annual personal income tax return, which means there is a federal and provincial component and the rate may vary by province. For example, in British Columbia for 2013, 50 per cent of the gain would be included in the determination of taxable income.

The two provinces focused on in this chapter are Ontario, of which Toronto is the capital city and most populated city and metropolitan area in Canada. British Columbia is the second province, of which Victoria is the capital. These two provinces have been selected as they are among the most progressive in Canada in fiscal management and, more specifically, in the administration and operation of recurrent property tax. Further, the valuation practices, bases of value and its administration are world-class, as are those generally across Canada. While Canada is one of the more progressive countries in the administration of the property tax, these provinces have encountered a number of challenges in the management of the assessment process and have introduced progressive reforms.

The chapter now examines the application and operation of recurrent property taxation in British Columbia and Ontario and, more specifically, looks at some of the challenges surrounding the financing of local government, the merits of the tax base and objection and review processes. It further examines the concessions and allowances expended, and highlights some of the strengths, which like the US system, result in a significantly high tax effort and revenues from this tax source compared with the tax effort of Australia.

British Columbia (BC)

Property transfer tax

At the provincial level of British Columbia, there is a property transfer tax which is determined on the fair market value of the land and improvements as of the date of registration. The tax is charged at a rate of 1 per cent for the first $200,000 and 2 per cent for the portion of the fair market value that is greater than $200,000. An example of the application of the tax follows.

A property is purchased for $500,000 and attracts the property transfer tax. The tax on this transaction is calculated as follows:

Purchase price	$500,000	
1% of the first	$200,000	= $2,000
2% of the remaining	$300,000	= $6,000
Total tax paid		**= $8,000**

A number of exemptions and concessions apply to the imposition of the property transfer tax, which are broken up into three broad categories as follows:

Family Exemptions

- Transfer of a principal residence
- Transfer of a recreational residence
- Transfer resulting from a marriage breakdown
- Transfer of a family farm involving individuals
- Transfer of a family farm to or from a family farm corporation

Other Exemptions

- Transfer to correct a conveyancing error
- Transfer to a registered charity
- Registration of an agreement for sale
- Transfer to or from joint tenants to tenants in common
- Transfer following bankruptcy
- Transfer for subdividing property
- Registration of multiple leases on the same property
- Transfer for company amalgamation
- Transfer for escheats, reverts or forfeits to or from the Crown
- Transfer under the Veterans Land Act (Canada)
- Transfer to minors from the Public Guardian and trustee
- Transfer to status Indian and Indian Band
- Transfer to or from trust companies or the Public Trustee

First Time Homebuyers

The First Time Home Buyers' Program is a provincial program to reduce or eliminate the amount of property transfer tax paid when someone purchases their first home. Provisions for qualifying for this grant are as follows:

To qualify for a full exemption, at the time the property is registered you must:

- be a Canadian citizen or permanent resident
- have lived in BC for 12 consecutive months immediately before the date you register the property, or have filed at least two income tax returns as a BC resident in the last six years
- have never owned an interest in a principal residence anywhere in the world at any time
- have never received a first time homebuyers' exemption or refund and the property must:
- be located in BC
- only be used as your principal residence
- have a fair market value of:
    - $425,000 or less if registered on or before February 18, 2014; or
    - $475,000 or less if registered on or after February 19, 2014
- be 0.5 hectares (1.24 acres) or smaller a partial exemption may apply if:
- it has a fair market value less than:
    - $450,000 if registered on or before February 18, 2014, or
    - $500,000 if registered on or after February 19, 2014
- is larger than 0.5 hectares
- has another building on the property other than the principal residence

Recurrent property tax

Responsibility for setting of property tax rates falls to both the provincial government level (School Tax portion) and municipal (local) level. The local governments (Surveyor of Taxes for rural properties) also collect levy portions for funding of the Assessment Authority (BC Assessment), and Municipal Finance Authority. The local government/taxing authority collects property tax revenue in the province of BC

Like the United States, a component of the property tax encompasses a school tax, which is one of the services earmarked for and included on the annual property tax notice. The tax is not determined on whether the taxpayer or their family use the public or private school system. The education system benefits all BC residents, including people without children in school. School tax is paid to share in the cost of providing education in BC. If a property is located within a municipality, school tax is paid to the municipal office. If the property is in a rural area, the school tax is paid to the province's Surveyor of Taxes. School tax is charged on every property in BC unless the property qualifies for an exemption.

Tax rates

Each year the province sets the *residential* school tax rate for each school district, which is based equally on:

1. the total number of residences in the district
2. the total residential assessed value in the district

Generally, these rates increase each year based on the previous year's provincial inflation rate. Each year the province sets the school tax rate for each *non-residential* property class. Generally, these rates increase each year based on the inflation rate plus new construction.

School tax revenue

Revenue from school tax represented approximately 33 per cent of the $5.412 billion spent by the Ministry of Education in 2013/14 as set out in the public accounts. School tax revenues from *residential property* represented about 13 per cent and *non-residential property* represented about 20 per cent of the education cost. If your property is located in a city, town, district or village, it is in a municipality. When you own property in a municipality, you will receive your property tax notice from your municipality. You pay your property taxes to your municipal office.

The municipality collects taxes for the services they provide and on behalf of other organisations to raise funds for their services. Some of the organisations the municipality may collect taxes on behalf of are:

- Provincial Government (school and policing)
- Regional Districts
- Regional Hospital Districts
- British Columbia Transit Authority
  - TransLink
  - BC Assessment
- Municipal Finance Authority

The provincial portion of the property tax is the school tax portion. The remainder of the property tax is determined by the tax rate set by the taxing authority. There are also portions collected to fund BC Assessment, the Municipal Finance Authority and potentially other groups (e.g., TransLink). These portions are remitted to the province and distributed accordingly.

Property tax applies to all classes of property. In BC, there are nine property classifications, which are set out in the Prescribed Classes of Property Regulation (BC Reg. 438/81):

Class 1 – Residential
Class 2 – Utilities
Class 3 – Supportive Housing
Class 4 – Major Industry
Class 5 – Light Industry
Class 6 – Business and Other

Class 7 – Managed Forest Land
Class 8 – Recreational Property/Non-profit Organisation
Class 9 – Farm

Where a property falls into two or more prescribed classes, the assessor shall determine the share of the actual value of the property attributable to each class and assess the property according to the proportion each share constitutes of the total actual value.

Class 3 – Supportive Housing, is a regulated class which requires designation of qualifying properties. The resulting assessment for qualified properties is an assessment of $1 for the land and $1 for the improvements, which provide tax relief.

Class 9 – Farm, requires certain income and reporting requirements, and also results in a regulated rate of assessment, which tends to be lower than market value. However, some taxing authorities may account for this by raising their mill rate (or tax rate) for this property class. Ski hills and port land are also valued according to designations in regulations.

The municipality/taxing authority has the full autonomy to set their tax rates. Ratios may be used to more fairly distribute the proportion of tax paid by each property type; however, ratios are not consistent across taxing authorities. The province (government) sets out the school tax rate, which is collected by the taxing authority on their behalf. The funds raised from school tax are remitted to the provincial government to help pay for the cost to provide education in BC.

Basis of value and valuation process

Taxable value is the value to which the appropriate tax rates are applied to determine the taxes payable; it is the actual value, less the value of any assessment or tax exemptions. *Actual value* is defined in section 19(1) of the Assessment Act, as – '"actual value" means the market value of the fee simple interest in land and improvements'.

Market and legislated values

For the determination of market value, the income approach is typically used to determine the assessment of certain property types which are income-producing properties and the direct comparison approach is used to assess other property to which the income method does not apply. As is the case in many other jurisdictions, direct comparison is used to assess most residential property. BC Assessment currently utilises a mass appraisal, computer-based model for the majority of valuations. The review and enhancement of mass appraisal tools assist in the valuation and assessment process.

Certain types of property do not trade in the real estate market, or do so only infrequently. Using market value is unsuitable for valuing such properties. A number of assessment appeals of major industrial properties in the early 1980s led to erosion in the stability of the property tax base. In response, government created legislated rates and manuals for stable, predictable and uniform valuation of properties that do not commonly trade in the real estate market. For example:

- Major industrial properties – in 1986, the Assessment Act was amended to authorise classification of major industrial properties, such as large mines, large sawmills and pulp mills, large cargo shipping facilities, etc., as a separate property class and to value the improvements at such properties based on legislated manuals. The land associated with such properties was, and continues to be, valued based on market value. In 1988, special depreciation rules were created for such properties that are shut down.
- Linear structures – in 1986, the Assessment Act was also amended to allow for legislated rates, established by BC Assessment, to be used to value continuous structures such as railways, pipelines, telecommunication improvements, etc. Rights of ways for such improvements are also valued using legislated rates. The year 1995 saw further amendments to the rules for valuation of railway property.
- Forestland – also in 1986, the Assessment Act was amended to allow for legislated rates to be used for the valuation of forestlands.
- Farmland – in 1984, land values for farms was established based on the productivity of land, depending on the location and type of agricultural activity.

More recently, legislative rates and manuals have been established for: dams, power plants and substations (1999); designated ski hills (2007); designated ports properties (2007); and designated supportive housing properties (2008).

Assessment authority

Property assessments are conducted by BC Assessment staff. BC Assessment is a Crown corporation. The Ministry of Community, Sport and Cultural Development is the overseeing Ministry of Government to which this authority reports. Each region throughout the province has an assessor who oversees the production of the annual assessment roll. The *valuers* may be one of a number of levels of staff – deputy assessors, senior appraisers, appraisers, property information collectors and appraisal assistants (admin staff).

The legislated makeup of the BC Assessment Authority is set out in the Assessment Authority Act. Private or contract appraisal staff are not used to determine property assessments for the provincial assessment roll. BC Assessment is responsible for producing an annual assessment roll, based on a valuation date of 1 July of the year prior to the taxation year. Other key dates on which assessments are based include 31 October, which is the physical condition and permitted use date. The roll reflects the physical condition and permitted use (zoning) of property as of this date, except where substantial damage or destruction of the buildings occurs between 31 October and 31 December.

Number of properties on the assessment roll (completed roll):

Roll year	*Number of properties*
2014	1,954,445
2013	1,935,426
2012	1,917,394
2011	1,902,875
2010	1,883,669

BC Assessment currently employs a total of 635 staff (including the appraisal, administrative and Head Office/Management). BC Assessment currently employs 262 working level appraisers, 54 per cent of which have a professional accreditation, and 135 other positions (senior appraisers, deputy assessors and assessors) that manage the working level staff as set out in Table 8.2.

Table 8.2 Appraiser to appraisal ratio – British Columbia

Total # of properties (2014 Assessment Roll)	*Working grade staff only*		*Working grade + mgmt appraisal staff*	
	# of staff	*Properties per appraiser*	*# of staff*	*Properties per appraiser*
1,954,445	262	7,460	397	4,923

BC Assessment currently utilises a mass appraisal, computer-based model for the majority of valuations. They are currently in the process of building a new system, which they expect to be ready for 2018. They hope to include more direct feeds and inputs from partners with the new system.

Complaints and reviews

There is no ability to appeal the property taxes for a given property. The appeal process exists at the property assessment (assessed value) level. The grounds for objecting to values are laid out in section 32(1) of the Assessment Act as follows:

Grounds of complaint to the completed assessment roll

32 (1) Subject to the requirements in section 33, a person may make a complaint against an individual entry in an assessment roll on any of the following grounds:

- (a) there is an error or omission respecting the name of a person in the assessment roll;
- (b) there is an error or omission respecting land or improvements, or both land and improvements, in the assessment roll;
- (c) land or improvements, or both land and improvements, are not assessed at actual value;
- (d) land or improvements, or both land and improvements, have been improperly classified;
- (e) an exemption has been improperly allowed or disallowed.

Property owners are encouraged to contact their local assessment office to discuss their concerns, prior to filing an appeal. There are processes in place for the assessor to make an amendment, without having to go through the appeal process (referred to as a *Bypass Agreement*), based on an error or omission that is brought to our attention, and where the change is agreed to by the owner. This is authorised under section 10(2) of the Assessment Act.

British Columbia has a two-tiered formal property assessment appeal process:

1. Property Assessment Review Panel (PARP) – Panels/process is administered by the Ministry of Community, Sport and Cultural Development, Property Assessment Services Branch. Property Assessment Review Panels (PARPs) are appointed by the Minister of Community, Sport and Cultural Development to review property assessments. PARP hearings take place between 1 February and 15 March each year. Approximately 75 panels are appointed throughout the province to hear property assessment complaints. An appeal (known in legislation as a "Notice of Complaint") must be filed with the assessor (the assessment office responsible for the property). More information on the PARP process can be found in Part 4 of the Assessment Act.
2. Property Assessment Appeal Board (PAAB or "the Board") – All Board members are appointed by the Cabinet of the Provincial government. The Board is led by the Chair – who reports to the Minister of Community and Rural Development. The Chair, two Vice-Chairs and Registrar manage the appeals, working with the parties to seek resolution through mutual agreement. If the appeal is not settled, it will be decided by one or more Board members who may be the Chair, Vice-Chair or a part-time member.

A person may proceed to further levels of appeal (i.e. BC Supreme Court and BC Court of Appeal) on questions of law. First Nations properties may have different appeal processes, depending on their contract with BC Assessment for assessment services.

Toronto, Ontario

Property transfer tax

Similar to British Columbia, Ontario also has a property transfer tax imposed by the province. The province of Ontario has further granted the local government of Toronto the right to impose an additional property transfer duty. This has, in effect, doubled the land transfer duty payable in Toronto compared with the rest of the province.

Recurrent property tax

Property taxes are determined by multiplying the assessed value of properties by municipal and education property tax rates:

Property tax = Tax rate × Assessed value

It is important to note that different classes of property exist in the property tax system, and are typically taxed at different rates. Each property is assigned to a property class based on its use and physical characteristics (Ministry of Finance 2013).

Ontario's current property assessment and tax system plays a fundamental role in supporting local municipal services, as well as the province's elementary and secondary school system. Property taxes raise approximately $24 billion per year in Ontario. The municipal portion of the tax raises over $17 billion and the education portion approximately $6.7 billion. In most provinces, the property tax is imposed by both the local and provincial levels, with the local level acting as the tax collector. In Ontario, the province sets the education component of the levy while the local authority sets the rate for its requirements.

For residential properties there is a uniform rate across the province. For business properties the business education tax rates vary considerably from municipality to municipality and by property type. This allocation is largely historical; however, the province has been attempting to harmonise the business education tax rate in the last few years. The education tax rate is added to the local tax rate and applied against the assessed value. The resulting tax is then collected by the local taxing authority. The education tax collected is remitted to the local school board. The example provided is what occurs in a single tier of local government. Where there are upper tiers local government, the upper tier sets its rate requirement and the lower tier local authorities collect the tax on their behalf.

There is no specific rate applicable to all property. Rates on residential property, depending on location and price, generally run from 0.5 per cent to 2 per cent, with business properties typically running three times that of residential. Farm property is taxed at 25 per cent of the residential rate. The property tax represents approximately 50 per cent of local government funding. The total property tax (including local and education tax) is approximately $24 billion.

Value and valuation

Municipal Property Assessment Corporation (MPAC) is an independent statutory authority which operates under and is governed by the Ontario Property Assessment Corporation Act, 1997.

- MPAC administers a uniform, province-wide property assessment system based on Current Value Assessment in accordance with the provisions of the Assessment Act. MPAC provides municipalities with a range of services, including the preparation of annual assessment rolls, which are used by municipalities to calculate property taxes.
- MPAC does not establish property taxes.
- MPAC currently assesses and classifies nearly five million properties, more than any other assessment jurisdiction in North America, with an estimated total value of $2.17 trillion dollars.
- MPAC employees are located in local offices across the province, in addition to a Head Office in Pickering and a Customer Contact Centre/Data Processing Facility in Toronto. MPAC is a non-share capital, not-for-profit corporation funded by all 444 municipalities in Ontario.
- MPAC is governed by a 15-member Board of Directors. Eight members of the Board are municipal representatives, five members represent property taxpayers and two members represent provincial interests. All members of the Board are appointed by the Minister of Finance.
- MPAC provides quality assessment services at an affordable cost. MPAC's cost-per-property of less than $37 is lower than comparable assessment authorities.
- MPAC has received international recognition and awards for the systems it uses to assess properties in Ontario.
- Representatives from countries around the world contact MPAC to learn about property assessment in Ontario.
- Ontario's province-wide Assessment Updates have exceeded international standards of accuracy.

First Nation property is based on the value of land, including any improvements on the land. The value is its market value and is updated on a four-year cycle. All increases in assessed value are phased-in in equal amounts over the four-year period. Any decrease in value from the previous cycle base year is immediately reflected on the roll. The *highest and best use* principle must be employed unless directed otherwise by the Assessment Act. There are only a limited number of current use based assessments, which include farmlands and managed forests.

MPAC is the body responsible for assessing all property in Ontario, including Toronto. MPAC rarely contracts for valuation assignments, but this has occurred for unique, specialised properties, such as airports and casinos. MPAC valuers generally work with outside experts to ensure knowledge transfer and to ensure scope of the work is appropriate and is completed according to the terms of the contract.

The valuation cycle is once every four years, in which approximately 5 million valuations are undertaken. A staff of 120 lead in the development of the valuation parameters to be applied in the valuation models. Another 600 or so review and adjust values to local and site-specific circumstances. As at the end of the last valuation cycle in 2012, the property types and numbers were as follows:

Table 8.3 Property type and numbers – Toronto

Property type	*Number of properties*
Residential	4,418,507
Multi-residential	16,083
Commercial	148,014
Industrial	76,721
Pipeline	1,505
Farm	208,958
Managed forests/ conservation	12,767
Special/Exempt	44,261
Total	**4,926,816**

Objections and appeals

The tax itself cannot generally be appealed against. There are tax appeals, but they are limited to the application of concessions when a business property has been damaged or vacant. The only grounds for objection are the classification and assessed value used in the assessment process. The first tier is administered by MPAC and is referred to as a *Request for Review*. This is a mandatory first step for all residential property owners. For business owners, this process is optional. The owner is required to clearly set out the basis for their objection. MPAC is required to formally respond to the request by the end of September. If the taxpayer is not satisfied, they have 90 days from the date of MPAC's response to file an appeal to the *Assessment Review Board*, which, under the auspices of the Provincial Attorney General, is an independent adjudicative body with quasi-judicial responsibilities to decide on the two issues set out above.

Reconsideration of assessment

Under the Assessment Act, for 2009 and subsequent taxation years, the owner of a property or a person who has received or would be entitled to receive a notice of assessment under this Act may request the assessment corporation to reconsider the following matters no later than March 31 of the taxation year in respect of which the request is made:

1. Any matter that could form the basis of an appeal under subsection 40 (1).
2. Any matter that could form the basis of an application under section 46. 2008, c. 19, Sched. A, s. 7 (1).

Appeal to assessment review board

40. (1) Any person, including a municipality, a school board or, in the case of land in non-municipal territory, the Minister, may appeal in writing to the Assessment Review Board:
   (a) on the basis that,
      (i) the current value of the person's land or another person's land is incorrect,
      (ii) the person or another person was wrongly placed on or omitted from the assessment roll,
      (iii) the person or another person was wrongly placed on or omitted from the roll in respect of school support,
      (iv) the classification of the person's land or another person's land is incorrect or
      (v) for land, portions of which are in different classes of real property, the determination of the share of the value of the land that is attributable to each class is incorrect; or
   (b) on such other basis as the Minister may prescribe. 2008, c. 7, Sched. A, s. 11.

Powers and functions of assessment review board

45. Upon an appeal, with respect to an assessment, the Assessment Review Board may review the assessment and, for the purpose of the review, has all the powers and functions of the assessment corporation in making an assessment, determination or decision under this Act, and any assessment, determination or decision made on review by the Assessment Review Board shall be deemed to be an assessment, determination or decision of the assessment corporation and has the same force and effect. 2008, c. 7, Sched. A, s. 14.

All property in Ontario and in most other provinces is subject to assessment with exemptions from taxation limited to government, places of worship, conservation lands and to lands used for philanthropic and social purposes.

9 United Kingdom – England and Ireland

Overview

Recurrent property tax dually operates in the United Kingdom as a Council Tax, applicable to domestic property, and as a Uniform Business Rate (UBR), also referred to as the National Non-Domestic Rate (NNDR). The UBR was introduced in England and Wales effective 1 April 1990 and subsequently introduced in Scotland effective 1 April 1995. It was levied at the same rate as England and Wales at the time it was introduced in Scotland to maintain harmonisation across the United Kingdom at the time (Plimmer 1998).

In broad terms, the amount of money raised from the council tax is similar to that raised from the Uniform Business Rate, with approximately 23 billion coming from each of the two taxes in 2013 (Audit Commission 2013 and HM Treasury/Dept Communities and Local Government). There are approximately two million non-domestic properties and 25 million domestic properties, so in relative terms, far more is paid by the occupiers of business property than by the owners of residential property. While the overall tax effort from recurrent property tax is high in the United Kingdom among the OECD nations, the revenue split from the UBR and Council Tax underlies the disquiet and high level of objections and appeals to the UBR.

England

Council Tax

The Council Tax replaced the Community Charge, better known as the Poll Tax, and took effect from 1 April 1993. In England, 326 councils collect council tax, comprising 201 district councils, 33 London boroughs, 36 metropolitan district councils and 56 unitary authorities. County councils do not collect council tax themselves (Audit Commission 2013).

This tax is levied on owners and occupiers of domestic property and is determined on two parts, each representing half of the tax revenue collected. Half of the tax payable has a personal element and half is determined on the property (Plimmer 1998). The personal component of the Council Tax is a carry-over from the former poll tax, as this component is determined on the basis that each household is occupied by two adult occupants. The rationale cited for the two occupants is explained by Plimmer (1998) to be that the most common number of adult occupants across all domestic dwellings in the United Kingdom was two (HMSO 1991:5). This would result in fewer adjustments and exemptions being sought from single occupant households and other concessionary occupants.

Concessions are provided for single-person households, and a number of other concessions exist, as listed under Schedule 1 of the 1992 Act. Further provisions apply to concessions for the personal component of the tax in both England and Wales. A summary of the main exemptions are as follows:

- Children under 18
- Persons on some apprentice schemes
- Persons 18 and 19 years old in full-time education
- Young people under 25 who get funding from the Skills Funding Agency or Young People's Learning Agency
- Student nurses
- For language assistance registered with the British Council
- People with severe mental disability
- Live-in carer who looks after someone who isn't their partner, spouse or child
- Diplomat

(United Kingdom Government Oct 2014)

The definition of *value* used to assess the value of property for classification into one of eight bands on which the tax is determined as follows:

> The value of any dwelling shall be taken to be the amount which, on the assumptions mentioned . . . below, the dwelling might reasonably have been expected to realise if it had been sold in the open market by a willing vendor on 1st April 1991.
>
> (Plimmer 1998:178)

The assumptions referred to in this definition are listed below:

(a) that the sale was with vacant possession;
(b) that the interest sold was freehold, or in the case of a flat, a lease for 99 years at a nominal rent;
(c) that the dwelling was sold free from any rent charge or other encumbrance;

(d) that the size, layout and character of the dwelling and the physical state of the locality were the same as at the date the valuation was made;
(e) that the dwelling was in a state of reasonable repair;
(f) in the case of a dwelling the owner or occupier of which is entitled to use common parts, that those parts were in a like state of repair and that the purchaser would be liable to contribute towards the cost of keeping them in such a state;
(g) in the case of a dwelling which contains fixtures to which this sub-paragraph applies, that the fixtures were not included in the dwelling. Such fixtures are those:
  i. which are designed to make the dwelling suitable for use by a physically disabled person; and
  ii. which add to the value of the dwelling;
(h) that the use of the dwelling would be permanently restricted to use as a private dwelling; and
(i) that the dwelling had no development value other than the value attributed to the permitted development.

(Plimmer 1998:179)

Council tax bands

In England and Wales, domestic property is allocated to one of the following value bands, on which the property's value is determined in line with the above definition of value. The rate in the pound applied to property within each band may vary across council areas. As noted in the following, the value bands in England have yet to be reviewed since the introduction of the tax in 1993. In contrast, Wales reviewed their value bands in 2005, which are determined on values as at 2003. The bands in England are so outdated that a majority of domestic property would now fall within the two highest value bands if domestic property were recalibrated under the existing value bands, hence once again rendering the council tax a de facto house tax in England. The sheer impact of Value Bracket Creep (VBC) means that most of the increase in tax revenue is manipulated by tax rate adjustment rather than adjustments to the value bands. To this end, the tax fails the principles of good tax design and, in particular, those of transparency and equity. The importance of regular valuation updates is by far the biggest issue in the determination of the value component of the Council Tax.

Council tax bands England

1993 to present
Band A Up to £40,000
Band B £40,001 to £52,000

Band C	£52,001 to £68,000
Band D	£68,001 to £88,000
Band E	£88,001 to £120,000
Band F	£120,001 to £160,000
Band G	£160,001 to £320,000
Band H	£320,001 and above

Council Tax Bands Wales

	1993 to 2004	2005 to present
Band A	Up to £40,000	Under £44,000
Band B	£40,001 to £52,000	£44,001 to £65,000
Band C	£52,001 to £68,000	£65,001 to £91,000
Band D	£68,001 to £88,000	£91,001 to £123,000
Band E	£88,001 to £120,000	£123,001 to £162,000
Band F	£120,001 to £160,000	£162,001 to £223,000
Band G	£160,001 to £320,000	£223,001 to £324,000
Band H	£320,001 and above	£324,001 to £424,000
Band I	Not applicable	£424,001 and above

Appeals to the council tax may be made on a number of bases which include the designation of the dwelling as a chargeable dwelling, the calculation of the tax payable and the imposition of a penalty for failure to supply information about the liable person. There are also rights of appeal to the valuation tribunal in the case of valuation matters; however, these are limited. There is no right of appeal against the inclusion of the dwelling in the wrong band unless some additional event takes place (Plimmer 1998).

Since the value re-banding in Wales, significant progress has been made in the transparency of the tax and the review and appeals process. A number of updated guidelines are available to taxpayers relevant to the updated value bands and assessments and how these impact their council tax to assist taxpayers. The Valuation Tribunal Service for Wales consists of four legally distinct and independent bodies which work together (and with others) to deliver their functions. 'The Valuation Tribunals have the duty to list, hear and determine appeals relating to: valuations for various Non-Domestic Rating (NDR) Lists (e.g. 1990, 1995, 2000 and 2005 revaluations); Valuations for Council Tax Banding List (e.g. 1993 and 2005 re-valuations); Liability for Council Tax (incl. Penalties) and valuations for Land Drainage Rates' (Welsh Assembly Government 2007:4).

Uniform Business Rate (UBR)

This tax is generally regarded as the second highest property outgoing for most companies after paying rent. The base of the tax is for non-residential/ commercial property income or rental value and is reassessed on a five yearly cycle. Rateable value is equivalent to the net rent (tenant assumed to be liable for repairs, insurance, etc.) at a fixed date – the antecedent valuation date. Revaluations have been conducted at five yearly intervals since 1990, but the government postponed the revaluation due in 2015 until 2017, which has caused a storm of protest as it means current rateable values – based on rental values in April 2008 – are way out of line with current rental values, which all went down during the recession and have not yet recovered.

The valuations are undertaken by government valuers/chartered surveyors in-house. All valuations are undertaken by the Valuation Office Agency (VOA), the central government agency; nothing is contracted out. However, when council tax was (hurriedly) introduced in 1993, about two-thirds of the work was contracted out to the private sector under direction of the VOA. Rental value information is collected each year or in the year prior to which the reassessment is to apply. Rental data for non-residential properties is collected continuously based on the VOA's lease register which shows when rents are due to be reviewed; but additional information, particularly data on accounts and costs, is collected prior to each revaluation. Sale prices for residential properties are also provided to the VOA, but are not used as there has been no council tax revaluation.

Rateable value is defined in the relevant legislation (Local Govt Finance Act 1988) as broadly the open market annual rental value as at a specified valuation date. Residential properties are based on banded capital values (Local Government Finance Act 1992). Precise definitions can be found in the two Acts mentioned.

Non-residential property are grouped by use and also by location; the rental values are looked at in these defined locations and uses and a rental value is determined and applied for all those properties in that location and use. The tax is determined on the actual rental value or, when the rental value is or appears to be out of line, a market rental value is used. Rateable value is always the open market rental value; if the actual rent paid is entirely in line with the definition of rateable value (unlikely), it may be used as the starting point for assessing rateable value.

The main critic of the UBR is that it is out of kilter with rents in some cases, that rates exceed market rents due to the recession and that the rental data used is as at 2008. The high tax rate, and its increase every year in line with inflation, further contributes to the challenges confronting businesses paying this tax.

The application of the UBR is simple and applies as follows:

Example: (rateable value) £100,000 × (tax rate) 0.482
= (UBR Bill) £48,200

However, the actual bill may be affected by a number of *adjustments*, e.g., transitional relief (to phase in changes due to a revaluation) or a variety of other supplements or reliefs.

A summary of the business rate system numbers is set out by HM Treasury and the Department for Communities and Local Government (2014:8) as follows:

- 326 billing authorities
- 1.787 million properties in England as at Sept 2013, valued separately at each revaluation
- £57 billion total rateable value of non-domestic properties in England
- 20% of properties in England account for 80% of the total rateable value as at March 2013
- Normally five years between revaluations
- Two years between the valuation date and the date the revaluation comes into effect
- Around £22 billion raised in England in 2012/13 to fund local services
- 2% cap to increases in business rates bills in 2014/15

The key public bodies involved in business rates in England, and their respective functions, are as follows:

- HM Treasury – overseas government tax policy including business rates.
- Department of Communities and Local Government (DCLG) – oversees business rates policy and legislation in England and the local government finance system.
- Billing Authorities – either councils or unitary authority responsible for calculation, collection and enforcement of business rate bills.
- Valuation Office Agency (VOA) – responsible for valuing non-domestic property and setting the properties rateable value.
- Valuation Tribunal for England – an independent and free tribunal service which reviews the appeals of business ratepayers when they disagree with the valuation officer's decision (HM Treasury and the Department for Communities and Local Government 2014).

At the time the Uniform Business Rate was introduced in 1990–91, the rate applicable was 34.8 pence in England and 36.8 pence in Wales. Scotland introduced the tax in 1995–96 at a rate of 43.2 pence, in line with England at that point in time. By 2013–14 the rate in England was 47.1 pence and 47.5 pence for London City, while in Wales it was 46.4 (Plimmer 1998; HM Treasury 2014).

Reviews and appeals

The right of appeal against council tax banding is very limited, but there are very few limitations on the right of appeal in rating cases.

The current system underpinning the UBR suffers from large numbers of speculative challenges aimed, in part, at testing the accuracy of the valuation.

As at the end of September 2013, almost 390,000 challenges had resolved on the 2010 rating list, but about three quarters of these resulted in no change to the rating list (Dept Community and Local Government 2013).

Currently, limited information is available in the public domain which allows the ratepayer to fully understand and check the accuracy of their rateable value before making a formal challenge. Whilst the Valuation Office Agency's web site includes details of the valuation itself and explanations of the valuation process, it does not provide all of the evidence on which the level of rateable value is based. Most assessments are based on comparisons with actual rents paid on the property or similar properties and it is the evidence of these rents which ultimately justify the rateable value.

In part, as a consequence of this, challenges are lodged by a ratepayer (or on their behalf by an agent) with little or no information about why the ratepayer believes the rateable value is wrong. Many challenges are made by rating agents on behalf of ratepayers and it has become normal practice for little or no evidence to be included with challenges. Most simply say, 'The rateable value is incorrect, excessive and bad in law and should be reduced to £1' (or a variation upon this).

Ireland

The Republic of Ireland has a dual property tax system in place: a non-domestic rating system for commercial properties similar to the United Kingdom and, recently introduced during 2013/14, a self-assessed banded capital value system for residential property called the *local property tax* (LPT). Given the similarities of the non-domestic rating system with those of the United Kingdom UBR, the following section will focus on the recently introduced LPT, which is defined as a tax payable on the market value of residential property in Ireland.

Liable Parties:

- Owners of Irish residential property, regardless of whether they live in Ireland or not
- Landlords, where the property is rented under a short-term lease (less than 20 years)
- Local authorities or social housing organisations that own and provide social housing
- Lessees who hold long-term leases of residential property (20 years or more)
- Holders of a life-interest in a residential property
- Persons with a long-term right of residence (for life or for 20 years or more) that entitles them to exclude any other person from the property
- Personal representatives of a deceased owner (e.g., executor/administrator of an estate)

- Trustees, where a property is held in a trust
- Where none of the above categories of liable persons apply, the person who occupies, or receives rent from, the property is the liable person.

(Irish Tax and Customs Revenue 2014)

The tax works under a self-assessment framework of four steps which follow:

1. Decide the current market value of the property.
2. Identify the valuation band and calculate the LPT due.
3. Determine if a deferral or exemption applies.
4. Complete and submit a return.

Value and valuation

The basis of the LPT is self-assessment. This means that it is the responsibility of the individual homeowner to determine the value of their property and select the appropriate tax band, as set out in Table 9.1. The LPT return letters received by homeowners contain an estimate of the LPT liability. Please note that this does not reflect an accurate valuation of the property and is simply an average estimate. It is the homeowner's responsibility to ensure that they correctly assess the value of their property. Irish Revenue (2013) provides property valuation guidance through their website which is available on www.revenue.ie. It further provides assistance through a Residential Property Price Register and also suggests the option of engaging a competent professional valuer.

Table 9.1 Value bands and tax 2013–14

A	*B*	*C*	*D*	*E*
Valuation band no	*Valuation band range €*	*Mid-point of valuation band €*	*LPT charge in 2013 (half year) €*	*LPT charge in 2014 (Full year) €*
01	0–100,000	50,000	45	90
02	100,001–150,000	125,000	112	225
03	150,001–200,000	175,000	157	315
04	200,001–250,000	225,000	202	405
05	250,001–300,000	275,000	247	495
06	300,001–350,000	325,000	292	585
07	350,001–400,000	375,000	337	675
08	400,001–450,000	425,000	382	765
09	450,001–500,000	475,000	427	855
10	500,001–550,000	525,000	472	945
11	550,001–600,000	575,000	517	1,035
12	600,001–650,000	625,000	562	1,125
13	650,001–700,000	675,000	607	1,215
14	700,001–750,000	725,000	652	1,305

Table 9.1 (continued)

A	B	C	D	E
Valuation band no	*Valuation band range €*	*Mid-point of valuation band €*	*LPT charge in 2013 (half year) €*	*LPT charge in 2014 (Full year) €*
15	750,001–800,000	775,000	697	1,395
16	800,001–850,000	825,000	742	1,485
17	850,001–900,000	875,000	787	1,575
18	900.001–950,000	925,000	832	1,665
19	950,001–1,000,000	975,000	877	1,755
20	Value greater than €1 million	Assessed on the actual value as follows: • At 0.18 per cent on the value up to €1 million • At 0.25 per cent on the portion above €1 million		

Source: Revenue Irish Tax and Customs 2013

All domestic property owners are required to determine the property value as at the same date of valuation, 1 May 2013. This value is used to assess their property tax for 2013, 2014, 2015 and 2016. Any changes to the property, either physical or otherwise, subsequent to the date of valuation are not taken into account for the years 2013 to 2016. In undertaking a self-assessment as at 1 May 2013, property owners are encouraged to consider some of the characteristics which make up the value of a property as follows:

- Location and proximity to local services
- Size of the property
- Layout of the property
- General condition of the property
- Property specification
- Decoration of the property
- Availability of parking
- Annual service charges (if applicable)
- Aspect of the garden (e.g., south facing)
- Proximity to any eyesores (e.g., electricity pylons, landfill, etc.)

(Society of Chartered Surveyors Ireland and Irish Tax Institute et al 2013)

Exemptions

Property purchased from a builder or developer, or property purchased by first-time buyers between 1 January 2013 and 31 October 2016 are exempt until the end of 2016.

Newly constructed but unsold residential property.

- Where ownership is vested in a public body or an approved charitable body and used to provide accommodation to people with special housing needs such as the elderly or people with disabilities

- Where a principal private residence is unoccupied by reason of long-term mental or physical infirmity
- Mobile home, vehicle or a vessel
- Property fully subject to commercial rates
- Houses in certain unfinished developments as prescribed by law
- Properties enjoying protection in other legislation – diplomatic or similar property

New exemptions will apply in the case of:

- New and previously unused properties that are purchased between 1 January 2013 and the end of 2016 will be exempt until the end of 2016;
- Second-hand property purchased by a first-time buyer between 1 January 2013 and 31 December 2013 will be exempt until the end of 2016.

Deferrals

In certain circumstances, a liable person may opt to defer payment of LPT. The following are some important points to note in relation to deferrals:

- A deferral is not an exemption.
- Interest of about 4 per cent per annum applies to deferred LPT.
- A claim for deferral must be made on the LPT return and the return must be filed with Revenue.
- The deferred LPT remains a charge on the property and will have to be paid to Revenue when the property is sold or transferred.

There are four separate categories of deferral available and full details of the conditions and procedures for each of these options are available on the Revenue website:

1. Income Threshold,
2. Personal Representative of a Deceased Person,
3. Personal Insolvency and
4. Hardship Grounds.

Appeals and reviews

If a taxpayer does not agree with the Notice of Estimate, it is a simple matter to displace it by making your own self-assessment and submitting your Return. If there is a disagreement between you and Revenue on matters relating to LPT, such as whether the property is residential, whether you are liable, matters to do with value, deferral, etc. that cannot be resolved, Revenue will issue a formal Notice of Assessment or a formal decision or determination. You may appeal to the Appeal Commissioners against those notices and this will be set out clearly on the notices.

Summary

While the property tax is well entrenched in the United Kingdom, much work is needed in reforming this tax over the next decade, particularly in England. This brings to the fore the major reforms needed in building the integrity of this tax and its revenues to bring it into the 21st century. The first issue for England is the split in revenues raised between domestic and non-domestic tax property revenue. This is closely aligned with the second issue in modernising the tax, that being its transition from an occupier tax to tax on economic rent incumbent on the owner. The final plank of reform is the need for a committed market valuation system with re-valuation undertaken at two or three yearly intervals.

The RICS (2013) undertook a review of the Uniform Business Rate and the impact of rates on empty property. This highlighted the need for reform; however, the reform required extends well beyond the finding of this study, which champions, among other points, increased relief from UBR for vacant non-domestic property. The impact of the global financial crisis affected many business hubs internationally and highlighted the shortcomings of taxing businesses which engage in retailing, trade and employing people as heavily as it does in England.

While a case may be made to reinstate the former tax abatement periods, tax deferment is a further alternative when property is vacant. If a business renting a property begins to slow, there is no parallel safety net to remove or reduce the UBR passed onto the operator of the business. Property investment is a business just like any other and should be treated no differently. Further, businesses which operate from premises they own, to which the property is an asset of the business, are not spared the UBR during recessions. Government must be careful in creating carve-outs for landlords with vacant property, of which the property is an asset of their investment business.

In England, the scheduled reassessment of the UBR due in 2013 was delayed to maintain artificially inflated rents in place from 2008. This firstly demonstrates that value is not the focus or arbiter of the tax, but that tax gouging has underpinned the delay, which has significantly damaged the integrity of this tax. Revenue robustness and maintenance is the overriding principle which drives the UBR in England. Further, less than two million non-domestic properties translate into approximately two million votes, in contrast to the domestic imposition of the council tax. Increased realignment on residential property determined on current market value is the first reform for England. The second is the removal of the fixed charge levy and moving the tax onto ownership.

The introduction of the self-assessment system in Ireland resulted from a lack of property information, including building improvements by the Irish taxing authority amid the need to introduce a tax on residential property at short notice. As revenues from central government stagnated following the global financial crisis, the ability to finance local government resulted in the need for a tax on residential property. The self-assessment system moves the valuation process to the property owner for determining the value of their property (land and improvements) and further avoids the cost of government administering and funding a full valuation contingency.

It is noted that there are 20 bands of value within which the taxpayers may nominate the value of their property. The total revenue collected and, more importantly, the way in which the revenue is adjusted annually is achieved by adjusting the rate applied to the various bands of value. After the first year of the tax, the tax moves from a value-based tax to a rate driven tax with no mechanism for adjusting values relative to the market until 2017. It is the present intention of taxing authorities to use owner-determined values as at 1 May 2013 for 2013 to 2016 tax years inclusive. It is imperative for the integrity of this tax and maintaining taxpayer confidence that values and the value bands are updated to account for bracket creep in values.

10 Denmark

Denmark is a constitutional monarchy with a parliamentary democracy comprising three branches of power: legislative, executive and judicial (Folketinget 2012). With a national population of approximately 5.6 million at 2014, one third of the Danish population lives in the four main cities of Copenhagen (capital), Aarhus, Odense and Aalborg. Denmark is characterised by universal welfare services which include free healthcare and free education. It has one of the world's most flexible labour markets, based on a flexi system, where employers can dismiss workers with relative ease. The Danish workforce has one of the world's highest levels of education and employment mobility, and it leads the world in the transition to a green growth economy (Denmark.DK 2014).

Denmark is a unitary structure, of which its sub-national government has undergone significant reform since 1970. The primary objective of reform is to bring services closer to the people and better define the roles and services provided by the tiers of government. This objective was further advanced in 2007 through the restructuring of municipal and regional government across Denmark, and it provides an important case study and possible model for sub-national government reform internationally. This review of Denmark firstly looks at the reorganisation of municipal and regional government, then progresses onto land and property tax reform as a source of public finance for sub-national government.

Reform of sub-national government

Up until 1970, Denmark consisted of more than 1,300 urban and rural municipalities, at which point the 1,389 municipalities amalgamated into 275 and the 24 counties merged into 14. In 2007, a further wave of reform was implemented by central government which abolished the counties which were replaced with five regions, with the 275 municipalities merging to 98. The restructuring of sub-national government as of 1 January 2007 resulted in two levels of government: Level 1, comprising the state, being central government, and Level 2, comprising regions and municipalities. It is noted that there is no subordination between regional and municipal government, as they each have different tasks and responsibilities. (Local Government, Denmark 2009).

The 2007 territorial reform followed two failed attempts and some successful experimentation. In 1995, an initial commission worked on the structure of metropolitan areas and suggested to merge the municipalities of greater Copenhagen. This project was mainly driven by civil servants and did not gain sufficient political support. It met with fierce resistance and was abandoned. In 1998, a new commission was created to consider the distribution of tasks, but did not have a mandate to propose a change to the administrative structure (its task was to show what could be gained by redistributing responsibilities within the existing structure) (Blöchliger & Vammalle 2012:76).

The Local Government Reform was the culmination of a four-year reform process. The primary goal of the merger was to improve the quality of municipal services by transferring new responsibilities from the county level to municipalities and increasing their size to ensure that they can assume these new responsibilities. The reduction of the number of municipalities implied a reduction of the number of directly elected officials, which could create major opposition. While the competences of the merged municipalities would be increased, of the 4,500 politicians at the municipal level, only 2,500 would remain after the mergers. Thus, the reform contained a provision stipulating that no public official would lose his job in the first year of implementation.

In general, politicians aligned with the position of their political parties, but in some cases local politicians from the coalition parties strongly opposed the reform. Most citizens were not opposed to the reform because it was presented as a way of guaranteeing the provision of effective public services in a decentralised way. Some special interest groups opposed the reform because they feared that service levels would be reduced when the responsibility was decentralised and that services would be cut in the more remote areas (Blöchliger & Vammalle 2012).

The Danish Local Government Reform bundled three reform elements; the first was a territorial reform (merging of municipalities); the second was a reallocation of tasks across levels of government; and the third was a financing and equalisation system of reform. This allowed compensating costs and benefits to carry over from one reform element to the other. For example, local politicians could be negatively affected by the territorial reform, as some positions would disappear, but those remaining in power would benefit from increased responsibilities. In the same way, a small municipality might have to bear the cost of a merger, but would benefit from increased equalisation payments (Jenson & Jacobson 2009).

One of the main elements of the reform was the increase of power and funding for the municipal level, which acquired more powers both from the dissolved county level and from the central government, while giving up only a few prerogatives. A number of tasks were transferred from the counties, leaving the municipalities responsible for handling most welfare tasks. Municipal responsibilities included:

- social services;
- child care;
- compulsory education and special education for adults;
- rehabilitation and long-term care for the elderly;
- preventive healthcare;
- nature and environmental planning;
- local business services and promotion of tourism;
- participation in regional transport companies;
- maintenance of the local road network;
- libraries; schools of music; local sports and cultural facilities

A responsibility for employment became shared with the central government. The new regions took over responsibility for healthcare from the counties, including hospitals and public health insurance covering general practitioners and specialists (Blöchliger & Vammalle 2012).

The central government was given a clearer role in overseeing efficiency in the provision of municipal and regional services. Employment services became a responsibility shared with municipalities, and responsibility for upper secondary schools was re-allocated to the central government. Tax collection was also transferred to the central government, as well as part of collective transport and road maintenance and nature and environmental planning. Finally, responsibility for culture was also transferred to the central government (Jenson & Jacobson in CEMR 2009).

An evolving finance path and equalisation system

The number of taxation levels was reduced from three to two, since the regions, unlike the counties, would no longer have the authority to impose taxes. Their revenues would consist of block grants and activity-based funding from the central government and the municipalities. In addition, in order to ensure that the local government reform did not result in changes in the distribution of the cost burden between the municipalities, a reform of the grant and equalisation system was carried out, which took into account the new distribution of tasks (Blöchliger & Vammalle 2012).

The municipal reform was thus presented by the government as a solution for reinforcing decentralisation. Voluntary horizontal cooperation between municipalities, rather than mergers, could have been an alternative solution for tackling the problem of size. But horizontal, voluntary cooperation agreements had existed for decades in Denmark, and were criticised for their lack of transparency and democratic control. These agreements covered only a few services (mainly garbage collection and treatment) and were not in force in all municipalities. That numerous voluntary cooperation agreements existed was interpreted as a sign that municipalities were too small, and this was used as an argument to promote the merging of municipalities. The reform actually reduced the number of collaborative arrangements, and some previous arrangements were even banned, including environmental issues.

In terms of the total cost of public services and their allocation across government levels, the reform was conceived as a zero-sum game, on the principle that 'the funds follow the tasks'. Apart from one-off transition costs, the reform was neutral with respect to the overall spending of the central and sub-central level. It did, however, result in higher grants, as on the one hand, municipalities receive new grants from the central government for the services it transferred to them, and on the other hand, the newly created regions have no taxing power and are funded by grants from the central government and the municipalities (Blöchliger & Vammalle 2012).

The Danish tax system is a leading world-class system, with high Optimal Operational Efficiency (OOE). It is well-resourced, with a detailed level of information accuracy and systems application accuracy and efficiency in the levying and collecting taxes. A summary of the revenue rate scale by tax as at 2014 follows:

- The corporate tax rate is 24.5 per cent and it is gradually being reduced to 22 per cent in 2016. There is progression in the rates for both resident and non-resident individuals and the marginal rate is app. 56 per cent.
- A special expat scheme may, under certain conditions, offer a flat rate of app. 32 per cent for up to five years.
- Value Added Tax is levied at 25 per cent.
- There is next to no social security payable for employers.

At the municipality level, the tax revenue collection is spread across several sources, of which income tax accounts for the largest portion at approximately 70 per cent. User payment charges and fees are the second largest source at 10 per cent, with the property tax as the third largest income source at 8 per cent. Grants from central government account for 7 per cent, with the rest of the tax revenue derived from loans, company tax and interest (Local Government Denmark 2009). Denmark, like other OECD economies, is faced with maintaining competitive income tax policy, particularly for its workforce. It has a high consumption tax rate (VAT) and labour taxes, with one of the higher top marginal tax rates for individuals in the OECD. In reforming its tax policy, the taxation of land and property is important, as it forms a relatively low proportion of total tax collected and as a percentage of GDP. It is a tax ripe in the reform of municipal and regional government.

Role of land and property taxation

Land and property taxation are under review in Denmark at present, with a new recurrent property tax system due in the 2015/16 tax year. The Tax

Ministry will scrap the current system used for assessing property value and instead ask an independent commission to come up with a more accurate model. The move comes after revelations that up to 75 per cent of property evaluations were either too high or too low (IPTI 2013b). The tax revenue from land and property tax has scope to be increased over the following decade, as the tax collected from this source as a percentage of GDP is low compared with other advanced OECD economies.

The commission reviewing this tax is expected to be made up of estate agents, accountants and others with knowledge of the property market. It is due to be ready for use when Skat, the tax agency, makes its next bi-annual property assessments in 2015 (IPTI 2013b). The review of the current system was founded on a number of arising issues which included assessments being too high and above market value, of which many of these dated back to 2003. Despite Denmark's tax system being one of the most advanced, much of the inaccuracy in the assessment of its property tax resulted from the adjustments made to the values subsequent to the tax freeze introduced in 2002.

Current land and property taxation

The economics of taxing land greatly assisted in addressing inflation and interest rates in Denmark between 1957 and 1960, a period when these were historically high. When this tax was repealed, by 1964 inflation rose to 8 per cent and land values skyrocketed from 17 billion DKK in 1960 to 320 billion Danish Kroner in 1981, a 19-fold increase, while prices in general rose only fourfold (Tholstrup 1980). From this point onwards inflation and land speculation increased until 1981, when a tax on land was reintroduced. The land value system was introduced in 1981, reformed in 1992 and a tax freeze introduced in 2002 (Muller 2003). From a tax point of view, it has demonstrated resistance not so much to the tax on land, but to the way in which this tax is applied, and specifically, to the main home residences (excluding the summerhouse) of the Danish population.

The biggest challenge for any government imposing a recurrent property tax is dealing with the main home or residence of the taxpayer. Fisher (1996) highlights that this has proven by far to be the biggest concern for government and taxpayers in any country. It has consumed much political debate and been the cause of significant disruption to the income from the property tax. In summary, the way this matter is addressed will greatly impact on the efficiency of the tax. While the Danish tax system is fluent and robust, this does not protect it from the simple economics which impact the price or value of housing, resulting in disproportionate growth in property values compared with increases in household income for most taxpayers. The dilemma for Denmark is twofold, as like in many Nordic countries, many homeowners also own a summerhouse, which is considered a second home.

Land tax revenue belongs to both the municipalities and regional governments of which the counties, now the regions, apply a fixed rate of 1.0 per cent and the municipalities may select a rate of between 0.6 (minimum) to 2.4 (maximum) per cent. Therefore, the combined land tax rates range between 1.6 to 3.4 per cent (Muller 2004). Land tax is levied by the municipalities and is payable in either two or four annual instalments and the portion of the tax collected for regional government is forwarded on by the municipalities. This better aligns this tax with the taxpayers' understanding of the purpose and use of this tax revenue to the services provided by the municipalities. In addition to the land tax bill, various municipal charges may apply specifically to roads, sewerage, water or district heating.

The service tax is the second recurrent tax imposed in Denmark, which is applied mainly in urban areas and was introduced in 1961. This tax is assessed on the capital value of the buildings, only it does not include or apply to the land component. In the case of private businesses, this tax is imposed by the municipalities, and approximately 29 per cent of municipalities impose this tax at the maximum rate of 0.7 per cent (Muller 2005). There is no service tax paid to the counties for private business buildings. For buildings owned by the central government, a service tax is paid to the counties at a rate of 0.375 and, in addition, this tax is paid to the municipalities at a maximum rate of 0.5 per cent.

The property value tax is the final recurrent tax imposed on property in Denmark and applies to all owner-occupied dwellings and summerhouses. Lefmann and Larsen (2000) highlight that this tax replaced a similar tax that operated for almost a century in Denmark, which was determined on the imputed rent from the house. This adjustment to assessing the tax on capital improved value resulted from perceptions of the imputed rent tax constituting a form of income tax. The basic rate is 1 per cent of the market value up to 2.6 million DKK, and for any value above this threshold the

Table 10.1 Recurrent taxes in Denmark

	Land tax	*Service tax*	*Property value tax*
Official Danish name	Grundskyld	Daekningsafgift	Ejendomsvaerdiskat
Year introduced	1926	1961	2000[(b)]
Coverage	All land	Buildings used for commerce, administration and manufacture	Owner-occupied dwellings and summerhouses
Basis of the tax	Land value (capital value)	Building values (capital value)	Property values (capital value)
Taxpayer	Owner	Owner	Owner
Beneficiary	Municipalities / Counties	Municipalities / Counties	Municipalities / Counties

Source: Muller, A. 2005: (b) this tax replaced a similar tax – income tax of imputed rent.

rate applied is 3 per cent. For taxpayers above 67 years of age, there is a reduction in the tax rate of up to 0.6 and 2.6 per cent. Further, for taxpayers who purchased the property before 1 July 1998 the rates are .02 per cent lower. There is a cap on the amount the property tax may increase by annually, of which the highest amount being 2,400 DKK; however, if this increase is less than 20 per cent, then the increase can be up to 20 per cent.

Operation of land and property value tax 2001/02 to 2014

This section sets out the complexity surrounding the assessment process, rules applied and adjustments made in the variation of land and property value tax in Denmark. It does highlight that a need for reform is required, both at the assessment and policy level. Property value tax is not part of income tax, but the calculated property value tax is collected with income taxes and is included on the tax card, which forms the basis for the tax that the employer withholds when paying wages. As a starting point, property value tax is 1 per cent of the part of the property value that does not exceed an amount of 3,040,000 DKK and 3 per cent of the rest. Property value tax is therefore progressive.

Two interim arrangements allow reductions in property value tax for property owners who took possession of their property before 1 July 2008. These owners firstly receive a reduction in the calculated property value tax to the value of 0.2 per cent of the property value. This means that the owners in question must pay 0.8 per cent of the property value tax on the part of the property value that does not exceed the progression limit of 3,040,000 DKK and 2.8 per cent on the rest.

Furthermore, these owners receive a reduction in the calculated property value tax of 0.4 per cent of the property value, although to a maximum of 1,200 DKK. The types of property that are entitled to these reductions are mainly the same types of property that were previously entitled to a standard allowance for things such as maintenance. This means that owners of owner-occupied apartments and owners of certain listed properties do not receive the reductions in question. The two reductions mentioned above are discontinued on change of ownership.

There is also a reduction for old-age pensioners. The reduction in the calculated property value tax is 0.4 per cent of the property value, although to a maximum of 6,000 DKK for homes and 2,000 DKK for summer cottages. In 2012, these reductions are regulated and reduced proportionate to income: single pensioners with an income over 174,600 DKK and married pensioners with an income over 268,000 DKK.

Tax freeze for property value tax

In 2002, a stop was put on any increases in property value tax. The basis for the freeze is the 2001 assessment with a 5 per cent supplement. However, if the 2002 assessment, or the assessment on 1 October in the income year, is lower than this figure, the lowest assessment is used as a calculation base.

The basis for property value tax can never exceed this ceiling. If property values fall, so that the current year's property assessment is under this ceiling, the property value tax is calculated based on the current year's property assessment. If property values later rise, the calculation base will be able to rise again, but only up to this fixed ceiling. According to the current rules, property values are taxed over a progression limit of 3 per cent. This limit has now been frozen to 3,040,000 DKK. This freeze also applies even if house prices fall.

Ceiling on buying and selling property

If a property is sold, the new owner assumes the previous owner's ceiling for the basis of property value tax. The new owner does not assume the special reductions that the previous owner may have had because the property was bought before 1 July 1998. If a property later undergoes a conversion or extension requiring planning permission, the calculation base for the ceiling is increased so the basis corresponds to the lowest of the following amounts:

1) the amount the changed property would have been assessed at in 2001 with a 5 per cent supplement;
2) the 2002 value of the changed property; or
3) the value of the changed property on 1 October of the income year. This also applies if the property's size changes or if the property changes in use. If the owner lives there both before and after the changed use of the property, the changed use does not come into force until the property is sold.

In practice, this means that the assessment authorities assess the property according to the rules that applied for the 2001 assessment, for the 2002 assessment and for the income year. They then apply the lowest assessment.

Limiting property value tax rises from year to year

The full extent of the limitation rules has been reduced due to the implementation of the ceiling over property value tax. There is still a need for these rules in cases where the value of the property had fallen below the ceiling for the evaluation basis only to rise again, or in such cases where the evaluation basis rose to conversions or extensions that required a building permit.

Visits to the Skat Ministry of Taxation and review of the various taxes during 2007 and 2010 resulted in a number of strengths and challenges emerging from the Danish land and property tax system. These are summarised in the following:

Strengths of the land and property tax system

1) Revaluation of land and property is undertaken every two years. (*However, these are not used; it is the 2002 values, which are being adjusted, that are being used*).
2) The centralised valuation process, which maintains consistency across the counties and across Denmark, is an outstanding attribute of the Danish property tax system.
3) The regular sales meetings between the valuers in selecting the best sales evidence and reaching consensus on the basis of determining updated values.
4) The tax is collected by the local authorities, of which the municipalities and regions are the beneficiaries.
5) The data collection sheet used when property is sold (Overdragelse af fast ejendom) is very detailed and comprehensive. The database of information on the individual property basis demonstrates the first step in the quality of the land and property tax system.
6) The level of information given to the taxpayer is detailed, which includes:
   a. information on how the values are determined;
   b. the values of other land, determined within the location of their land;
   c. the average age of improvements used in different locations for the added value of improvements;
   d. the information on improvements which allows accurate determinations of the added value of improvements to then calculate land values.
7) Whilst there is a level of computer-assisted mass appraisal (CAMA) used, the valuation process is very well-resourced and funded by professional staff, which has greatly contributed to the highest quality of values determined.

Challenges of the land and property tax system

The primary challenge confronting any tax system is first and foremost the understanding of how an assessment is determined and how this translates to a tax obligation. This factor is of particular relevance to land and property tax, as the base of the tax is determined by the taxing authority, with little or no input from the taxpayer. The assessment process is not simple in Denmark, which is evidenced from the myriad of rules and provisions as set out and explained in the previous section.

Once the assessment process of a tax system becomes overly complex, a shift away from the taxpayer and over-focus on the system used to generate the assessments will result in the demise of the tax base and integrity of the tax. This invariably became the case in Denmark, in which the base of the tax, the assessed value, became so removed from reality that it was impossible for government to defend the tax under the principles of good tax design.

With the above points made, it is of paramount importance that the reform of land and property tax in Denmark does not swing back too far the other way, in a non-market based assessment approach, as has been adopted in parts of the United Kingdom in the taxation of residential property. Some reference to current market value is needed in maintaining the integrity and progressivity of the tax. The use of bi-annual values is one particular strength of the Danish system; however, as is noted in the Tax Minister's speech in the following section, these became irrelevant as the base values from 2001/02 were still being used.

Options for reform in 2015/16 and beyond

The official reason for the 24 June order to reform the existing land and property tax system was largely driven by the tax freeze implemented during 2002. The whole tax and valuation apparatus became irrelevant as taxes for each subsequent tax year were adjusted off the 2002 values, which are set out in extracts from the Tax Minister's speech paper to the committee on taxes:

> *If we look at the ratio of home ownership, as it is known, that the assessments are used to calculate the property tax. But it is not the current property values that are used. It's the old values from 2001 and 2002. The current assessment has no practical significance, and the government therefore believes that these assessments can be terminated.*

On 24 June 2011, Parliament ordered the government to present a bill in the subsequent parliamentary session to abolish the current property assessment.

> *Parliament directs the government to present a bill in the upcoming parliamentary session to abolish the current property assessment system and thus provide a simpler basis for taxation of real estate.*

A summary of proposed changes as provided is as follows:

- *The general assessment of real property ceases.*
- *Establish a new model for assessment of newly constructed residential properties and the assessment of commercial properties.*
- *Municipal revenue from land tax, as well as contribution rates, remain.*

These points are further expanded on in the following envisaged provisions under which a number of proposed changes would operate as read out by the Minister for Taxation:

1. *The general assessments of home ownership ceases, so that the assessment per. 1st October 2011 will be the last. The amount as finally determined for property and land value assessment in 2011, remains standing.*
2. *For new residential properties there will be an 'assessment' of both the base area of the buildings at the level of the average assessment / evaluation in 2011 per. square meters of floor space and square feet of living space for the surrounding residential properties of the same type (single family, condominium, summer cottage) in the immediate area. This 'assessment' of such owner-occupied homes in the level that applies locally in the area where the property owner is listed, and there will thus be taken due account of individual factors such as amenity, etc.*
3. *If a dwelling is resized by extension, demolition or alteration of the basic size, it will be necessary to make a correction of 'evaluation'. It could happen in proportion to the area changes, the house is subjected to both land and building.*
4. *As mentioned, it would hardly be possible to find a schematic model that can accommodate all commercial properties. Therefore, the assessment of commercial properties untouched, as the resolution also proposes. Of course I would like to have the assessments of commercial properties simplified as much as possible, and we will try to make it.*

(J. Falk-Rasmussen & H. Muller, Skat Ministry of Taxation Denmark, personal communication, 25 November 2010)

Summary

Denmark, as a unitary structure of government, has undertaken a review of the operational tiers and has efficiently designated specific roles and tasks to these tiers. It has assigned the collection of taxes between central and local government, which fund the regional/county tier of government. This reform was undertaken in two steps, with the final structure put in place during 2007. Denmark, a comparatively high taxing country among the OECD members, has higher income taxes and relatively lower land and property taxes. It was identified that scope exists to reform its revenue base following the land tax freeze of 2002. Further, the base on which the tax is assessed is also under review following a number of difficulties which emerged in the way land value was determined.

The reform of the land and property tax system is a root and branch review, which is examining and reforming all aspects of this tax, including the amount of revenue raised, exemptions and concessions and the base on which the tax is assessed. The sequence of reform in Denmark has been exemplary as it commenced with addressing the structure and roles of government, followed by the reallocation of tax-raising powers, and finally it now has capacity to make meaningful reforms to its recurrent land tax system. It is highlighted that, among the key reforms, is the potential move to improved value on highest and best use for residential property, which is being considered as the key reform in maintaining value as the base of the tax. It will also consider the tax treatment of a great Nordic icon, the summerhouse, and determine the tax policy most appropriate for the ownership of second residences.

As a country with high relative rates of income tax and Value Added Tax compared with other OECD countries, the need to raise more revenue from capital, and in particular property from a recurrent property tax, is a reform priority. Denmark is well-placed to achieve this objective, as seen from firsthand experience in numerous visits to the Danish Ministry Skat over the past decade; the systems and taxpayer education programs are among the best in the world. The dedication to the tax principle of transparency is second to none and, with the introduction of reforms to enhance the principle of simplicity, Denmark should regain the mantle of once again becoming the leading property tax system in the world.

11 New Zealand

Overview

New Zealand is a constitutional monarchy with a parliamentary democracy, although its constitution is not codified. Elizabeth II is the Queen of New Zealand and the head of state. The Queen is represented by the Governor-General, whom is appointed on the advice of the prime minister. New Zealand has a unitary structure of government, with its subnational government primarily devolved to local government; local government does not have constitutional status. As at 2013, New Zealand had a population of approximately 4,450,000 and, as experienced in other OECD countries, New Zealand is impacted by an aging population.

The demographic change of the New Zealand population indicates that by the late 2020s, the 65+ age group will outnumber those in the 0–14 age group. In 2013, the 65+ age group numbered some 600,000 people, while there were 900,000 children. The projections also indicate that population growth will slow as the population ages, and there is a 1 in 3 chance that deaths will outnumber births in 2061. Figure 11.1 sets out the age distribution of the aging population mix as at 2012 and its project over the following 50 years, based on projections made by Statistics New Zealand. As demonstrated in Australia, this will impact tax revenues from labour and income and will turn focus to other tax sources, such as consumption and capital.

New Zealand's tax system and tax mix

As a country with a unitary structure of government, New Zealand tax laws are established under the Constitution through acts of parliament. Tax raising powers are the domain of central government, with local government created as an operational arm of central government under local government legislation.

New Zealand has a simple tax system with predictability and equity being among its key attributes. The prime sources of tax revenue are derived from income tax and the goods and services tax at the central level of government

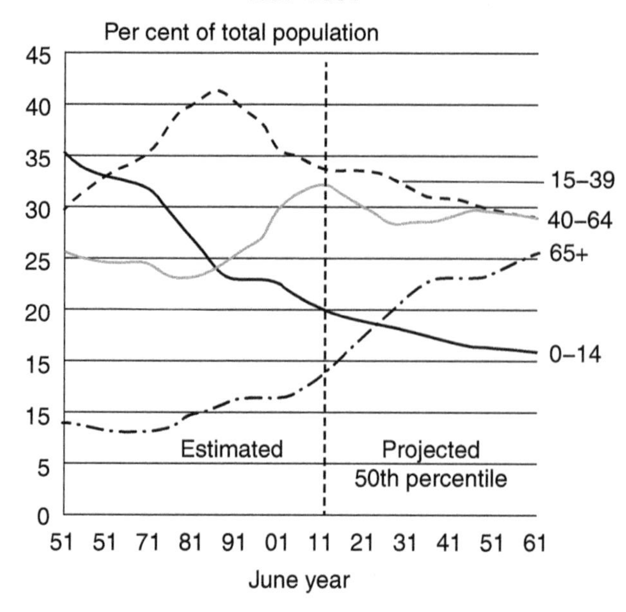

Figure 11.1 Age distribution in New Zealand

Source: Statistics New Zealand

and the property tax (council rating) at the local government level. It has no inheritance tax, very limited application of capital gains tax, no local or state taxes apart from rates, no payroll tax, no social security tax and no health care tax.

In summary, tax rates in New Zealand in 2014 are as follows:

Company tax rate:	28%
Personal income:	33% from $70,000
	30% $48,001 to $48,000
	17.5% $14,001 to $48,000
	10.5% $0 to $14,000
Goods and services tax:	15% GST applies.

There are some exemptions to which a zero-rate applies. New Zealand does not levy stamp duty; this tax was abolished in 1999.

Inland Revenue has an internal appeals process which is binding on the government. Taxpayers have a right of appeal to the courts in the case that an unsuccessful appeal results from the Inlands internal appeals process on taxation matters.

The current tax system in New Zealand was reformed following a review during 2010. The review of the tax system, known as the Tax System for New Zealand's Future, made a number of recommendations. Among the analysis of the existing system, the following points were identified:

1) There was a 'major hole in the tax base concerning capital'.
2) The tax burden is disproportionately borne by PAYEE taxpayers, as many with wealth can restructure affairs through trusts and companies to shelter income from taxes.
3) International competition for capital and labour from Australia will impact on the sustainability of corporate and personal tax rates.

(Centre for Accounting, Governance and Taxation Research 2010)

The recommendations were made by the reviewing panel, of which the following recommendations were adopted and have resulted in New Zealand ranking the second most competitive tax system in the OECD (Tax Foundation 2014:5):

1) The company, top personal and trust tax rates should be aligned to improve the system's integrity.
2) New Zealand's company tax rate needs to be competitive with other countries' company tax rates, particularly that in Australia.
3) The top personal tax rates of 38 and 33 per cent should be reduced as part of an alignment strategy and to better position the tax system for growth.
4) The most comprehensive option for base-broadening, with respect to the taxation of capital, is to introduce a comprehensive capital gains tax (CGT).
5) The other approach to base broadening is to identify gaps in the current system where income, in the broadest sense, is being derived and systematically under-taxed (such as returns from residential rental properties) and apply a more targeted approach.
6) Increasing the GST rate to 15 per cent would have merit on efficiency grounds because it would result in reducing the taxation bias against saving and investment.

(Centre for Accounting, Governance and Taxation Research 2010)

As noted in the review of the current tax system in New Zealand, many of these reforms have been adopted, most notably the increase in the GST rate, decrease in the corporate tax rate and reduction in the top tax rate for individual taxpayers. The recommendation to improve tax from returns of residential property is now in process and it will take some time to work through the application of this reform. The tax from residential property may be applied in one of two forms; the first is on the capital gains on disposal and the second is recurrently on the annual return or imputed rent from property.

The second tax exists in the form of local government rates, and presents the task that, if further tax on residential property were to apply, how would this be distinguished from it representing an increase in current local rates? The simple answer is that entry and exit taxes on property are not as theoretically or economically pure as a recurrent tax on land or property; however, these are less visible and dealt with once at the transaction of the property rather than on the holding of the property. In addressing both the palatability and operational dimensions of a recurrent tax, a review of the evolution of the tax and the sphere of government that would apply it in New Zealand is needed.

Local government

The first local government body in New Zealand was established in 1842 in Wellington by the Municipal Corporations Ordinance. In 1852, the New Zealand Constitution Act established the six provinces, which were each responsible for setting up their own local government, with the result that 'by 1867, 21 municipal local government units had been constituted under no fewer than 14 separate provincial ordinances'. When the provinces were abolished in 1876, central government assumed responsibility for local government and 'established the system which was, broadly speaking, in place until the 1989 reorganisation' (New Zealand Parliamentary Library 2014).

The way local government is managed in New Zealand is unique and a model for other unitary structures, and indeed the states of federated structures of government, to observe in the management of their local government structure. Following World War II, Derby (2012) defines the importance of the Local Government Commission, which was established in 1946 as an independent review board with powers to recommend local government restructuring and amalgamation.

Both the 1989 reform and the Auckland amalgamation were unusual because there was no provision for those affected to vote on whether or not they wished for amalgamation to occur. In 2012, a change to legislation removed this requirement, replacing it with an optional poll which will only occur if a petition signed by 10 per cent of electors in any of the affected council districts requests it. As at 2011, Derby (2012) states that there were 67 territorial authorities, 12 city councils, 54 district councils and one unitary authority. There are currently four proposals for reorganisation under the new legislation being considered by the Local Government Commission (LGC), all of which are now amalgamation proposals as follows:

- Northland: application for the Far North District Council to become a unitary authority. Submitted by the Far North District Council.

- Hawke's Bay: application for the amalgamation of the Central Hawke's Bay, Wairoa, Napier and Hastings councils.
- Wellington: application for the amalgamation of the Greater Wellington, Kapiti Coast, Porirua,
- Wellington, Hutt, Upper Hutt, South Wairarapa, Carterton and Masterton councils. Submitted by the Greater Wellington Regional Council.
- Wairarapa: application for the amalgamation of the South Wairarapa, Carterton and Masterton councils. Submitted by all three councils, this proposal is being considered alongside the Wellington proposal due to the overlap.

(New Zealand Parliamentary Library 2014)

New Zealand councils rely on a single form of tax: property tax. The power to levy a property tax is contained in the Local Government (Rating) Act 2002. In addition to property tax revenue, councils receive income from a range of additional sources including:

- Sale of goods and services, including swimming pool charges
- Regulatory fees, such as parking fines and infringements
- Interest from investments, including Council Controlled Trading Organisations (CCTOs)
- Grants and subsidies which encompass a share of road taxes and charges, as well as development contributions.

(LGNZ 2014)

While rates are the sole source of tax for local government, they may be applied differently and on a range of different bases which include:

- general rates – based on the land, capital or rental value of a property (with the exception of Auckland Council which is required to use capital value)
- targeted rates –calculated on the basis of a feature of a property and used to fund a specific service, such as funding the cost of a sea wall to halt erosion based on the frontage size of affected sections
- uniform annual general charges (UAGC) – a standard cost per property, not related to property value (only 30 per cent of rates can be made up of UAGCs).

Local government in New Zealand fares well, with 90 per cent of its revenue being own source with revenue from rates representing just over 60 per cent of its own source income. This equates to approximately 2.3 per cent of household expenditure spent on rates, which is just over 50 per cent more than Australians expend on rates and land taxes collectively. New Zealand does not levy stamp duty on land transactions and has a notably more advanced tax effort from recurrent property tax than Australia. New Zealand stands as a glowing example of what Australia needs to do to improve its property tax.

While it has a flexible and adaptive property tax structure, recurrent land and property tax policy (local rates) have endured a number of reforms, which are a work in progress in maintaining New Zealand's high tax efficiency standard ratio among the OECD nations. The following section is a review of the property tax (local rates) in New Zealand and examines the strengths, opportunities and challenges in the operation, application and potential expansion of revenue from this tax over the next decade. It demonstrates why New Zealand's tax system, and in particular its property tax rates, are high under the principles of good tax design.

Land and property taxation

A national land tax was introduced in New Zealand in 1878, under the provisions of the Land Tax Act 1878 (NZ), and was repealed in 1992 (New Zealand Local Government Rates Inquiry Panel 2007). The introduction of land tax in New Zealand had an impact on the colonies of Australia, which had not yet introduced a tax on land; however, Australia followed New Zealand's lead soon after. This tax was levied on the unimproved capital value of land, of which New Zealand established the first centralised valuation authority in Australasia on which the tax on land was assessed (Herps 1988).

The land tax was originally a major source of national revenue, yielding over 70 per cent of total land and income tax revenue in 1895. The tax 'also served a social purpose in acting as an inducement to the breaking up of unduly large land holdings' (New Zealand Tax Review Committee 1967). While income tax became the tax of choice during the 20th century, little incentive existed for reforming a recurrent tax on land, and the GST was seen as a preferred option to reforming a national land tax. This did, however, pave the way for local government to increase its tax revenue from this source, which it has progressively done since the mid-1990s, following the relinquishment of a land tax by national government.

Up to and including the 1988 income year, land tax was generally 2 per cent of the value of the land less a special exemption of $175,000. It therefore was not payable by any taxpayer whose total interest in property has a value of less than $175,000. The exemption reduced by $1 for every $1 by which the land value exceeded $175,000. Thus, no exemption was allowable when the land value exceeded $350,000. For the year ending on 31 March 1989, legislation reduced the rate of land tax to 1.5 per cent from 2 per cent of the land value. From 31 March 1990, the rate of land tax was set at 1 per cent of the land value (Simpson & Figgis 1998:31–32).

In 1991–92, the rate was lowered to 0.5 per cent, and the tax was abolished in the following year. Simpson and Figgis (1998:32) highlight that the

1990 New Zealand Budget Papers estimated that the net cost to revenue of abolition was estimated in 1991–92 to be $200 million, approximately 1 per cent of total tax revenue. The estimate was net because allowance was made for the deductibility of land tax against income tax liability (Simpson & Figgis 1998). The land tax base was broadened by abolishing many exemptions and applying the tax to virtually all commercial and industrial land uses. The exemption of land owned by public and local authorities was removed and the special exemption of $175,000 was repealed. The 1989 budget changes technically simplified the rate scale by removing the threshold, making the rate scale proportional. At the same time, the removal of the threshold broadened the tax base by catching in the tax net those who had previously been exempted (Simpson & Figgis 1998).

As set out in Table 2.4, recurrent land tax, which now operates in the form of local government rating, as at 2010 represented 2.1 per cent of GDP, being a marginal increase over the 2 per cent of GDP raised from both local rates and the former national land tax which co-existed in 1965. Since the abolishment of the national land tax from 1992 in New Zealand, local government has progressively captured this revenue through increases in local rates. Collection of land or property tax revenue (depending on the basis of value used) by local government is far more palatable and likely to be acceptable to taxpayers rather than a national land tax.

The Local Government (Rating) Act 2002 (NZ) invests local authorities with powers to charge rates 'in order to promote the purposes of the [Local Government] Act'. The four types of rates that may be charged are: a general rate, chargeable against all rateable land; a fixed amount, universal annual general charge (UAGC), payable in respect of each rateable unit; a targeted rate for particular activities identified in a local authority's funding impact statement, such as waste removal; and a targeted rate for water supplied. The aggregate of targeted rates (excluding the water rate) and UAGCs may not exceed 30 per cent of a local authority's total rates revenue. Differentiated rates may be charged for different categories of land.

Bases of value

Up until 1985, land value was the preferred base on which to assess the property tax in New Zealand; however, by the 2006–2007 fiscal year, capital value had become the tax base for the majority of local authorities (Franzsen, cited in Dye & England 2009:37). The rationale for the transition to CIV in the capital cities of New Zealand was due to the limited transaction of land, and despite the transition, the use of land value as the base of the property tax remains strong in regional government (New Zealand Local Government Rates Inquiry Panel 2007:7–8).

Land value is determined on the market value of land, which is re-assessed annually. The efficiency of land value used for the rating of land outside the capital cities of New Zealand is effective. However, the efficiency of land value based on highest and best use has come under scrutiny. In contrast to land value being determined on highest and best use, it is suggested that land value be determined on the existing use of land (New Zealand Local Government Rates Inquiry Panel 2007:7–8).

New Zealand now has a well-developed rating system in which local government has the option of adopting one of three bases of value for the rating of property (McCluskey et al 2006:381). Four of the main cities of New Zealand (Auckland, Wellington, Christchurch and Hamilton) all utilise a capital or annual value rating system (McCluskey et al 2006:381). Improved value is said to be the best means to achieving equity between the ratepayers, based on their ability to pay (Mander 1982). The property tax is applied at the local government level in New Zealand, to which the local government determines a rate in the dollar, which is applied to the CIV to determine the tax assessment.

The bases of values (land, capital and annual) are determined under the Rating Valuations Act 1998 under *section 2 – Interpretations* as follows:

Annual value, in relation to any rating unit, means the greater of –

(a) the rent at which the unit would let from year to year, reduced by –
  (i) 20 per cent in the case of houses, buildings and other perishable property; and
  (ii) 10 per cent in the case of land and other hereditaments.
(b) 5 per cent of the capital value of the fee simple of the unit.

Capital value of land means, subject to sections 20 and 21, the sum that the owner's estate or interest in the land, if unencumbered by any mortgage or other charge, might be expected to realise at the time of valuation, if offered for sale on such reasonable terms and conditions as a bona fide seller might be expected to require.

Land value, in relation to any land, and subject to sections 20 and 21, means the sum that the owner's estate or interest in the land, if unencumbered by any mortgage or other charge, might be expected to realise at the time of valuation if –

(a) offered for sale on such reasonable terms and conditions as a bona fide seller might be expected to impose; and
(b) no improvements had been made on the land.

Section 20 – Trees and minerals, and *section 21 – Leases*, as set out in the above definitions of value, are addressed in these sections and how they are to be accounted for in the determination of the various bases of value.

Section 9 – Valuation roll intervals: the Act sets out the frequency of the valuations needed for the determination of rates. In summary, valuations must be undertaken no less frequently than three times yearly by the Valuer-General, or delegated valuer.

Objections to valuations

Part 4 – Objections – sections 32–40, set out the provisions for objections to values.

The first step of the review process is set out under section 34, which prescribes that the review of a value which is the subject of an objection is to be made by a valuer, to which the result of that review is to be provided to the objector under section 35.

Any person dissatisfied with the outcome of the review undertaken under section 34 may require the value to be referred to the Land Valuation Tribunal within 20 working days, as set out under section 36 of the Act. Under section 37, the Valuer-General may object to any valuation and join the proceedings, or object to any valuation determined under review of section 34.

Under section 38, in the presence of at least a District Court Judge and a registered valuer, who is a member of a Tribunal, it is necessary to constitute a sitting of the Tribunal for the purpose of hearing an objection made under the Act, to which the onus of proof on any objection rests with the objector.

Summary

New Zealand is a unitary structure of government which, in its early foundation as a colony, became a federated structure through the formation of provinces, which were subsequently abolished. The structure of government today is Commonwealth (Central) and local government, with operational boards for the provisions of services. The tax system is ranked second most competitive in the OECD by the Tax Foundation for its mix and efficiency, and has avoided the use and reliance on property transaction taxes, creating an important and leading benchmark for other OECD countries. The local tier of government has gone through a number of restructures and amalgamations, creating a stronger tier of sub-national government in the provision of services and local infrastructure to the community.

Rates remain the sole and dominant source of tax revenue for local government across New Zealand. Subsequent to abolishing a national land tax in 1992, local government has progressively increased rates revenue to 60 per cent of its own source revenue and maintained recurrent land tax in the form of local rates at just above 2 per cent of GDP. The bases on which this tax is determined are flexible and local government has the option of adopting alternate bases of value on which to assess their local rates. Robust and well-defined objection and appeal provisions are available for the review of this tax and the valuations used to assess it.

New Zealand, as part of the ongoing review of their tax system, is looking at the potential to increase recurrent tax revenue from land or property in one form or another. The opportunity exists to improve tax revenue from this source in line with some of the larger, advanced OECD economies, which also require further refinement in both maximising and equalising revenue across the various classes of land and property. What remains to be determined is how this is to be achieved and applied across the taxpaying community in New Zealand and, most importantly, by which tier of government.

Part 4

Reforming land value taxation and fiscal reform of sub-national government in Australia

Introduction

Parts 1 to 3 were a review of the broader tax system, followed by the operation of land tax across Australia. A detailed state-by-state account of the operation and application of land tax followed, with several international case studies which gave insight into application of this tax abroad. Part 4 explores some of the options for reforming land taxes gained from the insights of the international case studies, as well as from some of the factors which work well across the jurisdictions of Australia. This part further provides a number of options worthy of consideration in transitioning from inefficient mobility taxes and recurrent land tax limited to a narrow number of property owners, to a land tax which is efficient and neutral and better contributes to the tax effort from this source.

The reform of land and property taxes is part of a broader agenda which traverses all three tiers of government and, in particular, the Commonwealth. Reform is of particular importance for the Commonwealth, which in part has not been able to operationalise the reforms needed to modernise Australia's Federation and centralised tax system. Through the nation's early development, the redistribution of tax revenue has been vital in sustaining the evolving state-based structure that has defined Australia's geographic landscape. Now in the 21st century, the fast-evolving phenomenon of urbanisation, and the transitioning away from industry to a service-based economy supported by a progressively aging workforce, requires revision of tax policy in Australia.

It was emphasised in Chapter 1 that taxes on labour are counterproductive and the focus of tax reform internationally has been on value-added or consumption-based taxes over the past half century. Australia embraced this tax reform in the past 15 years when it replaced the states' wholesale sales tax with the goods and services tax. During this same period, operationalising tax reform stalled during the abundance of tax revenue generated by the mining and resources boom. It was not until the Global Financial Crisis onwards that resource-dependent countries like Australia required a tax system that accounted for alternate sources of tax revenue. As mining moves from the set-up phase to the extraction and exportation phase, the Mineral Resources Rent Tax (MRRT) is in principle an important tax source in retaining and sharing some of the wealth exported beyond Australia's boarders.

The states may also put their stamp on natural resources generated through the exportation of food by foreign-owned land and business entities which export produce from Australia. Rather than the states and Commonwealth jousting between mining royalties imposed by the states and a MRRT imposed by the Commonwealth, this dichotomy must be agreed upon by these two tiers of government. The states may alternatively turn their focus to the taxation of primary production land which is foreign-owned and whose produce is shipped abroad for non-domestic consumption. While much focus has centred on foreign-owned primary production land, Australia's relative consumption of its primary production needs to be better focused away from limiting foreign ownership of land to properly taxing land that is foreign-owned.

The states and Commonwealth were caught short when the global financial crisis set in during 2008. Tax reform policies were needed to make-up declining tax revenues generated from the mining boom and the reduction in tax revenues from the GST, a revenue that flat-lined, resulting from reductions in consumption and a declining economy. It was during 2008 that Australia's Future Tax System review was established to make recommendations to address revenue shortfalls and further identify contributions that sub-national government must make in addressing vertical fiscal imbalance. To this end, while vertical fiscal imbalance is cited as the problem, the solution sits in all tiers of government.

With the need to keep income taxes internationally competitive along with the levelling out of consumption tax revenue, we now enter the important focus on land and property tax reform in Australia. Part 4 primarily focuses on options for reforming sub-national government's largest tax source, land taxation, which currently comprises the amalgam of state land taxes and local government rating. This tax source was demonstrated in Chapter 1 to be well below the level collected by advanced OECD economies, including New Zealand, in which the tax effort from recurrent land tax significantly outshines that of Australia. Part 4 further addresses the need to reduce and cap the imposition of stamp duty, which impacts housing mobility and hence affordability.

While most of the objectives and recommendations of tax reviews comprise a critique of the issues, in contrast, the primary contribution of Part 4 is to fill the void of transitioning change at the coalface of sub-national government across Australia. Operationalising the recommendations made in the review of Australia's Future Tax System (2010), the Productivity Commissions into Housing Affordability (2004) and the various inquiries into the states' tax systems requires more granulation in grounding the recommendations for change. Such granulation includes reforms to tax policy and the harmonisation of some taxation, valuation and administrative processes which aim to improve the efficiency of a tax which operates across two tiers of government.

Reform is not about centralising land taxation, but decentralising the tax from state to local government, and ensuring such divestment is strictly governed by and in line with the principles of good tax design. This requires, as far as possible, the greatest level of simplicity, transparency, consistency and operational efficiency across the administrative processes and legislative provisions which govern the tax. With these principles in place, further tax reforms may be achieved between the Commonwealth and states. In achieving this objective, the Commonwealth grappled with how it might divest some of its tax base to sub-national government in addressing the recalibration of tax revenue collected across the three tiers of government. In conjunction with tax reform is the need to align the roles and functions of the various tiers of government in Australia, as has been undertaken internationally in Denmark and New Zealand, and which is ongoing in Canada.

Chapters 12 and 13 are progressively organised to commence with the review of the limitations of state land tax and the shift needed in transferring this tax source to local government by maximising the tax effort from all land uses through the council rating system. While scope exists for land tax to coexist at both state and local government levels, these taxes should be applied with a more succinct rationale and purpose for taxpayers to understand. This is followed by the reforms to the operation and application of this tax, which include changes to the base of the tax, allocation of the tax burden across the categories of land and the development of a capacity to pay model which replaces the need for the capping and pegging of rate revenue. Further addressed are the important reforms in the transitioning from transaction taxes to recurrent taxes on land, resulting from improved revenues collected from land tax.

In conclusion, reforms to own source revenues and defining the role of sub-national government show that land tax revenue sources are not necessarily defined by geographic boundaries. By synchronising tax effort across the states, opportunity exists for the sharing of revenues from other tax bases that are collected and administered by the Commonwealth. This then provides a more transparent way for the collection and distribution of tax revenue to sub-national government as it is collected by local government, which allocates the tax to the services and infrastructure provided by state and local government.

12 Reforming recurrent land tax under the status quo

This chapter focuses on reforming land tax as it currently operates across two tiers of government and addressing factors which influence the base of the tax, how the tax is assessed and the much neglected void in taxpayer understanding. It further provides options for improving tax effort and reforming some of the outdated restraints that have impacted revenues. The first step in reforming land tax is to understand the limitations and challenges confronting the tax, which has, in part, been discussed in Parts 1 to 3 of the book and which are implicit in a number of the international jurisdictions. In exploring these factors further, it is apt to consult participants involved in the administration of this tax and consider the breadth of options for reform.

Two focus groups were used to review the operation of land tax and explore the options needed for reform. Focus Group 1 had a strong legal and educational representation, balanced with two practicing valuers with rating and taxing experience. Focus Group 2 comprised two legal practitioners and one law graduate with a strong contingent of valuers with extensive experience in rating and taxing matters. An independent and experienced focus group facilitator introduced the topic and maintained the direction of each group. A summary of the composition of each focus group is set out in Table 12.1, in which each participant is assigned a code. The link to the detailed transcript from these focus groups is available via the reference list under Mangioni (2013b).

A summary of the results compiled from the simulations and surveys, which are set out in Chapter 4, was used in a brief presentation at the beginning of each focus group by the author to provide context to the discussion. Given the optimum length of a focus group of between 90 and 120 minutes as defined by Krueger and Casey (2009), the two focus groups were first led in open discussion on the general relevance of land tax, which acted as an important icebreaker. This discussion tended to focus on the rationale and economics of taxing land and its use as a basis for a general-purpose revenue tax. Following the icebreaker phase, a presentation of the research problem in determining land value was presented to each of the focus groups.

Table 12.1 Focus group composition

Focus Group 1	*Focus Group 2*
Property Solicitor (PS)	Barrister / Valuer (BV)
Property / Construction Solicitor (CS)	Solicitor / Valuer (SV)
Valuer (V1)	Property Law Graduate (PG)
Valuer (V2)	Valuer (V3)
Valuer / Educationalist (VE1)	Valuer (V4)
Valuer / Educationalist (VE2)	Valuer (V5)
Researcher (R)	Valuer (V6)
Independent Facilitator (FAC)	Researcher (R)
	Independent Facilitator (FAC)

This chapter proceeds by reviewing the matters discussed in the focus groups and subsequent research undertaken into the financing of sub-national government. Where relevant, reference is made to international best practices, as set out in the international case studies, which support the suggested reforms needed in the rating and taxing of land.

What is land tax in a modern Australian tax system?

The answer to the question as to what land tax is, is controversial and vexing and may be broken down into a number of sub-questions, which are covered in these final two chapters. It may be viewed that these questions would be addressed at the beginning of the book rather than in the final chapters. As a tax that has been imposed by different tiers of government over its 130-year history and used for a variety of purposes in Australia, it is apt to address this question now at the point of its reform. This is important in driving reform rather than allowing this question to remain contested.

The responses to the question as to what land tax is and the complex diversity of answers given have consumed much of the debate. Bird, Slack and Tassonyi (2012:7) epitomise the complexity of this question in the following:

> An old story tells of a group of blind men, each of whom touches a different part of an elephant, and their resulting disagreement about how best to describe the beast. Many analyses of property taxes are like this story. What one sees is some combination of what one is looking for and what is there, and how one interprets the relationship between theory and what seems to be reality offers yet more room for interpretation.

The way this tax is perceived is highly dependent on the local context in which it is applied. In the United States and parts of Canada, one of the designated purposes of the property tax is the assignment of part of the revenue collected from this source to education. This purpose was evident in the review of this tax in Chapter 8 in the United States and Canada. In the case of the review of Ontario, Canada, Bird, Slack and Tassonyi (2012) emphasise that, without understanding education finance and the changing structure of local government over the past decade, it is not possible to fully understand property tax reform. The context in which the property tax is designated will impact how it will function, how it may be reformed and what it actually stands for.

It is important to highlight the conundrum which confronts the assignment of tax revenue in any country. It is not uncommon for taxpayers to interpret and attempt to relate tax policy and the use of revenue generated back to their own circumstances. In the case of education, it may be reasonable for any taxpayer who has paid for their own education or paid for their offspring's education to argue that they should not pay the component of the tax designated to education; however, they contribute to a broader educated society and does not relate to the specificity of their own individual circumstances. This principle is well articulated by C. Fair (B.C. Assessment, personal communication, IAAO Conference, Sacramento, 26 August 2014) as follows:

> The tax isn't determined on whether the taxpayer or their family use the public or private school system. The education system benefits all B.C. residents including people without children in school. School tax is paid to share in the cost of providing education in B.C.

Likewise, in contrast to education, if land tax were to be designated to infrastructure, a similar argument may be mounted that a particular infrastructure project would not be used by the individual taxpayer. The fact that something is not used is not the test of relevance; instead, the option to use and the benefit the infrastructure project provides to the community is the primary consideration and test of relevance.

This matter was raised in the focus groups, which exhibited a clear lack of consensus as to what land tax and local rates stand for in Australia. Differing views became apparent on matters including the level of government imposing the tax, the differing roles of the states and local government and the various services and functions of these tiers. A level of discussion evolved as to land taxes being linked to infrastructure, to which a counter argument arose that there was little perceived linkage between the land tax and infrastructure, as infrastructure was largely bourn by developers through developer contributions.

Discussion in one focus group highlighted that trans-local government infrastructure was not funded by developers but by the community through taxes and financed, in some cases, through Private Public Partnerships. This brings to the fore the distinction between local infrastructure versus infrastructure which traverses several local government areas. At the local level, power, sewer, water amplification and reticulation are paid for by the taxpayer, while at the broader level, transportation, roads and rail to and from employment hubs are paid for by the taxpayer. Further, the terms *finance* and *funding* in the debate over infrastructure are loosely interchanged when referring to its provision. The Productivity Commission (2014) rightly points out that funding, which is that component needed in the development of infrastructure, ultimately comes from either taxes or user charges. The fact is that taxes are a key component in the funding of infrastructure, which is sufficient to move the debate forward as to how tax revenue may be raised and assigned to this objective in Australia.

The focus groups debated land tax with views ranging from its abolition and replacement with a higher rate of goods and services tax, which in turn was strongly opposed by participants who believe that a larger tax contribution should be levied from land as a store of wealth. A further level of discussion arose, which identified the competition between state and local government raising the same tax in different forms, one as a land tax and the other as local rates. This brought forward the question as to why the same tax is imposed twice by two different tiers of government, of which the primary focus of debate centred on the imposition of state land tax. Some may still be lost in trying to reconcile the proposition that state land tax and local government rates are the one generic, recurrent tax spread across two tiers of government in Australia. This is understandable, as these taxes are presented as two different taxes, with state land tax viewed as a consolidated revenue tax, and council rates viewed as more closely aligned to the services and infrastructure. At present, neither the current split imposition of these taxes, or the combined revenue generated from them, adequately contributes to tax effort in Australia.

Most apparent from the interaction of the focus groups is that attempts to reform a tax in which there is little consensus as to its objective, and which is subject to a variety of interpretations and points of view, easily sidelines the more important debate as to how this tax is to be reformed. This void of understanding paves the way for the tax to be more clearly defined in Australia, which is in a unique position to reform this tax. It is in a strong position to achieve a structured reform which provides sustainable tax revenue while reforming less efficient transaction taxes on land. It is also in a prime position to define the fiscal roles of each tier of government and remove the duplication of administering dual land tax systems.

While the question as to what land tax constitutes was keenly debated, in contrast to earmarking tax revenue to education, as is the case in the United States and Canada, it is clear that infrastructure is a distinct target to which land tax revenue could be earmarked to in Australia. It is further made clear that the states are responsible for most of the infrastructure across Australia (Mangioni 2014c) and a rationale exists for a component of land tax revenues to be earmarked for this important purpose.

Land tax reform under the status quo

The reform of existing and the introduction of new taxes are treated with great caution by government in the current political environment. The ability to sell the merits of and the need for tax reform has been impacted by using this important reform to scare rather than educate the taxpaying public. Alternatives to reforming the tax system have resulted in governments pruning programs and reducing funding projects, a practice which has a place among the options in managing any economy. This aside, the need to address both the total tax collected and most challenging, defining which taxes will be the subject of increases, has thwarted reform in Australia since the introduction of the GST. It was discussed in Chapters 1 and 2 that the total recurrent land tax collected from state and local government in Australia is one of the sources from which the increase in tax revenue must be achieved.

There are two approaches in undertaking this reform in Australia. The first is to reshape tax policy under the current state and local government arrangements, and the second approach is to address this reform at a much broader and innovative level. The remainder of this chapter addresses the first option, while Chapter 13, the final chapter, addresses the latter option. How this tax can be reformed to improve the efficiency of current revenue collected and what elements of state land tax and local rating can be transposed to a modern tax system are the questions to be answered. In answering these questions, a review of the operation and elements of the current system is needed.

It is apt to first set out the differences and factors impacting the reform of a dual land tax system operating across the tiers of sub-national government in Australia, as set out in Table 12.2. It is clear from this table that recurrent land tax, in addition to being spread across two tiers, is assessed on several different bases of value, has a number of mechanisms used to adjust revenues, and impacts tax effort through the use of a variety of differentials, concessions and exemptions. Each of these factors aims to mitigate the shortfalls of the other in the absence of an operating environment which enhances the tax revenue that might be collected, and hence the economic robustness of this tax source. These factors highlight the need for reform of this tax in improving the robustness of revenue from this source.

The first item addressed in Table 12.2 is the limitation of increases applicable to each tax. One of the strengths of the current state land tax system is that changes in revenue are primarily driven by the movement in value on which the tax is assessed. This works well in the case of state land tax, to which the principal place of residence is exempt. If the current state land tax regimes were to be expanded to include the principal place of residence, few homeowners could afford to pay this tax in its present form and at the level it is imposed. This suggests that the rate of the tax, and not solely the base of the tax, is the determinant of whether the broadening of this tax is achieved.

Table 12.2 Comparison between state land tax and local government rating in Australia

	State land tax	*Local rates*
Revenue increase limitations	N/a	Rate pegging NSW, rate capping Victoria and Ministerial oversight in other states.
Base of the tax	Land/Site/Unimproved Value	Land/Site/Capital Improved/Net Annual Value/Base Amount Per Property/Minimum Rates
Valuation cycle	Annual/Biennial/Up to five yearly in different states	
Tax imposition	Tax revenue is directly related to value and subject to a tax free threshold in each state.	Categorisation of land uses and rating differentials
Revenue break-up	Revenue: approximately 25 per cent from residential land and approximately 75 per cent from non-residential land; varies from state to state.	Varies across local government areas of Australia; collectively higher percentage of rate revenue is derived from residential use land.
Sources of tax expenditure	Residence/Threshold/Primary Production/Not For Profits/Government Land Not Leased	Not For Profits/Some Government Land
Summary	Australia's tax effort from local rates and state land tax combined is less than two thirds of New Zealand's, less than half of each the United States, Canada and the United Kingdom, which only levy a tax at the local government level.	

In contrast to the broadening of tax revenues, the increases in local government rates are moderated in New South Wales by rate pegging and, soon to commence in Victoria, through the use of rate capping. Unless this point is addressed in detail within the context of a reform that measures the taxpayers' capacity to pay, real tax reform is limited. This point is addressed in Chapter 13, in which it is demonstrated that increasing tax from this source is to be achieved at the local government level. What the states have focused on to date is moderating increases in rate revenue. What local government has yet to grasp is the ability to fully demonstrate to the states is the taxpayers' capacity-to-pay, resulting from increases well beyond the increases currently being sought by local government.

We now ground this sentiment by first referring back to the examples at the beginning of Chapter 2, which set out the assessment of state land tax and local government rating in New South Wales. These examples are merely templates, and the sections and chapters which followed in Part 2 provided the legal operational framework which govern these taxes across the states and local government jurisdictions of Australia. In order to reform land taxes, it is important that their applications are unpacked so elements which work well and contribute in achieving the principles of good tax design are maintained and strengthened while those that do not are removed and replaced by more progressive processes.

The remainder of this chapter addresses the three key points of the options used to assess the tax, the mechanisms used to adjust revenue, and the evolving phenomenon now impacting the first two points: the fast-emerging cities and urban hubs of Australia where most land tax is raised. Stemming from the review of these points, in unison with defining the purpose of land tax, the way forward emerges for refining land tax under the status quo of a dual tax system. What becomes apparent is that these reforms have far more potential to raise revenue with broader structural reforms and with all tiers of government cooperating. This is, in contrast to retaining taxes in silos, built around the now outdated operational tiers of government. It is particularly the case for state and local government in the application and administration of land tax.

Bases of assessing land taxes

The base on which land tax is assessed and the methods used to adjust revenue across Australia are the subject of much debate and depend on the level of government imposing the tax. In order to improve taxpayer confidence in state and local government land taxes, the first and most important step is to build taxpayer education and improve transparency of how these taxes operate. This point mystifies many tax administrators and politicians, of which the determination of the base of the tax raises more questions than it answers.

It was pointed out in Chapter 2 that several different bases of value currently exist in assessing state land tax and local government rates. Further noted was the option in some states of Australia to assess local government rates on more than one basis of value. The analysis of recurrent land taxation across Australia, and its application internationally, supports the view that there is no one correct or precise basis of value on which this tax should be assessed. In the various jurisdictions examined, and within their respective local contexts, some bases have evolved to meet the needs of the built environments in which the tax operates. In others jurisdictions, dual bases of value may operate effectively, as was demonstrated in New Zealand, Victoria and South Australia for local government rating purposes.

In evolving urban and regional centres, land value as the base of the tax still operates effectively on two accounts. The first account is the availability of vacant land transactions, which provides the requisite evidence for the determination of land value in a simple and transparent manner. The second account is that land value is neutral, in that it does not discourage the development of land as it is taxed the same regardless of whether it is developed or not. Land value is still used as the base of local government rating in regional New Zealand, with capital improved value used in the four main cities. It was further highlighted that in Victoria and South Australia the option exists for local government rates to be determined on land/site value, improved value or net annual rental value.

Tasmania has recently reviewed its rating valuation system, which has resulted in the recommendation of the transition from gross rental value to site value in line with its state land tax system. This brings Tasmania in line with New South Wales and Queensland, which also levy state land tax and local rates on the same basis of value. This makes common sense in a dual state land tax and local government rating environment. Why pay for the manufacture of two bases of value when you can run your rating and taxing systems off the same tax base? From a cost-centred perspective, this is a perfectly logical and understandable conclusion to have reached. What has been overlooked is the bigger picture of how the whole recurrent land tax system should be reformed and how the tax effort from this source is to be maximised and applied in accounting for the differences of the built environment in which the tax is applied.

Before further discussing the bases of value and which base might best work in the assessment of local government rates, it is first apt to qualify that state land tax and local government rating are primarily an ad valorem tax. That is, value is the primary rationale for the assessment of these two taxes across Australia and, likewise, for the assessment of this tax abroad. This point was demonstrated in the following case, in which the court acknowledged the importance of the ad valorem component in *Sutton v. Blue Mountains City Council* (1977) 40 LGRA 51. It was argued by the ratepayer that: 'council set their minimum rates so high and their ad valorem rate in the dollar so low, that all ratepayers paid the minimum and the ad valorem applied to no one.'

A review of the minimum rates paid by the applicant shows that the General Rate, Water Rate, Local Sewerage Rate and Library Rate in the years 1976 and 1977 each comprised a minimum rate with no application of an ad valorem component. The percentage of ratepayers in the Blue Mountains paying the minimum rates ranged from 76.2 to 97.1 per cent. The Court did not give any specific indication of what would be an acceptable 'cut-off' point in determining the ad valorem percentage; however, in upholding the objection and appeal of the ratepayer, Holland, J. stated at paragraph 66:

> The problem in the point of view that I have expressed is not in saying that the minimum rating power is limited but in postulating where the limit lies. I think that the answer to this problem is that it is a matter of degree in which some cases will be considered to be clearly below and some clearly above the line and that there would be an area of boarder line cases which would be difficult to decide and on which minds might differ.

The objection in this case clearly opposed rates being assessed on a straight benefits received basis with all ratepayers levying the same rate regardless of the value of their land. This kind of activity often results from pressure driven by ratepayers affronted by rates being imposed on value. They view equality representing the total rate revenue derived from residential property by dividing total rate revenue by the number of properties to arrive at the same rates paid by all owners. Following the revision of the NSW Local Government Act 1993, sections 499 and 500 of the Act allow local government to raise up to 50 per cent of total rate revenue of a base amount per property. The other 50 per cent of the rate revenue is raised as an ad valorem tax on the land value.

The initial rationale for the imposition of sections 499 and 500 was to address variations between large residences and land holdings versus smaller residences and land holdings located within the same local government area. The view that drove base amounts per property was that residents drove on the same roads and used the same facilities provided by local government, had one garbage bin per residence, and hence should not be paying rates solely on the value of land. The imposition of minimum rates and the use of section 500 allowed some local governments to temper the differences in rates between disparate residences and rates imposed on land value.

In Sutton's case, the Blue Mountains predominantly comprised single-dwelling houses, in 2010 and beyond, some local government areas with 50 per cent of residents living in medium- and high-density housing. This has resulted in some local government areas with a majority of rates being raised off minimum rates, which brings forward potential challenges to those noted in the Sutton case. Minimum rates are used to ensure that property with land value which is considerably lower than others makes a minimum rate contribution. As rating on minimum rates increases, this creates a challenge for many local government areas in the capital cities of the states that solely use land value to assess local government rating and state land tax.

As will be demonstrated in the following section, the use of minimum rates, rather than being the exception in many local government areas, has become the norm. This is due to the evolving change in housing structure in the capital cities of Australia, a point to be expanded on next. Of specific imperative for many local governments which impose minimum rates, is potential challenges from residents whose rates are determined on ad valorem. This brings forward the tax principle of equity, to which potentially gives rise to challenges from residents in houses who pay rates on ad valorem, in contrast to residential unit owners who pay the minimum rate.

Table 12.3 sets out the emerging differences in rating policy since the Sutton case and how the evolving city phenomenon now raises the question as to which basis of value best meets the needs of rating authorities in locations where most of the current and potential rate revenue is to be

Table 12.3 Comparative analysis and rating restructure relative to the Sutton Principle

	Sutton case 1976/77	*Local Government Act 1993 NSW*	*Local Government 2010 onwards*
Rating structure	76.2 and 97.1 per cent of rate revenue was derived from minimum rates	Sections 499 and 500 allow up to 50 per cent base amount. Section 548 allows for minimum rates.	Applications made to exceed the 50 per cent base amount in parts of Sydney.
Relevant date	1976 and 1977	1993	2010 onwards
Residential profile	Primarily single-dwelling housing	N/a	Over 75 per cent medium- and high-density housing in some LGAs
Taxing rationale	To ensure minimum rates became the rule with ad valorem being the exception.	Create a balance between base and ad valorem rating. A base and minimum rate cannot exist together (s548(7)).	To even the rate relativity between residential housing types resulting from the diminishing relativity of LV to CIV in some LGAs.
Commentary	In challenging the rates over two years, 1976 and 1977, it was determined that the rationale for the rate structure was to remove the ad valorem component for rating property within Blue Mountains City LGA.	The Local Government Act was reviewed in 1993, in which provisions were made to restrict rate revenue derived from a base amount to no more than 50 per cent, with the balance derived from an ad valorem component of no less than 50 per cent.	Local government is seeking to establish a more even imposition of rating across its increasingly diverse housing type. In doing so, it seeks to ensure the consistent application of the ad valorem component of its rating.

Source: *Sutton v. Blue Mountains City Council* (1977) 40 LGRA 51 and Local Government Act 1993 NSW

raised. From a state land tax perspective, the primary reason this matter has not yet become highly contested results from the principal place of residence exemption and the investor threshold, in which approximately 25 per cent of state land tax revenue is derived from residential property across Australia. Further, land taxpayers have no idea who pays this tax and which property is subject to the tax. This is predominantly driven by the fact that land taxpayers do not occupy the property on which state land tax is imposed and hence are not aware which of the surrounding property or owners are subject to this tax.

The emerging value and rating phenomenon

The juxtaposition between residential houses and units raised in the previous section has been exemplified over the past 20 years, with the development of high-density housing located across local government areas of the capital cities of Australia, and in particular Sydney and Brisbane. To demonstrate the impact on the assessment of rate distribution resulting from the transition in housing and using the different methods of calculating rates in a hypothetical scenario, Table 12.4 sets out the rates for three housing types, each with different land values. This is then followed by an analysis of land versus improved value and the variability which is fast-emerging between the diverging housing types across local government areas, which further impacts the rates payable, using the same three properties in Table 12.4.

A brief introduction and background is first needed to provide a context to the following example, showing the variability in the way rates may be calculated for the same property as well as a comparison of how rates may be calculated across the types of property. It is further useful to distinguish the following comparison with that made at the beginning of Part 2, in which an example was used to demonstrate the differences between local government rates versus state land tax paid by Frances in that hypothetical scenario. In contrast to juxtaposing state land tax with local government rates, the following analysis juxtaposes the breadth of variability that exists in one state for the calculation of local government rates. It demonstrates the necessity

Table 12.4 Methods available for applying local government rating in NSW

Rating scenarios	*Suburb A Land value $100,000*	*Location B Land value $500,000*	*Location C Land value $900,000*	*Total rate revenue*
Option 1 Ad valorem	$400	$2,000	$3,600	**$300,000**
Option 2 Ad valorem and minimums	$550	$1,950	$3,510	**$300,500**
Option 3 Base amount and ad valorem	$960	$2,000	$3,040	**$300,000**

of aligning the processes for calculating rates across local government as a seamless tier of government rather than siloed local government areas.

Kerrigan is a hypothetical local government area in New South Wales which comprises three suburbs, which are referred to as locations A, B and C. Location A comprises 50 strata units, Location B comprises 50 terraced houses, and Location C comprises 50 large detached houses. Kerrigan Council has resolved that it needs to raise approximately $300,000 in general rate revenue next year and has three options available for raising this revenue as follows:

Option 1: Rates determined with an ad valorem rate of 0.004 cents in the dollar applied to the land value.

Option 2: Rates determined with an ad valorem rate of 0.0039 cents in the dollar applied to land value with minimum rates of $550.

Option 3: Rates determined on 35 per cent of the total rate revenue as a based amount per property and 65 per cent as an ad valorem tax at 0.0026 cents in the dollar.

What sits behind the rationale for the differences of Option 1 and Option 3 in Table 12.4 is the tax base design which, in a residential application, worked well in an evolving city environment where consistency in the uniformity of land uses and diversity of housing was not too disparate. This point has gained significant momentum over the past 20 years, as more Australians progressively migrate from living in houses to living in medium- and high-density housing. A number of local governments struggle with equalising rate revenue across the classes of residential property

As pointed out in Chapter 2, in the states of New South Wales and Queensland, land and site value are both the base for state land tax and local government rating. In each of these states, the author has been asked, on occasion, why the land or site value of high-value residential units is a fraction of the capital improved value. This factor is questioned by tax administrators in the assessment of state land tax and rating officers in the assessment of local government rates. While a detailed analysis of the differences between land or site value and improved value is beyond the scope of this research, Table 12.5 provides a summary example of the different ranges that exist between land and site value compared with the improved value for three categories of residential housing.

Table 12.5 Indicative percentage of land to improved value

Housing type	*Ratio range*
High-rise units	10–20 per cent
Medium-density units	20–30 per cent
Single-dwelling houses	35–45 per cent

What is increasingly evolving across Sydney and Brisbane is that the land value for units represents a smaller proportion of the capital improved value relative to houses. This difference is further distinguishable between medium-density and high-density units. As land becomes more densely developed, two factors are identified; firstly the land component is smaller per unit, and secondly, the ability to measure the underlying value of higher density housing in highly urbanised locations is diminished through the absence of vacant land sales, a matter to be expanded next. This issue largely escapes scrutiny in the application of state land tax due to the land tax free threshold, which is set out in Table 7.1 using average values and demonstrates the number of residential properties an investor could hold before being subject to land tax across the various states of Australia.

Melbourne, while having experienced housing with higher levels of density than Brisbane and Sydney over this same period, is not the subject of the same differences as Sydney and Brisbane due to the flexibility of the bases of value it has to choose from in the administration of its council rates. This now brings to the fore the important, but often overlooked, rationale as to which bases of value are best suited in the assessment of local government rates. Again, the answer rests with the built environment within the location in which the tax is assessed and finally the simplicity and transparency of determining the basis of value used to assess local government rates.

The use of land values to assess state land tax and local government rates in highly urbanised locations where most revenue is raised meets two challenges and prompts debate for options in addition to land or site value. The first reason is governed by the fact that the more densely developed land becomes, the higher the added value of improvements will be as a proportion of the total value as the land component reduces. The second is a valuation constraint governed by the availability of either vacant land sales or land transactions derived from improved property purchased and subsequent demolition of improvements, to which the price paid for the property represents land value.

The NSW Ombudsman (2005:8) highlighted the propensity for conservative land values resulting from the absence of market evidence to support increases in land values, as follows:

> While the practice of applying a conservative component factor in areas where there is little sales evidence is defensible as a valuation practice, the major concern of practices such as this being applied in some valuation districts but not others is that it undermines the consistency of the valuation process across the State.

Since the NSW Ombudsman findings of 2005, considerable improvements and advancements have been made in the mass appraisal system used to value land in NSW. These and similar practices are used in other states for the assessment of rates and taxes where land is the base of the tax. Despite these improvements, this brings to the fore the questions as to whether one basis of value, i.e. land or improved value, is arguably more consistently determined than the other, and whether the exercise of determining land value, as set out in Chapter 4, remains a sustainable base to assess recurrent land tax in a modern tax system. It was demonstrated in Chapter 4 that the standard deviation in the valuation of residential houses, for two of the three houses valued, was within an acceptable margin of error in the initial simulation. While not specifically measured in the case of medium-density housing, this result would be amplified for high-density housing, where land was fully developed and an absence of vacant land sales existed.

In discussions with valuing, rating and taxing authorities in Australia and internationally, the most common debate surrounds which is the most suitable base to assess recurrent land tax in Australia. The answer to this question varies; however, over the past decade it has moved towards the direction of Capital Improved Value, particularly for residential property. Much opposition exists for the adoption of this base in the case of business use property. Arguments exist that this base would include a tax on goodwill and business elements of value that are included in land or site value, but excluded from land value. This was elaborated on in Chapter 3 under licensed premises, shopping centres and special uses land, from which the deduction of the value of improvements does not solely equate to land value.

Which basis of value for assessing an efficient land tax

In recapping briefly on the bases of value that were examined in Chapter 3, land (or Site) Value put to highest and best use values, land (improved or not) based on the sale value of comparable vacant land. Mangioni and Warren (2014) state that as vacant land is sold in the market it will reflect its potential highest and best use, resulting in a base definition which is efficient. With Capital Improved Value (CIV), in practice different concepts find application depending on how improved value is determined. If CIV and ARV are determined by what improvements are permitted on the land, rather than what is actually on the land, then the landholder can take no action to influence their land tax assessment. This results in the land's capital improved value being highest and best use and economically efficient.

> Highest and best use of a land as improved relates to the use that should be made of an improved land in light of the existing improvements and the ideal improvement described at the conclusion of the analysis of highest and best use as though vacant.
>
> (Australian Property Institute 2007:240)

Mangioni and Warren (2014:468) further point out that,

> in contrast, if CIV (or ARV) is determined by the existing use of the land (CIVEU), regardless of whether the improvements are maximally productive or highest and best use, taxing this base will distort land use since it is non-neutral in its impact on improvements to that land. Since CIVHBU and LVHBU both reflect the highest and best use of land, taxing either base will not distort land use and will therefore be neutral (and therefore economically efficient) in its impact on any decision on improvements to that land.

The factor determining the prevalence of capital improved value existing use (CIVEU) appears to be its ability to be simply estimated from market land sales of comparable properties, a process which is outwardly transparent to taxpayers (and so less subject to disputation). The cost of this approach is higher than capital improved value highest and best use (CIVHBU), where the later can be determined by a benchmark value being established within a location that is most indicative of surrounding similar uses. However, valuing existing use capital improved value (or what is actually on land) might not be so simple if comparable sales are not available and, more importantly, any estimate which reflects existing use will not be economically efficient. The cost of CIVEU is further prohibitive as records of improvements of individual property are necessary and tracking changes to including the aging of improvements further adds to the complexity and cost of the task.

Table 12.6 highlights the differences of the impact on neutrality between Land Value (LV) and Capital Improved Value Highest and Best Use (CIVHBU) versus that of Capital Improved Value Existing Use (CIVEU). The overriding importance is the neutrality of the tax base, of which the taxpayer is unable to influence.

In contrast, Mangioni (2013) states that adopting CIVHBU requires the definition of some *Physically Defined Standard State* (PDSS) which, when

Table 12.6 Applying the bases of value

Valuation base	*Land / Site Value (LVHBU)*	*Capital Improved Value / Annual Rental Value (CIVHBU)*	*Capital Improved Value/ Annual Rental Value (CIVEU)*
Applied Tax / Value Principle	Highest and Best Use Vacant land	Highest and Best Use Improved land	Existing Use (Under developed/ obsolete use or improvements)
Outcome	Neutral	Neutral	Non-neutral

Source: Mangioni & Warren 2014

assessed for tax, reflects the indicative highest and best use of that land within a location. If highest and best use is defined as being 'land with improvements which are highest and best use', the Physically Defined Standard State for each piece of land would not need to consider the age, condition and scale of improvements by location on improved land (which is its existing use). The challenge is how to give practical meaning to a Physically Defined Standard State. In contrast to LVHBU, where the challenge is to identify (now non-existent) vacant land sales in urban environments, CIVHBU is more readily identifiable as being the sale of land recently improved.

The question is whether Capital Improved Value determined on highest and best use can be more broadly implemented and, most importantly, clearly demonstrated to be determined on either highest and best use or a similar proxy which maintains neutrality. In response to McCluskey et al (2010) the characterisation of Capital Improved Value, while the term market value is used to describe the base on which capital improved value is assessed, it is not always theoretical market value that is used. The successful adoption of CIVHBU, through the implementation of a Physically Defined Standard State definition, depends crucially on the valuation process being simple and transparent and exhibiting integrity. However, how any valuation process impacts on the integrity of land tax design, and therefore on taxpayer acceptance, is not well researched despite its pivotal role in linking theoretical concepts with practice.

In the research carried out by Mangioni (2013) in Chapter 4, where land value was determined in the absence of vacant land sales in a residential and retail simulation, the valuation process commenced with the analysis of improved sales. What then proceeded was that valuers then determined the land value by deducting the added value of improvements from the sale price to arrive at the land value. It was shown in the revised simulations that, when sales of property which are CIVHBU were the starting point of the valuation process (sales analysis process), a more consistent land value was achieved across the population of valuers who valued the same land at the same date of valuation. To this end, the use of Capital Improved Value to assess state land tax and local government rating in highly urbanised locations has significant merit.

A basis of value that is understood by the taxpayer will be more accepted, as it removes one layer of valuation mechanics, that is, the stripping back of improved sales in highly urbanised locations, where most of the states land tax and local government rates are raised. The valuation approach used in New Zealand is land value and improved value for rating purposes. South Australia and Victoria each use site value for state land tax and improved value and annual rental value for rating purposes. The cost of producing more than one basis of value in the two-tiered tax system which operates in Australia is a much lower priority to that of producing a basis of value that better fits and operates within the built environment in which land tax evolved. This brings to the fore the matter of cost and the need for annual valuations which are addressed in the following section.

Valuation frequency

The valuation of land is the largest cost in administering state land tax across Australia and accounts for over 40 per cent of the total cost once factoring the time and cost of objections, appeals and court cases against the values used to assess the tax. Across the states, the valuation cycles differ for state land tax and local rates across Australia, as does the frequency of valuations for the different bases of value. The valuation frequencies within Australia and internationally, along with the valuation authorities, are set out in Table 12.7 and vary across the taxing jurisdictions. The question as to how frequently land values should be reassessed for rating and taxing purposes in Australia was asked of valuers by Mangioni (2013) in a survey and discussed in the focus groups referred to in this chapter. Valuers were given four options of selecting either one, two or three years, or another option for the valuation frequency of land values in Australia, of which the mean response resulted in 2.3 years. It was clearly stated that annual valuations are unnecessary. The potential volatility of values has been combated through the use of averaging of values over a three-year period in NSW, QLD and ACT.

There is no need for annual valuation of the tax base in any jurisdiction of Australia. Currently with so little residential property attracting state land tax in Australia, the money and resources spent on producing these valuations could be better spent on advancing the valuation base on which state

Table 12.7 Valuation/Assessment frequency

Australia			
State/Ter	*Base*	*Frequency*	*Valuation provider*
NSW	Land Value	Annual	Contractor
VIC	Capital Value Site Value Assessed Annual Value	Two yearly (Biennial)	Contractor or Valuer-General
QLD	Site Value	Annually	Valuer-General
WA	Gross Rental Value Unimproved Value	Three to five years and annually	Valuer-General
SA	Capital Value Site Value Assessed Annual Value	Annually	Contractor or Valuer-General
NT	Unimproved Capital Value	Three yearly	Contract valuers
ACT	Land Value	Annually	Contract valuers

Table 21.7 (continued)

International			
Jurisdiction	*Base*	*Frequency*	*Assessment provider*
New Zealand	Site Value (regional) Improved Value (cities)	Up to three yearly	Contract valuers
Denmark	Land Value Improved Value Income	Biennial	In-house government valuers
Canada	Improved Value	Yearly (B.C.) Four yearly (Ontario)	Government and contract assessors
United States	Improved Value Improved Value	Annually (California) Annually (New York)	Government assessors Government assessors
England	Improved Value Income	1993 (last valuation) Five yearly	Government assessors Government assessors

land tax does apply. At present, only two states, NSW and Qld, rate and tax on the same base, while Tasmania considers a move to the same base for each of their recurrent land taxes. In South Australia, values are produced for all three bases of value annually, the cost of which is under pressure at the present. The need to revalue multiple, or indeed single, bases of value annually is highly questionable. The current system in NSW is a classic example of the expense of revaluing one basis of value annually, of which over 80 per cent of the land values are not used on an annual basis. For local government rating purposes, values are updated every three to four years, while for the narrow number of land tax-liable residential properties, the land value of every residential property is undertaken annually.

As highlighted in the literature, in the case of countries and jurisdictions that have a land value tax, very few undertake annual valuations. In the case of Australia, the focus group participants resolved that annual valuations were not needed and placed too much demand on the valuation process. Three years was determined to be the maximum period required between valuations. Given that the date of valuation each year in NSW is 1 July, and that sales would occur up to six or more months either side of that date, the need to differentiate the market from one year to the next is challenging. The need to analyse data required time and, in the case of many of the sales that valuers relied upon as potential redevelopment sites, the improvements remained on the land for months, years and, in some cases, indefinitely.

It was pointed out that a two- to three-year interval between valuation cycles would allow valuers to analyse the market at the same date of valuation. One valuer in the focus groups highlighted that the issue raised by the Ombudsman (2005), of valuers not having a basis in accounting for time between the sale date and date of valuation, would be overcome and differences for time would be more easily assessed with a longer valuation cycle. Assessments may still apply to value; however, where the value is determined less frequent than annually and in alternate years where existing values are used, alternate mechanisms may be used to adjust the tax revenue collected. The use of indexes for alternate years between valuation cycles is a more than acceptable approach.

In contrast, the approach used abroad in the United States and Canada, is one in which budgets are set each year and the rate in the dollar applicable to tax assessments is adjusted to raise the requisite tax revenue needed to administer local government commitments. However, in Australia the factor that government solely relies on for state land tax revenue is adjustment of values determined from the market place and used to assess the tax. Rather than defining the financial commitment of the tax to be raised, the states rely on as much land tax as is forecast, as their revenue shortfall from own source taxes is a fraction of the revenue needed to administer their tier of government. The determination of local government rates in Australia is more closely aligned with those of the United States, Canada, New Zealand and Denmark where some rationale exists for the imposition of local government rates.

Summary

It has been demonstrated that the dual land tax and rating system applied across state and local government of Australia served the initial purpose of funding all three tiers of government at the time of Federation. The states ceased taxing land soon after Australia Federated, leaving this tax to Commonwealth and local government to collect. In the late 1940s the Commonwealth vacated the imposition of this tax with the states, once again imposing land tax soon after surrendering their income taxes to the Commonwealth. The fractured imposition of this tax across sub-national government has impacted the tax effort from this source and leaves Australia as the only advanced OECD economy that imposes a dual recurrent land tax across two tiers of government.

The basis of value on which recurrent land taxes are assessed must be revised and broadened in some states and, in particular, within the urban hubs where a majority of recurrent land tax is raised. In the case of residential property, the transition to Capital Improved Value as the base of rating residential property is of high importance in reforming the rating system under the status quo. For states where valuers are directly employed by government to carry out these valuations, it is a zero sum gain, as their costs remain fixed. In the case of states where contract valuers are used, more pressure is being brought to bear on valuing authorities to further reduce valuation costs. This is understandable, as less than 20 per cent of residential property attracts state land tax. To this end, valuing every residential parcel of land on an annual cycle in some states is redundant.

In the case of local government rating, broadening the options on which rates may be assessed will address the vast disparities that have emerged in the rating of land in states that solely retain land value as the base of assessing rates. The cost savings from undertaking annual valuations would be far more usefully assigned to strengthen local government rating assessments, particularly in the states of New South Wales and Queensland. Such a move would further improve the operational efficiency of local government rating, which could be undertaken on a two or three yearly cycle. It is noted in Table 12.7 that the United States and British Columbia in Canada use annual valuations; however, their recurrent property tax system has been fully divested to local government and is administered with far greater efficiency and revenue robustness compared with Australia.

By far the most compelling rationale for change in Australia is noted in neighbouring New Zealand, where a dual land value and improved value system operates, in which values are assessed on a three yearly cycle. The tax is administered at the local government level and the level of revenue collected is something Australia should aspire to. This is dealt with in more detail in the last chapter, in which much more is needed in reforming Australia's land tax system.

13 Realigning Australia's tax system

Blue sky reform

Introduction

The reform of Australia's tax system requires more than a root and branch review of individual taxes, but an inter-government approach of coordination and sharing tax bases through more innovative tax design. Such coordination will improve the effort from recurrent land tax while harmonising the processes and administration costs by collecting land tax at the local government level. The last chapter highlighted the broadening of options for the taxation of land on more than one basis of value and provided options for local governments to choose the basis of value on which they assess the tax, as noted in the states of Victoria and South Australia. The use of dual bases of value was also observed internationally in New Zealand, Denmark and the United Kingdom.

It was highlighted in Chapters 1 and 12 that the states worked together in the late 1800s with the outcome of a Federated Australia in 1901, in which tax reform was one of the key drivers for strengthening Australia's fiscal powers. Over a century on, the next phase of reform is needed to modernise the tax system and maximise revenue from land tax, a key source of revenue for sub-national government across Australia. The Commonwealth plays an important role in participating in the harmonious divestment of the tax system to sub-national government. The objective of this chapter is to provide a framework for the participation of all three tiers of government in the realignment of land tax which is to be divested to and administered by local government across Australia.

In the realignment phase, we commence at the point the Commonwealth took control of income tax revenue from the states. This period marked an important point for Commonwealth, state and local government relations, as it moved the reliance from local government on its legislative parent, the states, to its financial parent, the Commonwealth. At the point when the Commonwealth took control of income tax, local and state government became the fiscal subordinates of the Commonwealth. This was particularly pertinent at that time, as there was no Goods and Services Tax. Revenue from income tax was and still remains the largest tax source in the Australian federation. As the shift from taxing income moves to taxing consumption, the need to balance this shift by taxing capital in the form of land is important in maintaining an efficient and balanced tax system.

At the time of Federation, the states worked together to form the Commonwealth, an objective which was achieved with great success and served Australia for over 110 years. It is now time for the Commonwealth to respond by assisting sub-national government in its reform to a modern federated structure. This participation is best achieved through a leadership role with financial incentives for both state and local government in driving reform. State land tax reform cannot be achieved in just one or two states, but requires a broader approach to ensure all states participate in lifting the tax effort from this source across Australia. We now consider land tax within the context of, firstly, reforming less efficient taxes such as conveyance stamp duty, which taxes mobility. This is followed by the reforms needed to improve revenue from recurrent land tax.

It is impossible and unnecessary for any government to justify the relevance of any tax in minutia; however, streets, suburbs and local government areas are not gated communities and perceived geographic boundaries should not dictate their tax effort. This applies in both the application of land tax and local government rates. From a financial perspective of overriding importance is the need to ensure that tax effort is being met consistently and that local communities and governments are taxing land in a fluent, transparent and efficient manner.

Property tax reform and housing mobility

As was highlighted in Chapter 1, during the 1970s and 80s a groundswell emerged for the removal of death duties, of which the bequeathing of property was the largest asset contributing to death duty revenues. The lost tax revenue from the removal of this tax was front-loaded onto the purchase of property in the form of conveyance stamp duty. The cost of purchasing property has become prohibitive, and it is sometimes argued that removal or reduction of stamp duty would be of limited benefit, as in high demand areas the price of housing would adjust to include the impost of the stamp duty foregone. The primary case mounted for the reduction of stamp duty is that, for middle and outer ring housing of cities, stamp duty impacts mobility, while in parts of the inner ring of cities, its imposition still impacts the decision to move, but to a marginally lesser degree.

Conveyance stamp duty as a mobility tax impacts decisions to trade up and down in the property market, resulting in inefficient utilisation of property in the capital cities of Australia. Table 13.1 highlights the recent trends in house and unit holding patterns over the past 10 years in the capital cities of Australia. It is clear that more property is being held for longer periods, of which transaction costs are the main factor impacting this trend. While advocates for the retention of stamp duty will argue that there is little evidence to suggest the removal of stamp duty on the purchase of property will increase affordability, what is shown in Table 13.1 is that stamp duty does impact housing mobility, which contributes directly to housing affordability.

While it may have been politically popular to replace death duties with stamp duty, it is now time to evolve tax policy, firstly, by reducing each of the brackets used to assess stamp duty across Australia and index the thresholds to address bracket creep. The revenue foregone from this move is to be replaced by imposing stamp duty on property inherited. At present, not only is there no inheritance tax on the transfer of assets bequeathed in a will, but the beneficiaries are also exempt from paying stamp duty on the transfer of property as an exemption exists in the transfer of assets which pass through a will. It is estimated that over the next 30 years the tax foregone on stamp duty from property bequeathed will account for a sizable per cent of conveyance stamp duty revenue per annum. The broadening of the stamp duty net to include bequeaths could easily reduce the amount of stamp duty paid by all property purchasers.

The intergenerational wealth transfer in property is being subsidised by new entrants to the property market who, in many cases, do not qualify for stamp duty relief. The salient factor for government, as is the case in all tax reform, is the introduction and transitioning of tax reform. The reintroduction of transfer duties must not be stopped by headlines that death duties have been reintroduced; this is an equitable tax imposed on beneficiaries and brings them in line with every other home purchaser and investor in Australia. Essentially, it is the have not's subsidising the haves, as Australia lags behind the world with no inheritance tax at present, but expends billions of dollars in stamp duty exemptions granted to beneficiaries. This tax is imposed in the United Kingdom and across Europe. Stamp duty concessions are further provided to select groups of first home buyers, as shown in Table 13.2; however, as noted, these concessions are increasingly applicable to new dwellings only, which emphasises the objective of stimulating the sale of new accommodation and hence supporting the construction of new housing.

In looking at the international case studies, New Zealand does not impose conveyance stamp duty. In the two provinces of Canada reviewed, stamp duty is imposed at 2 per cent in British Columbia and 2 per cent in Ontario, of which an additional 2 per cent is imposed by the City of Toronto. In the United Kingdom and other parts of Europe, inheritance tax is paid. In all of these countries, recurrent property tax is the more efficient tax of choice and is collected at a rate significantly higher than that of Australia. The Commonwealth, in contributing to tax reform, must further adjust grant revenue to the states, including the distribution of the Goods and Services Tax based on the states reform of this mobility tax on housing.

Table 13.1 Holding periods of housing across the capital cities of Australia

1	*2*	*3*	*4*	*5*	*6*	*7*
2012	*2012 ave cost of housing*	*Stamp duty*	*Ave hold period house 2002*	*Ave hold period house 2010*	*Ave hold period unit 2002*	*Ave hold period unit 2010*
Sydney	$580,000	$21,590	6.60	10.10	5.70	8.20
Melbourne	$470,900	$20,224	8.40	10.80	6.80	9.2
Brisbane	$437,000	$13,720	7.20	9.20	6.10	7.7
Adelaide	$386,000	$15,630	4.50	8.10	4.20	7.9
Perth	$491,000	$17,338	5.60	8.20	5.00	7.7
Hobart	$320,000	$10,735	6.10	8.70	6.10	8.1

Sources: RP Data for holding periods, stamp duty scales for each state on average cost of housing

Table 13.2 Concessions for pensioners and first homebuyers

State	*Legislation*	*Provision*	*Housing type*
Victoria	Duties Act 2000 – Division 5	Pensioner and first homeowner exemptions and concessions	New
New South Wales	Duties Act 1997 – Pt 8 Div 1	First new home	New
Queensland	Duties Act 2001 – Div 3	Concessions for homes and first homes	New
Western Australia	Duties Act 2008 – Div 3	First homeowners concession	Not specified
South Australia	Stamp Duties Act 1923	s71c Concessional rates of duty in respect of purchase of first home	Not specified
Tasmania	Duties Act 2001	s36G Exemptions and concessions	New from 2014

Source: State Duties / Stamp Duties Acts

Recurrent land tax reform

The second plank of reform lies in local government fully utilising their tax raising potential in a modern federation. The largest own source revenue generated by local government is rates, followed by grants in second place, with user charges ranked third. This generalisation varies considerably between city and regional local governments across Australia. Within the cities there is unmet rate revenue raising capacity, while in some regional and rural local government areas, grant revenue remains the main source of income. As a financial seamless tier of government, local government should have capacity to fund itself and stream surplus revenue up to the states. This scenario will be set out in more detail in the sections that follow; however, under the present structure of central/local government funding, local government is currently defined by reference to geographic boundaries, rather than its revenue raising capacity as an operational tier of government.

The question to be raised in reforming the tax transfer system, is how sub-national government can contribute further to the tax effort in Australia from the existing taxes currently under its control. Secondly and more importantly, how these tax bases can be integrated and collected by local government is the specific challenge to be addressed in reforming Australia's tax system and defining sub-national government's fiscal realignment. To understand the options for the integration of taxes by the tiers of government, and in particular, taxes which are currently the domain of central government, it is apt to look back and observe the Commonwealth/local government funding relationship over the past 40 years. This relationship gained momentum following the fiscal realignment of the state and Commonwealth income tax collection arrangement of the late 1940s. It was further followed by the rapid expansion of local infrastructure needed to support the population growth of Australia's cities since the 1970s.

Centralist government policy in funding local government post-1970

Over the past 40 years, several funding approaches have been used by central government in supporting local government. The policies adopted depend on the perceived role and functions of local government, which has experienced ongoing urbanisation and re-urbanisation within the capital cities. The factor which most impacts central government's approach to funding local governments is its perceived relationship with the states. Table 13.3 is a summary of the funding policies of the Commonwealth with local government since 1972. Central government recognises the emerging importance of local government, has sought to ensure adequate funding and has gone as far as supporting local government programs and, further, the constitutional recognition of this tier. While constitutional recognition is a potential pathway for local government, it does not address the fiscal reform and financing of Australia's federation and, in particular, sub-national government. The question of determining Australia's fiscal capacity and how this is to be achieved is an important prelude to reorganising the tiers of government.

At the peak of the Commonwealth's financial support for local government, the Local Government (Personal Income Tax Sharing) Act 1976 (Cth) was introduced to directly fund local government. The Oakes Inquiry (1990) highlights that, between 1980/81 and 1985/86, the guaranteed share of income tax money granted to local government by the Commonwealth reached a high of 2 per cent. The Hawke/Keating Government abandoned this system, adopting a 'macroeconomic policy of fiscal restraint' and by 1988/89 Commonwealth income tax receipts represented 1.32 per cent of grant revenue provided to local government across Australia and grant revenue having declined in real terms (Oakes Inquiry 1990:58).

During the transition of the Whitlam to Fraser Government (Commonwealth), the state government in NSW introduced rate pegging in 1977. 'Its introduction was seen as a response to the economic conditions of the time including spiralling cost-push inflation. However its use in NSW has no parallel in any other State' (Local Government Association of NSW 2003:3). When rate pegging was introduced, there was little resistance from local government in New South Wales, as Commonwealth grants were increasing during the Whitlam and Fraser Governments.

It was not until the Commonwealth funding progressively diminished during the 1980s and 90s that more local governments across Australia became financially unsustainable. Victoria has similar statutory provisions to NSW, which are referred to as rate capping. In contrast to NSW, which limited the increase in total rate revenue generated at the local government level, it is not yet clear how rate capping limits are to be imposed in Victoria. Rate capping, at this point, may be more closely aligned with those provisions used in California, under Proposition 13, and New York, under Maximum Permissible Assessment (MPA) provisions. A similar land tax freeze was introduced in Denmark in 2002, which limited the increase in local government rates to 2 per cent per annum.

In reflecting on the various policies of earlier central governments, it was in the Whitlam and Fraser governments that innovative thought was shown for the potential divestment of income tax revenue to local government. The measures adopted under the Personal Income Tax Sharing Act were progressive and, in fact, ahead of their time. The measures were simple by designating 2 per cent of personal income tax receipts received by the Commonwealth as income assigned to local government. Following the Whitlam and Frazer era of the 1970s, Hawke and Keating, who had engineered the containment of wages growth through the introduction of the Wages Accord, changed the funding of local government and introduced Financial Assistance Grants (FAGs) to local government, which was allocated on a per capita basis. This was viewed positively by local government, as the allocation formula was determined on a per capita basis which was seen as being transparent and simple.

The Howard Government further extended funding to local government through the introduction of the Roads to Recovery Act 2000 (Cth) (Twomey 2013). During a period of great fiscal buoyancy in Australia, this grant earmarked revenue to local government for its largest capital expense at the time of introduction: road construction and maintenance. As Australia's tax revenue collections began to slow during the GFC, the Rudd and Gillard Governments once again saw the future limitations in being able to sustain the level of funding needed by sub-national government. To this end, the question of constitutional recognition of local government again resurfaced and gained momentum during the two terms of the Rudd/Gillard Governments.

During the Abbott/Turnbull Government's first term, in which tax revenues have not grown as anticipated, the Commission of Audit and the review of the tax system are examining government spending and revenue policy. Among the matters being examined are VFI and the revenues raised by central government, and how the Commonwealth could alternatively assist sub-national government in the more efficient assignment of existing taxes. This brings forward questions of how a modern Australian Federation and its tax system should operate. The grant system of financing sub-national government has worked well to date; however, opportunity exists for local government to improve its own revenue raising capacities from recurrent land tax. The way the Commonwealth funds local government into the future will set an important tone of this reform direction.

Table 13.3 Commonwealth funding of local government

Government	*Ethos*	*Approach*
Whitlam Government 1972–75	Supported regionalism and the road funding for local government.	Local government is the answer to centralism in Australia.
Fraser Government 1975–83	Moved away from regionalism and funded local government direct through the states, using a fixed percentage.	Introduced Local Government (Personal Income Tax Sharing) Act 1976 (Cth). By 1980, 2 per cent of income tax revenue was received by local government.
Hawke/Keating Governments 1983–96	Commissioned national inquiry into Local Government Finance 1985.	Removed fixed percentage revenue from income tax revenue and introduced Financial Assistance Grants (FAGs) distributed on a per capita basis.
Howard Government 1996–2007	Bypassed s96 of granting money to local government via the states.	Introduced the Roads to Recovery Act 2000 (Cth).
Rudd/Gillard Governments 2007–13	Build stronger links between the Commonwealth and local governments. Provide for constitutional recognition of local government.	Australian Council of local government formed in 2008.
Abbott/Turnbull 2013–Present	Low-key approach to this subject to date during first term of office.	Freeze on the indexation of Financial Assistance Grants (FAGs) to local government.

A modern Australian tax framework: top-down and bottom-up reform

This section presents a conceptual framework of land tax reform in the transition from Australia's current framework to one that better aligns land tax with the taxpayer's capacity to pay. The objective is to expand the role of the Commonwealth support of local government as local government transitions to improve its tax collection effort from recurrent land tax, e.g. local government rates. This is achieved through the development of a local government rating system which assesses local government rates on the amalgam of two capacity-to-pay indicators. The first indicator is the value of the property and the second is an overlay of the owner's income. This provides a framework for local government to defer outstanding rate revenue until the property is disposed of.

It was noted in Chapter 1 that Vertical Fiscal Imbalance (VFI) is the problem that some reformists state must be addressed in Australia. When viewed from a sub-national government perspective, it is understandable why VFI is the rationale that emerges as the need to reform the tax system in Australia. What is of overriding importance, however, is not VFI in isolation, but the issues of tax effort, tax mix and the tax capacity of Australia as a whole, which are the underlying drivers of reform. While the transfer system is a key matter for reform identified by AFTS (2010), the underlying issue which resonates is a symptom of the structural framework of the tax system. In the case of recurrent land taxation currently spread across state and local government, it operates in a dysfunctional and disjointed manner which significantly impacts the revenue raising effort from this source.

What has fundamentally changed between the current and proposed future framework, as set out in Figure 13.1, is first and foremost, breaking the flow of revenue from upper to lower tiers of government and repositioning the tax raising powers of sub-national tiers and, in particular, local government. This moves the nexus of a top-down 20th century tax funding approach which worked well during the initial urbanising phase of Australia's development. Empowering the lower tiers of government and improving their tax raising capacity is where much of the tax reform is needed. Over time, the dependency on central government grants should be progressively reduced as sub-national government is given and willingly assumes greater tax raising responsibility in a modern federation.

The current tax system, as pointed out in Chapter 1, depicts the Commonwealth's collection of over 80 per cent of total tax revenue, to which it grants revenue to sub-national government via the Commonwealth and State Grants Commission. Located at the bottom of the current tax framework in Figure 13.1 is the direct funding of local government by the Commonwealth through the Roads to Recovery Grants and Financial Assistance Grants, of which the latter passes through the State Grants Commissions.

Setting this point aside for the moment, the current tax system is a heavily oriented top-down, master–servant relationship which worked well initially; however, now it leaves sub-national government reliant on the Commonwealth acting as the primary tax-collecting tier of government. Given the vast responsibilities of the states, their energies are best served providing and managing the large important portfolios of health, education and transport infrastructure, among other matters.

The future tax collection and transfer framework in Figure 13.1, in contrast, represents a far more decentralised system in which central government collects a reduced share of tax revenue, while sub-national government increases its share of the total tax collected. This latter approach streamlines the Grant Commission's role, as firstly, local government should become progressively, fiscally self-sustainable as a tier of government and independent of the need for funding from the Commonwealth. The reduction of this funding results in more revenue available from the Commonwealth to the states. Further, the surplus local government revenue is collected for the states by local government. In the international case studies, this approach of the lower tier of government collecting revenue on behalf of regional or state government applies in Denmark (local and regional government), while in Canada and the United States a similar arrangement operates between (local and state/provincial government).

This bottom-up approach of local government acting in part as tax collection agency for the states is an approach that should be progressively transitioned in Australia. Before the proposed approach is elaborated on further, it is first apt to recognise a number of important steps needed in the reform of Australia's federated structure, which includes unpacking the current structure from both the service and delivery side as well as the income and fiscal sides of the equation. While traditional thinking is to first define which tier of government is responsible for the various services provided, this will temporarily be set aside. This is not because it is unimportant; however, it is beyond the scope of this research to address this matter in tandem with the core matter of realigning recurrent land tax revenue at the sub-national government level. Some may argue that these two points are not severable, particularly at the local government level. The debate on reform may drift towards services in one community compared to those in an adjoining community; however, the emphasis remains on fiscal realignment and management in this research.

It is tempting to enter the debate on resourcing and services and the role and functions of the tiers of government. A short comment is warranted in understanding that each existing tier of government in Australia plays and will continue to play an important role in a modern Australian federation. Both state and local government are key participants in addressing one of the fastest-emerging issues confronting Australia, that is, the provision of infrastructure. At the local level, local government is responsible for the provision, maintenance and replacement of infrastructure, while at the trans-local government level, it is the states that provide and maintain infrastructure that spans local government. Further, the states play a crucial role in providing oversight of health, education, power, water and most transportation infrastructure across their geographic expanses.

While much more could be added to the role and functions of the various tiers of government, it is the review of recurrent land tax and the role it plays in reforming the funding of sub-national government that is the subject of this research, in conjunction with tax reforms which strengthen

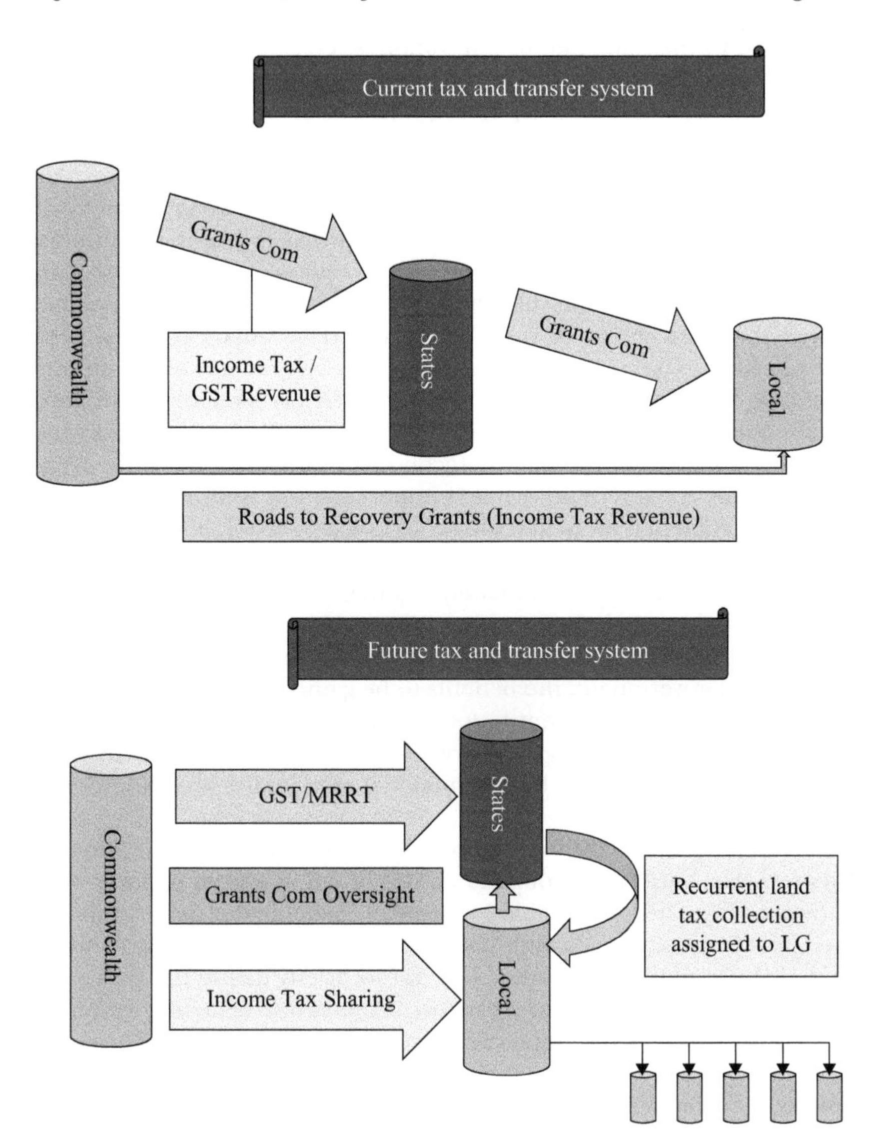

Figure 13.1 Current and future tax collection concepts

the framework and underpin the decentralisation of the tax system. It is the improvement in tax effort from land tax that makes the contribution to the broader tax reform agenda.

Which tier of government should levy recurrent land tax?

The previous section set out a modern tax framework, which now requires the fiscal reform of sub-national government and, specifically, in the imposition of recurrent land tax. As discussed in Chapter 2, land tax in Australia is imposed by the states and its imposition is referred to as land tax within the state's tax legislation. It was further highlighted in Chapters 5 and 6 that, resulting from the exemptions granted to the principal place of residence, the investor threshold and qualifying primary production land, state land tax is imposed on approximately 20 per cent of property in Australia. In contrast, local government imposes land tax in the form of rates, with very few exemptions and tax expenditure. However, local government rates may be the subject of rating differentials, the imposition of ad valorem and base amounts, and are the subject of rate pegging or capping, which neutralises the strict impact of value as the sole determinant of assessments. The question to be answered is which tier of government is best positioned to impose land tax in Australia, which is the primary focus of this section.

This question must be splintered into two parts; the first part asks which tier should collect the tax and the second part asks which tier is the recipient of the tax revenue. This divide is important, as it aptly addresses which tier of government is more likely to be accepted by taxpayers to collect land tax and, indeed, the progressive increase in its imposition. Once collected, the relevance of which tier ultimately receives the revenue is less important to the taxpayer, as their focus turns to where the tax revenue is spent. With Australia's languishing land tax effort currently spread across two tiers of sub-national government, the benefits to be gained from the restructuring of land tax and its divestment to local government are significant. In support of a bottom-up reform to Australia's federation and the restructure of sub-national government, the international case studies are an important reference.

In referring back to the international case studies in Part 3, the five countries reviewed have been grouped into two broad structures of government known as federated and unitary structures. Federated structures include the United States, Canada and Australia, while England, New Zealand and Denmark are unitary structures. It is apt to briefly define the differences between federated and unitary structures. A *unitary structure* of government, or a unitary state, is a sovereign state governed as a single entity. The central government is sole authority, and the administrative divisions exercise only powers that the central government has delegated them. Administrative units are created and abolished, and their powers may be broadened and narrowed by central government.

In contrast to unitary states, under a *federated structure*, power is shared between federal and state tiers of government. The states themselves are unitary in nature in their relationship and administration of local government. In a federation, the component states are in some sense sovereign, insofar as certain powers are reserved to them that may not be exercised by the central government. However, in some countries these points of separation are blurred and, in the case of Australia, were never fully partitioned between federal and state government. Both health and education are prime examples in which both state and federal government in Australia have designated portfolios and ministries. In contrast, matters of defence, customs and foreign policy are clearly defined solely within the federal domain, while town planning, policing, housing and most infrastructure are the domain of the states.

In each of the countries reviewed in Part 3, it is clear that local government is the tax collection agency for higher tiers of government. No country or jurisdiction apart from Australia imposes and collects a recurrent land or property tax at the state or provincial level; rather, this tax operates at the local tier of government. This does not mean that the provinces, states or higher tiers of government do not receive tax revenue from recurrent land or property tax. In the unitary structured countries of England and Denmark, the revenue from this tax is shared between local and central government. In the United States and Canada, it is shared between the states/provinces and local governments.

The tax is levied at the tier of government that it is most acceptable to the taxpayer, with greater tax effort being achieved abroad compared with Australia in every instance. Of particular note in the case studies are the United States and Canada, which, like Australia, are federated structures of government, the property tax being a recurrent tax on the capital improved value. In these jurisdictions, part of the property tax collected is retained by local government, with a component collected on behalf of the states/provinces. The component of the tax collected on behalf of the states is defined and earmarked to education. However, there is no direct link between the local taxing authority, the property tax and education programs, as education is primarily administered by state and provincial government in the United States and Canada.

The overriding rationale for the acceptance of an earmarked tax on education is that the tax is collected by the local taxing authorities who are seen by the taxpayer's to be the tier of government that best aligns with the services that impacts the value of their property. There is little or no evidence that suggests that the tax collected from the education component of the property tax is redistributed back to the local authority that collects it at the same proportion collected from specific locations. If land tax were to be collected solely by local government in Australia, there is no reason why this tax could not be earmarked to state infrastructure projects which may cross a number of local government areas, or specifically be targeted to infrastructure in defined local areas as needed and prioritised.

As was discussed in the review of Denmark, the newly formed counties, which took on the larger projects, have no taxing powers at all and are reliant on the tax collected by local government and the grants allocated by central government. The importance of defining the roles of the various tiers of government in Denmark has resulted in a much superior structure of government which has reduced duplication of services and roles across the tiers of government. It is noted that Denmark is a unitary structure; however, the counties and local government are a model for the relationship between the states and local government in Australia.

In England a similar process applies to Denmark, in which the property tax and uniform business rate is collected solely by local government, with the business rate revenue paid to central government, which is then redistributed back to local government on a needs basis. While this is the model in England, increasingly more own source revenue is kept by local government. This is determined on their needs, of which the maintenance of existing and the provision of new local infrastructure are the targets for retention of increases in tax revenue. This is a relevant rationale in the reform of land tax in Australia, in that the component of revenue collected for the states can be redistributed back to some local governments for infrastructure projects.

In progressing the effort from recurrent land tax further, the rivalry for this tax source between the tiers of sub-national government in Australia must first be addressed. Figure 13.2 sets out a simple process for sub-national government to work together efficiently in the collection and administration of recurrent land tax across Australia. In summary, local government is by far the most suitable tier of sub-national government to collect and administer this tax; it is also the most acceptable of the tiers of sub-national government to taxpayers. Further, defining the term *infrastructure* will be essential in providing context for taxpayers to broadly understand what land taxes are being allocated to. This is regardless of whether the infrastructure is provided by state or local government.

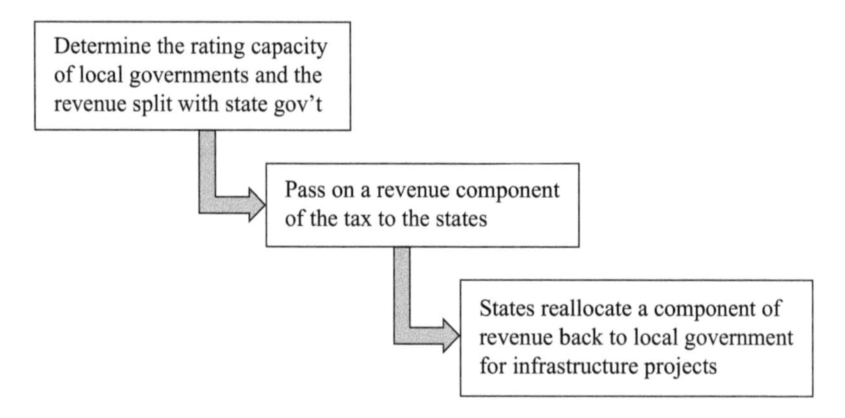

Figure 13.2 Land tax collection and reallocation model

The focus groups in Chapter 12 have highlighted the lack of nexus between a state land tax and the services and infrastructure provided by the states. The international cases clearly show that land tax operates successfully at the local government level in both federated and unitary structures of government. Until this restructure is addressed in Australia, sub-national government is unable to contribute to the tax reform agenda by improving the tax effort from this source. In assisting sub-national government in achieving tax reform, the realignment of the tax system is needed across tiers of government, which includes the transition of the states' hold on its revenue from recurrent land tax.

The Commonwealth is reluctant to tax the income derived from the capital gains of housing; the states cannot impose a state land tax, but impose rate pegging and capping on local government; and at the same time sub-national government marches towards a fiscal cliff. The reform of this tax needs to be addressed at the most fundamental level, and the spheres of sub-national government must zoom out and look at reform at the macro as well as the micro levels. At present, the states are enjoying the revenue windfalls from conveyance stamp duty. However, once this abates, the states, like the Commonwealth who enjoyed the mining boom revenues, will again be confronted with the need for fiscal reform.

In putting to rest the potential impost of state land tax on the principal place of residence, we refer to the experience of this impost between 1998 and 2004 in New South Wales. The 1998 experience was the expansion of state land tax to include high-value residential land under the provisions of the Premium Property Tax Act 1998. This was a well-intended attempt to expand revenue from recurrent land tax collected by the state government of New South Wales. The tax lasted five years and resulted in a High Court challenge, which brought to the fore the limitations of firstly taxing land as the base. It secondly demonstrated that taxing the principal place of residence at the state level was naïve and out of touch with the taxpaying public.

Land tax reform: building revenue within a capacity-to-pay framework

In Chapter 2, the capacity-to-pay principle was briefly introduced in the context of understanding the relationship of the taxpayer's ability to meet their tax commitment. It was identified that there was no specific definition or application of this principle as it may be determined on either wealth (assets) or income of the taxpayer. What remains unattended in its application to land tax, which is currently assessed on the value of land, is the potential for this tax to be assessed on the measures of value with an income dimension of the taxpayer. This is possible where the tax is so structured that the taxpayer's ability to pay reflects their income in unison with the value used in the assessment of a tax liability.

Land tax and, to a lesser degree, council rates are determined on the value of land or, in some states, improved value which reflects the taxpayer's stored wealth, which is assumed to reflect their capacity to pay. It was highlighted in Chapter 2 that over the lifetime of a taxpayer the relativity between income and stored wealth will vary significantly, and that a lag exists between the level of a taxpayer's income and the value of their assets over the holding period. In the earlier years, income is higher in line with earning capacity of the taxpayer, to the point where property is purchased, after which time income starts to decline relative to the value of the property for two reasons. The first reason is that for a PAYEE income earner, which comprises the majority of taxpayers in Australia, income increases between 2.5 to 3 per cent per annum compound, while in the capital cities of Australia the value of real estate increases by as much as 7.5 to 8 per cent per annum compound over a 10-year time horizon. In essence, property values in many cases grow at between two and a half to three times faster than a taxpayer's income.

Secondly, when many homeowners move into later years of work and then into retirement, income plateaus and then reduces further while the value of their residence continues to grow. The offset for owners in retirement generally is that the mortgage debt on property has reduced or has been paid off when income reduces in retirement. However, increasingly where the mortgage has not been paid off it is serviced with retirement income, or paid off using lump sum retirement sources. At the local government level, a rating subsidy and offset for the residence of aged homeowners applies in some states.

The variability in taxpayer circumstances, of which income earning capacity and the point the property owner is at in their income earning lifecycle, impact the policy design and transition needed to reform local government rating across Australia. The challenge of any tax reform is the transition, which does not constitute an objection to paying higher recurrent land tax in the form of rates. It is therefore incumbent on government to establish at what level local government rates should progressively be set. Before proceeding further on this point, it is important to highlight the players in the local government rating regulation process upon which these decisions fall.

Who's who in the regulation zoo?

It was highlighted earlier in this chapter that a freeze on the indexing of Financial Assistance Grants (FAGs) was recently introduced by the Commonwealth. This marks an important first step in reforming local government funding in Australia. The measure begins the shift of responsibility to local government to ensure that revenue is being raised in line within ratepayers' capacities to pay, and brings forward the prospect of each local government assessing their income revenue from rates accordingly. In NSW, the Independent Pricing and Regulatory Tribunal (IPART) has the responsibility of assessing rate variation applications made by local government above the rate increase peg set each year. The criteria used to assess such applications follows:

> IPART will assess applications for minimum rates above the statutory limit against the following set of criteria (in addition to any other matters which IPART considers relevant):
>
> 1. the rationale for increasing minimum rates above the statutory amount,
> 2. the impact on ratepayers, including the level of the proposed minimum rates and the number and proportion of ratepayers that will be on the minimum rates, by rating category or sub-category and
> 3. the consultation the council has undertaken to obtain the community's views on the proposal.
>
> It is the council's responsibility to provide enough evidence in its application to justify the minimum rates increase. Where applicable, councils should make reference to the relevant parts of their Integrated Planning and Reporting documentation to demonstrate how the criteria have been met.
>
> (NSW Office of Local Government 2014)

While IPART has the responsibility of assessing rate variation agreements, it has no power or authority to compel any of the 152 local governments in NSW to seek variations above the rate peg. Further, it does not have any capacity to compel local government to show why those that do not exceed the peg are raising revenue at an optimum level. The balancing act on determining the optimum level will fall to the State Grants Commission and how the tax effort of local government is assessed in the distribution of Financial Assistance Grants (FAGs). At present, in NSW the Grants Commission standard formula for distribution according to expenses of local government is as follows:

> Expenditure allowances are calculated for 20 functions or areas of expenditure. An additional allowance is calculated for councils outside the Sydney area that recognises their isolation. The general formula for the calculation of expenditure allowances is:
>
> **Allowance = No. of Units x Standard Cost x Disability**
>
> Where: **No. of Units** is the measure of use of the function for the Council. For most functions the number of units is the council's population. For others it may be the number of properties or the length of local roads.
>
> The **Standard Cost** is the average of annual average net expenditure, per unit, by all councils in the state, averaged over five years.

> The **Disability** is the measure of the extent of relative disadvantage a council faces in providing a standard service because of issues beyond its control. For each function the characteristics likely to influence the cost are identified and measured. The measure is then related to the potential additional costs to councils.
>
> (NSW Local Grants Commission Annual Report 2014:19)

Victoria is currently formulating a framework for capping increases in local government rates. The Essential Services Commission, a similar government instrumentality to IPART, is responsible for developing a framework which will govern increases in rates. This regime will commence from the 2016/17 taxing year, allowing a short continuation of the current status. The role of the Commission will be delicate in the initial phase, as it determines the benchmark for assessing the base level of rates, which is a crucial requisite on which to assess any applications for increases in rates to be assessed. This brings forward the question as to what that optimum level might be within, and indeed across, local governments. While it is incumbent on local government to justify increases beyond the rate cap, it is incumbent on the regulatory authority to articulate the tolerances under which it draws its conclusion to either support or reject a rate variation application. In essence, the Grants Commission and IPART / ESC are at opposite ends of the regulatory and funding process.

It is further worth noting that important caveats underpin the limitations of some local governments applying for increases in rates to regulatory bodies. The cost of preparing applications for rate variations is well beyond the affordability of many rural and some regional councils. The inherent risks of this factor are that it is not always possible to distinguish between councils which cannot afford to apply, those that simply do not need to apply and those which choose not to apply. In the amalgam of these circumstances the freeze on the indexation of Financial Assistance Grants and the way this revenue is distributed are of high importance for local government with limited capacity to improve rate revenues. The primary objective is to improve the tax effort from land tax using a sound platform of the principle of capacity-to-pay within the tax reform agenda, a matter still in its infancy in Australia.

It is clear that if local government were to assume the role of becoming the sole imposing and collecting tier of recurrent land tax through the rating system, much reform would need to be undertaken on the imposition of the tax on residential property. The revenue take from non-residential and business use property contributes the majority share of revenue collected from state land tax. The rates in the dollar applied in the assessment of local government rates are in the main currently higher than those applied to residential land. The principal place of residence exemption from state land tax and residential investment property, which receives the investor threshold, would require the greatest attention in the transition of recurrent land tax to local government across Australia.

Benchmarking capacity-to-pay

The phenomenon of the movement in values on which land tax is assessed relative to income is more commonly benchmarked internationally and has been the prime reason for the limitation of property tax increases abroad. In parts of the United States, Canada, Denmark and the United Kingdom, increases have been regulated and are now being carefully scrutinised by tax authorities in New Zealand. To this end, the progressive phase of monitoring and regulating local government rates in Australia is not new internationally. While this is the case, the property tax represents a larger proportion of the property owner's income for the tax paid on their residence in all advanced OECD economies.

In the United States and Canada, land tax, known as the property tax, on the home accounts for over 3 per cent of GDP and on average accounts for approximately 3.25 per cent of the taxpayer's income in the United States and closer to 3.5 per cent in parts of Canada. These statistics are not well developed in Australia; however, an initial review shows local government rates to be in a considerably lower range of 1.0 to 1.6 per cent of taxpayer income. While these averages are broad indicators, the range of the relativity between local property tax and household income in the United States is broad and ranges between 0.6 to 9.5 per cent, with the median being at 3.25 per cent of the taxpayer's income. In contrast, in Canada it was noted in Toronto that 3.5 per cent is the upper tolerable limit of rates as a percentage of the taxpayer's income.

The question now arises as to how to improve recurrent land tax revenue across Australia in a way that embraces a capacity-to-pay framework, while meeting the principles of good tax design. Like any tax reform, the challenge is in the progressive transition from an existing to a new framework. The objective is for land tax revenue across Australia to progressively reach 3 per cent of GDP and represent a greater percentage of taxpayer's income. In the period designated as retirement years, this percentage will increase; however, national uniform provision must be made for any outstanding local government rates to be paid on disposal of the property, which may be triggered by either disposal, which constitutes sale, or on death of the ratepayer. This is where the Commonwealth must assist local government in financing this reform in the later and retirement years of taxpayers.

When New South Wales removed the land tax free threshold in 2005 for one year, concerns were raised about property investment migration to other states which had maintained their land tax free thresholds. Land tax is a deductible expense for income-producing property and property used for business purposes in many circumstances. This tax deduction provides a means for the Commonwealth to indirectly divest revenue to sub-national government by allowing land tax and local government rates as a deductible expense. The removal of the threshold would provide a means of contributing to moderating the rate in the dollar, on which the recurrent tax is collected on behalf of the states. Such a move would require a rating differential for residential property used as an owner's residence compared with residential investment property if the collection of land tax were to be divested to local government.

It would be the role of the states to remove the threshold prior to divesting collection and administration of this tax to local government. The importance of this initial reform is that the removal of the threshold is uniform across Australia as the tax is divested to local government. This then brings Australia in line with the United States and Canada which, like Australia, are federations, as well as neighbouring New Zealand, which is a unitary structure. The transition of state land tax to local government is long overdue in Australia, in which this tax was divested to local government in the United States and Canada progressively during the 20th century. In New Zealand, this tax became the sole domain of local government in 1991.

In summary, New York City, with a population of 8.5 million people, collects recurrent property tax revenue of approximately 85 per cent of the total combined revenue collected from state land tax and local government rating across the whole of Australia. Likewise, the province of Ontario, Canada, with a population of approximately 13 million people, collects 110 per cent of the total recurrent land tax revenue collected across Australia. Once Australia makes the much-needed structural reforms which follow, we can start addressing vertical fiscal imbalance, improve tax effort and address the fiscal reform of sub-national government in Australia.

Figure 13.3 sets out the outdated structure of the land tax system in Australia followed by the broader tax reform framework needed across sub-national government to transition change. As discussed in Chapters 12 and 13, the summary of changes are:

1) Divestment of recurrent land tax to local government from the states, and a component of the revenue collected by local government streamed back up to the states for infrastructure projects;
2) The progressive increase in recurrent land tax revenue to represent 3 per cent of GDP and approximately 9–10 per cent of total tax collected over the following 5- to 10-year period;
3) The option for the assessment of recurrent taxation on either land/site value or capital improved value as decided by the relevant local government on either a two- or three-year valuation cycle;
4) Uniform options to quarantine the component of local government rates above the 3 per cent of taxpayers' income until disposal of the property, with Commonwealth support for local government until outstanding rates are paid or recovered;

5) The progressive reduction of stamp duty rates to half of the present rates over a five-year period, with stamp duty imposed on all property bequeathed through wills. A means test and provisions are to be made available to defer such payment for a defined period as determined by the states.

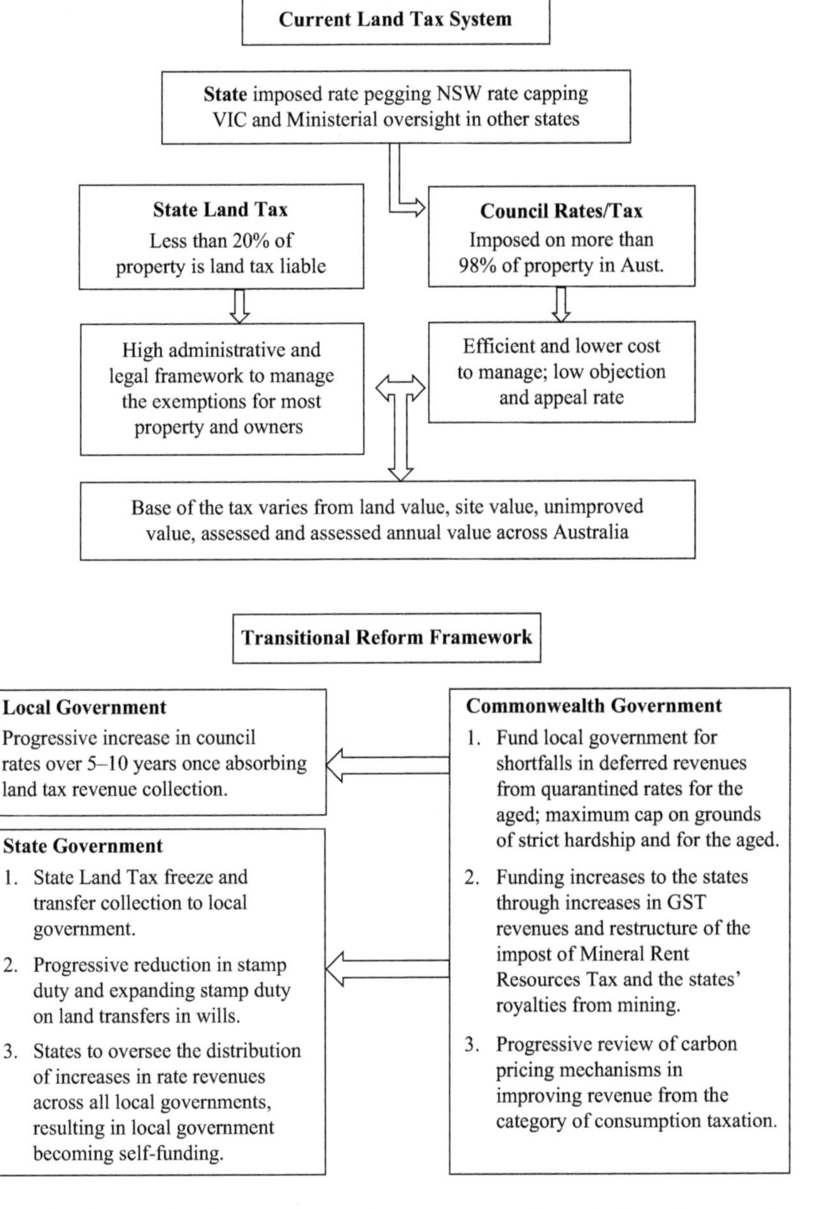

Figure 13.3 Current land tax system and proposed transitional reform framework

Improving economic and operational efficiency in land tax reform

There are three key benefits of assigning the collection and administration of land tax to local government in contrast to the current two-tier system. Local government issues a rate notice for each property within its area; hence the operational efficiency of a single tier collecting the tax is simple. Local government issues four rate notice instalments in most states each year and has a high percentage collection ratio. The data matching between owner details, property description and built attributes is far more accurate at the local government level. While state land tax is a taxpayer-focused tax, driven by exemptions and thresholds, local government rating is a property-focused tax, which is efficiently imposed and collected, and is far more tolerantly accepted by taxpayers than state land tax.

At present, only two states, NSW and Qld, rate and tax on the same base, while Tasmania looks to move to the same base for each of their recurrent land taxes. In South Australia, annual values are produced for all three bases of value annually, the cost of which is under pressure at present. The need to revalue multiple bases or indeed a single basis of value annually is highly questionable. The current system in NSW is a classic example of the questionable expense of revaluing one basis of value annually of which 80 per cent of the land values are never used in most years. For rating purposes, values are updated every three to four years for council rating purposes, while for the narrow number of land tax-liable residential properties, the land value of every residential property is undertaken annually.

There are major factors arising in the imposition of this tax at both the state and local government level in the application of land tax to non-residential property going forward. This matter will be addressed in a subsequent release to this publication. Suffice to say that the tax principle of revenue efficiency is of high importance and one that government must attend to in sustaining the longevity of this tax. This publication clearly addresses the first step of reform through the taxation of residential property. Queensland has made good progress in achieving reform and better shoring up of land tax revenue from non-residential property. However, more is needed in Australia and ongoing research into practices used abroad in the taxation of non-residential property. This will be one of the focuses of the subsequent publication of recurrent land tax and its reform.

Summary of reforms – present and future

1. Operational Efficiency: Collection and administration by one tier of government
2. Economic Efficiency: Base of the tax (land and improved value on a two- to three-year valuation cycle)
3. Revenue Efficiency: Robustness of revenue (codification of the determination of the base)

Conclusion

The current challenge of land tax reform in Australia is that it is ramped up in times of revenue shortfalls and becomes a classic austerity tax. Rather than reforming the tax in buoyant economic times, it is a tax reform of last resort for the reasons set out in the earlier chapters. Its high visibility and public perception, which has not been helped by some novice commentators who ride the wave of selling taxation moderation rather than reform, has resulted in land tax being left in the too hard basket. Reforming the most important plank of this tax, its application to all residential property, requires more than a simple broadening of the tax net, to the coordination of sub-national government and the way this tax reform is presented to the taxpaying public. The insatiable desire to protect property owners in the housing market by not raising the required recurrent tax revenue in lieu of taxing transactions to a prohibitive level is one of the greatest challenges confronting sub-national government today.

It is imperative that those presiding over housing and tax policy disaggregate discussion on tax revenue derived from property transactions from revenue derived from recurrent land taxation. It has been demonstrated that Australia has capacity to improve its tax effort, of which sub-national government plays an equally important role with central government in this reform. The Commonwealth has the dual role of firstly, financially incentivising the states by waking them from their tax reform slumber, and secondly, supporting sub-national government as it navigates these necessary reforms. However, as highlighted above, without broader reform of the tax system, the states and, to a lesser degree, local government are complacent in remaining the fiscal dependents of the Commonwealth. Without a net fiscal gain for the states, the tax reform agenda remains the problem of the Commonwealth, which is charged with raising further tax revenue to fund their fiscal dependents who argue for revenue shortfalls to be addressed by the Commonwealth each year.

At some point, the tax reform must be addressed once the trimming of government spending reaches its limits. It will have been incumbent on government to assess how tax reform will be operationalised in advance. It has been highlighted that at the Commonwealth level the aim to reduce taxes on income, including companies which employ the Australian workforce, will need to be funded from other sources, of which the Goods and Services Tax will contribute to that reform objective. At the state and local level, the transition from transaction taxes to a recurrent taxation of land collected by local government will greatly assist in reducing the need for the Commonwealth to fund local government. This, in turn, will allow the Commonwealth to reallocate that revenue to the states. The Commonwealth and states, in turn, will need to build the administrative capacity of local government as the prime tax collection agency for sub-national government in line with practices of the more advanced OECD federations of Canada and the United States.

At the operational level of a reformed recurrent land tax system will be the need to build on the challenges which currently confront the economic and revenue efficiency of land tax across Australia. Many of the technical details of strengthening the base on which the tax is assessed are imperative, as is defining the capacity to pay this tax, which will largely impact the success and longevity of reform. More than sufficient evidence exists to show that reform of the tax base is needed, and that options are available from experiences and case studies included in this research. Building a platform for gauging the capacity to pay, rather than sole tax policies which engage rate capping and pegging, must be avoided. The divestment of land tax collection and administration to local government in Australia for the states and for their own source revenues is the first necessary step, as achieved in other OECD countries.

Once these matters are addressed, taxing authorities are in a better position to gauge how revenue from this source may be better predicted and moderated. Bolt-on tax reforms, such as states imposing levies comprising fixed amounts per property, would be a most unsatisfactory approach to reforming land tax. Whether land or improved value is used to advance the tax effort from this source, shying away from value as the dominant base on which this tax is imposed must be avoided as all costs. This forces the tiers of government to work collectively, regardless of political persuasion, in defining the capacity-to-pay framework for property owners. In contrast to the costs outlaid by the states in administering the massive tax exemptions afforded to the principal place of residence and tax free investor thresholds, it is time to modernise recurrent land tax to address what the tax should apply to rather than what it exempts.

These two tax carve-outs which dominate the state's land tax legislation, in relation to the principal place of residence and the threshold, is where the reform of this tax predominantly sits. Rather than being a taxpayer-focused tax, applied to a narrow number of properties, it is time for the tax, as its name suggests, to apply to all property as do local government rates, with very few exceptions. Once local government, as a financially seamless tier of government, has become fiscally self-sufficient, this tier along with the Commonwealth can turn their attention to assisting the states. To this end, the fiscal reform of sub-national government is one of the key pillars of reforming Australia's tax system. Realigning fiscal responsibility is an overdue phase of Australia's federation, which remains a work in progress.

Where to go from here

This edition provides a review of Australia's tax system, the operation of land tax around Australia, and several international case studies. The final part of the book gives insight into the reforms needed through a reorganised land tax system at the sub-national government level to bring Australia's tax into the 21st century. Over the following 18–24 months, the author is further researching ways to improve land and property taxes and the tax efforts of sub-national government. A detailed critique and ranking of the operation of this tax on a state-by-state and local government basis will be provided in a forthcoming publication. This will include a critique of land tax reforms which are to be put to the tax paying public at the local government level.

As part of reforming the federation, harmonisation of land tax across the tiers of government is needed, as is the fast-emerging importance of revenues generated through utility charges imposed on property. As property emerges as a conduit for the collection of tax revenue, the way revenues are imposed on land and utility charges across the tiers of government will require a succinct and coordinated rationale. To this end, the cost-of-living pressures must be better understood by government as it embarks on revenue realignment from these sources. The direction of these charges and how they are managed and imposed on the tax paying public, will be further articulated in future editions of this research.

What has yet to evolve in Australia is the understanding that, while land and property have traditionally been administered and organised geographically, sub-national government has siloed its revenue raising capacity along these same geographic boundaries. The ability for local government, in particular, to achieve financial self-sufficiency is at odds under its current revenue constraints, which are not solely impacted by the states. Realigning and improving the revenue raising capacity of local government as a financially seamless tier of government is a more important reform than realigning geographic boundaries, which is seen to be achieved through physical amalgamations as the first step. This important matter is a further focus of the author's ongoing research.

While the above are important aspirations, one of the most constructive and useful discussions over the years of researching land tax occurred during a visit to the OECD in Paris during 2012. This discussion led to the insightful comment that went along the lines of:

> *Do not rely on any one political party or persuasion to achieve tax reform. This will come from like-minded people within the various parties and persuasions who band together, some crossing the floor with colleagues to get the job of reform done.*
>
> (H. Bloechliger, *Economist OECD*, Paris, personal communication, 3 May 2012)

This resonates the idea that anyone involved in tax reform will usually default to the impact that reform will have on their own circumstances and those who champion them. However, those that see the importance of reform will see beyond their own backyard or balcony.

References

Access Economics. (2010). *Valuation and local government rating in Tasmania: A robust framework for the future.* Access Economics Barton, ACT.

ACT Government, Treasury Directorate. (2012). *ACT Taxation Review.* Canberra, ACT: Author. Retrieved from http://www.treasury.act.gov.au/documents/ACT%20Taxation%20Review/ACT%20Taxation%20Review%20May%202012.pdf

Alonso, W. (1964). *Location and land use.* Cambridge, MA: Harvard Press.

Arnold, J. (2008). Do tax structures affect aggregate economic growth? Empirical Evidence from a panel of OECD countries. Economics Department Working Papers No. 643. Paris: OECD.

Arnott, R., & Petrova, R. (2002). *The property tax as a tax on value: Deadweight loss.* National Bureau of Economic Research.

Asprey, K.W. (1975). *Taxation Review Committee.* Canberra: Australian Government Publishing Services.

Audit Commission. (2013). *Council tax collection.* London: City of London Corporation.

Australian Bureau of Statistics. (2013). *Total tax revenue* (ABS Cat. No. 5506.0). Retrieved from http://www.abs.gov.au/

Australian Bureau of Statistics. (2012). *Population projections Australia* (ABS cat. No. 3222). Author.

Australian Property Institute. (2000). *Professional practice.* Canberra, ACT: Author.

Australian Property Institute. (2004). *Professional practice.* Canberra, ACT: Author.

Australian Property Institute. (2007). *Appraisal of real estate,* ed. R. Reed. Canberra, ACT: Author.

Australia's Future Tax System. (2008). *Consultation paper – Commonwealth of Australia.* Barton, ACT: Author.

Australia's Future Tax System. (2010). *Final report – Commonwealth of Australia.* Barton, ACT: Author.

Bahl, R. (2002). Property taxes in South Africa. In M.E. Bell & J.H. Bowman (Eds.), *Fiscal decentralisation, revenue assignment and the case for the property tax.* Cambridge, MA: Lincoln Institute of Land Policy.

Bahl, R. (2009). *Property tax reform in developing and transition countries.* Washington, DC: United States Agency for International Development.

Balconi, M., Pozzali, A., & Viale, R. (2007). The 'codification debate' revisited: A conceptual framework to analyse the role of tacit knowledge in economics. *Industrial and Corporate Change* 16(5), 823–849.

Bird, R.M., & Slack, E. (Eds.). (2004). *International Handbook of Land and Property Taxation.* Northampton, MA: Edward and Elgar Publishing Ltd.

Bird, R.M., Slack, E., & Tassonyi, A. (2012). *A tale of two taxes: Property tax reform in Ontario.* Cambridge, MA: Lincoln Institute of Land Policy.

Blochliger, H., & Vammalle, C. (2012). Reforming fiscal federalism and local government – beyond the zero sum game. OECD.

Brennan, F. (1971). *Canberra in crisis.* Canberra: Dalton Publishing.

Business Council of Australia. (2013). Action plan for building prosperity – Ch 01 Tax, fiscal policy and the federation. Business Council action_plan_booklet_1_tax_fiscal_federation_final_31-7-2013 (1).

Carlson, R.H. (2004, September). *A brief history of property tax.* Paper presented at the IAAO Conference on Assessment Administration, Boston, Massachusetts.

Centre for Accounting, Governance and Taxation Research. (2010). *Reflections on a tax system for the future.* Parkville: University of Melbourne.

Chancellor, J. (2013, April 29). Victorian first home buyers grant jumps from $7000 to $10,000 but only for new houses and apartments. *Property Observer.*

Commonwealth Australia. (2014). *History of Australia's government.* Retrieved from Australia.gov.au

Comrie, J. (2013). In our hands, strengthening local government revenue for the 21st century. Working paper for the Australian Centre of Excellence for Local Government, University of Technology, Sydney.

Constitution of California, State Constitution. Retrieved from www.leginfo.ca.gov/const-toc.html

Daly, M.T. (1982). *Sydney boom Sydney bust.* Sydney: Allen and Unwin.

Daw, C.A. (2002, February). Land taxation: An ancient concept. *Australian Property Journal* 1, 20–25.

Denmark.DK (2014). Quick facts and society. The official website of Denmark. Retrieved from Denmark.dk.

Department of Communities and Local Government. (2014). Administration of Business Rates in England. Discussion Paper. Retrieved from https://www.gov.uk/government/uploads/system/uploads/attachment_data/file/308504/PU1623_administration_of_business_rates_discussion_paper.pdf

Department of Premier and Cabinet Tasmania. (2013). *Valuation and local government rating review, local government division.* Hobart, Tasmania: Author.

Derby, M. (2012). *Local and regional government – Early forms of local government.* Wellington, New Zealand: NZ Parliamentary Library.

Dotzour, M.G., Grissom, T. V., & Liu, C. H. (1990). Highest and best use: The evolving paradigm. *The Journal of Real Estate Research* [17] 5(1), 17–32.

Dye, R.F., & England, W.E. (Eds.). (2009). *Land value taxation: Theory, evidence and practice.* Cambridge, MA: Lincoln Institute of Land Policy.

Folketinget (2012). Review of the Danish Parliament, Copenhagen. Retrieved from http://www.cbs.dk/en/library/databases/folketinget-danish-parliament-home-page

Franzsen, R.C.D. (2009). The international experience. In R.F. Dye & W.E. England (Eds.). (2009). *Land value taxation: Theory, evidence and practice.* Cambridge, MA: Lincoln Institute of Land Policy.

French, N., & Gabrielli, L. (2007). Market value and depreciated replacement cost: Contradictory or complementary? *Journal of Property Investment & Finance* 25(5), 515–524.

Gaffney, M. (1975, November). *The many faces of site value taxation.* Paper presented at the twenty-seventh Annual Conference of the Canadian Tax Foundation, Quebec City.

Gaffney, M. (1995, November). *Proposition 13: What happens when a State radically slashes its property tax.* Paper presented at the Jerome Levy Institute, Excerpts from a paper, "Big plans to stir the blood and steer the course."

George, H. (1879). *Progress and poverty*. New York: Cosimo.

Hansards. (1973) Canberra, ACT: Commonwealth House of Representatives.

Hendy, P.W., & Warburton, A.O. (2006). *International comparison of Australia's taxes*. Canberra ACT: Commonwealth of Australia.

Herps, D. (1988, May). Land value taxation in Australia and its potential for reforming our chaotic tax system. The Walsh Memorial Bequest Address delivered at Macquarie University School of Economics, Sydney.

HM Treasury. (2014, June). Administration of business rates in England. Discussion paper. Retrieved from https://www.gov.uk/government/uploads/system/uploads/attachment_data/file/308504/PU1623_administration_of_business_rates_discussion_paper.pdf

Hudson, M. (2001). The land-residual vs. building residual methods of real estate valuation. *Some Prefatory Remarks to the N.Y.U. Real Estate Institute discussion.*

Hudson, M. (2008). Henry George's political critics. *American Journal of Economics and Sociology* 67(1).

Hughes, M.A. (2006). *Why so little Georgism in America: Using the Pennsylvania case files to understand the slow, uneven progress of land value taxation.* Cambridge, MA: Lincoln Institute.

Hyam, A. (2004). *The law affecting the valuation of land in Australia* (3rd ed.). Leichhardt, NSW: The Federation Press.

IPART. (2008). Review of state taxation, independent pricing and regulatory tribunal for New South Wales. Sydney: Independent Pricing and Regulatory Tribunal of New South Wales.

IPTI. (2013a). Xtracts, update on Australia. Toronto, ON: International Property Tax Institute.

IPTI. (2013b). Xtracts, update on Denmark. Toronto, ON: International Property Tax Institute.

Irish Tax and Customs Revenue. (2014). Property tax update. Retrieved from www.revenue.ie

Irvine, J. (2009, July). First-home incentive is deterring more buyers. *Sydney Morning Herald.*

Jenson, P.R., & Jacobson, E. (2009). Local and regional government in Denmark. Paper presented at the CEMR Conference, Copenhagen, Denmark.

Kelly, S. (2003, December). Self-provision in retirement? Forecasting future household wealth. Paper presented at the International Micro simulation Conference on Population Aging and Health, Canberra, ACT.

Kupke, V., & Murano, W. (2002). The implications of changes in the labour market for the homeownership aspirations, housing opportunities and characteristics of first home buyers. *AHURI* (Final Report No. 18), 1–56.

Krueger, R.A., & Casey, M.A. (2009). *Focus groups: A practical guide for applied research.* Sage Publications.

Lefmann, O., & Larsen, K.K. (2000). Denmark. In R.V. Andelson (Ed.), *Land value taxation around the world.* Blackwell Publishers.

Local Government Association of NSW. (2003). NSW local government rate determination model. Sydney, NSW: Author.

Local Government Denmark. (2009). The Danish Local Government System. Retrieved from http://www.kl.dk/English/Local-Government-Reform/

Local Government National Reports. (2008/09).

Local Government Rates Inquiry Panel. (2007). *Funding local government: Report of the Local Government Rates Inquiry.* Wellington, New Zealand: New Zealand Government.

Mander, M.R. (1982). New Zealand rating valuation system. *The Valuer* 16(3), 239–241.

Mangioni, V. (2011). Transparency in the valuation of land for land tax purposes in New South Wales. *eJournal of Tax Research* 9(3), 140–152.

Mangioni, V. (2013). Codifying value in land value taxation. *Australian and New Zealand Property Journal* 4(3), 248–257.

Mangioni, V. (2014a, December). Land value taxation and the valuation of land in Australia. *Nordic Journal of Surveying and Real Estate Research* (2).

Mangioni, V. (2014b). Refining principles of compensation in land acquisition for urban renewal. Proceedings from the PRRES Conference. Pacific Rim Real Estate Society, 1–13.

Mangioni, V., & Warren, N.A. (2014). Redefining the land tax base in highly urbanised locations. *Australian Tax Forum* 29(3), 455–476.

McAlester, V., & McAlester, L. (1997). *A field guide to American houses.* New York: Random House.

McCluskey, W.J., Bell, M.E., & Lim, L.J. (2010). Rental value versus capital value – alternate bases for the property tax. In R. Bahl, J. Martinez-Vazquez, & J.M. Youngman (Eds.), *Challenging the conventional wisdom of the property tax.* Cambridge, MA: Lincoln Institute of Land Policy.

McCluskey, W.J., Grimes, A., Aitkin, A., Kerr, S., & Timmins, J. (2006). Rating systems in New Zealand: An empirical investigation into local choice. *Journal of Real Estate Literature* 14(3), 381–397.

McLean, I. (2004, June). Land tax: Options for reform. Nuffield College Politics Working Paper 2004-W7. Paper presented at the Oxford seminar on New Politics of Ownership: Financing the Citizens Stake, Oxford, England.

Meade, J.E. (1978). *The structure and reform of direct taxation.* London: Unwin and Allen, and Institute for Fiscal Studies.

Ministry of Finance. (2013). Ontario economic outlook and fiscal review – background papers. Retrieved from http://www.fin.gov.on.ca/en/budget/fallstatement/2013/paper_all.pdf

Muller, A. (2002). *Valuation of land and buildings for recurrent property tax.* World Bank – Tax Policy and Administration Division.

Muller, A. (2003, June). Importance of recurrent property tax in public finance, tax policy & fiscal decentralisation. Presented at the international conference on Property and Land Tax Reform, Tallinn, Estonia.

Muller, A. (2005). Development of the Danish Valuation System. Presented at an OECD Seminar about Property Tax Reforms and Valuation, Vienna, 19–21 September.

Musgrave, R.A., & Musgrave, P.B. (1976). *Public finance in theory and practice.* Tokyo: McGraw-Hill, Kogakusha.

National Archives of Australia. (2009). *NSW Dept of Lands, valuation procedures manual.* Retrieved from http://naa.gov.au/collection/explore/cabinet/by-year/1974.aspx

National Commission of Audit. (2014). *Towards responsible government: The report of the national commission of audit.* Sydney, NSW: Commonwealth of Australia.

New Zealand Local Government Rates Inquiry. (2007). *Funding local government.* Wellington, NZ: Author.

New Zealand Parliamentary Library. (2014). Local government amalgamation. Retrieved from http://www.parliament.nz/en-nz/parl-support/research-papers/00PLLawC51141/local-government-amalgamation

New Zealand Tax Review Committee. (1967). *Taxation in New Zealand: report of the Taxation Review Committee* (L. N. Ross, chairman). Wellington, New Zealand: Author.

Nile, F. (1998). Report on the inquiry into changes in land tax in New South Wales. Parliament of New South Wales, Legislative Council49.

NSW Local Grants Commission. (2013/14). *Annual report 2013/14.* Nowra, NSW: Author. Retrieved from http://www.olg.nsw.gov.au/sites/default/files/Grants 2013-14AR.pdf

NSW Office of Local Government. (2014). *Guidelines for the preparation of an application to increase minimum rates above the statutory limit 2015/2016.* Sydney, NSW: Author. Retrieved from https://www.olg.nsw.gov.au/news/14-26-special-rate-and-minimum-rate-variation-guidelines-and-process-201516

NSW Ombudsman. (2005, October). *Improving the quality of land valuations issued by the Valuer-General.* Sydney, NSW: Author.

NSW Treasury and Ministry for Police and Emergency Services. (2012). *Funding our emergency services: A discussion paper.* Sydney, NSW: Author.

Oakes Inquiry. (1990). *Report on the committee of inquiry into local government rating and other revenue powers and resources.* Sydney, NSW: Parliament of NSW.

Oates, W. E., & Schwab, R. M. (1997). The impact of urban land taxation: The Pittsburgh experience. *National Tax Journal* 50(1), 1–21.

OECD. (2010). *Organisation for Economic Cooperation and Development Revenue Statistics 1965–2010,* Table 22–23.

OECD. (2011). Tax statistics. Retrieved from http://www.oecd-ilibrary.org/taxation/total-tax-revenue_20758510-table2

Office of State Revenue. (2009). First home owners boost fact sheet. Retrieved from http://www.finance.wa.gov.au/cms/uploadedFiles/_State_Revenue/FHOG/First_Home_Owners_Boost_fact_sheet(1).pdf

O'Keefe, J.A.B. (1974). *The legal concept and principles of land value.* Wellington, New Zealand: Butterworths.

Pearson, L. (1994). *Local government in NSW.* Leichardt, NSW: The Federation Press.

Plimmer, F. (1998). *Rating law and valuation,* Harlow, England: Pearson Education Limited.

Prest, A. R. (1983). *Some issues in Australian land taxation.* Canberra, ACT: Centre for Federal Financial Relations.

Productivity Commission. (2004). *First home ownership.* (Report No. 28). Melbourne, VIC: Commonwealth of Australia.

Productivity Commission. (2014). *Public infrastructure.* Melbourne, VIC: Commonwealth of Australia.

Proposition 13. (1978). The Jarvis-Gann property tax initiative, California Assembly May 1978.

Revenue Irish Tax and Customs. (2013). *Your guide to local property tax & how to pay and file.* Author.

RICS. (2007). The potential for the property tax in the 2004 accession countries of central and eastern Europe. *RICS Research Issues Paper* 7(17).

RICS. (2013, February). Empty property rates. *RICS Research.* Retrieved from http://www.rics.org/au/knowledge/research/research-reports/empty-property-rates/

Rost, R. O., & Collins, H. G. (1993). *Land valuation & compensation.* Canberra, ACT: Australian Institute of Valuers.

Sanderson, P. (2012). CHOVA Commonwealth heads of valuation agencies conference. *Beyond the Horizon,* 18–21.

Simpson, R., & Figgis, H. (1998). Land tax in New South Wales. (Briefing paper no. 6/98). Sydney, NSW: NSW Parliamentary Library.

Smith, S. (2005). *Land Tax: An Update.* Sydney, NSW: NSW Parliamentary Library.

Smith, G. (2009). Australia's tax reforms: Past and future, draft paper. Retrieved from http://www.victoria.ac.nz/sacl/centres-and-institutes/cagtr/pdf/smith.pdf

Society of Chartered Surveyors Ireland and Irish Tax Institute. (2013). Local property tax – a public information guide. Retrieved from http://taxinstitute.ie/portals/0/Tax%20Policy/LPT/SCSI%20ITI%20LPT%20Public%20Guide%20Web%2009APR13.pdf

State Business Tax Review Committee. (2001). *Review of state business taxes.* Melbourne, VIC: Department of Treasury and Finance.

Statistics New Zealand. (2012). *Population statistics 2012.* Wellington: New Zealand Government.

Tax Foundation. (2014). Washington, DC: International Tax Competitive Index.

Tomson, A. (2005). Mass valuation theory and practice in transitional countries – Estonia, Latvia and Lithuanian cases. *Journal of Property Tax Assessment and Administration* 2(1), 43–55.

United Kingdom Govt. (2014). Council tax update from the Valuation Office Agency. Retrieved from http://www.gov.uk/browse/tax

Walton, J. (1999). *Report into the operation of the Valuation of Land Act 1916, Parliamentary Report.* Sydney: Author.

Warren, N.A. (2004). *TAX: Facts fiction and reform.* Sydney, NSW: Australian Tax Research Foundation.

Warren, N.A. (2006). *Benchmarking Australia's intergovernmental fiscal arrangements – final report.* NSW Government.

Welsh Assembly Government. (2007). *Review of valuation tribunals in Wales.* Retrieved from http://ajtc.justice.gov.uk/docs/RTOW_English_t.pdf

Whipple, R.T.M. (1986). *Commercial rent reviews: Law and valuation practice.* Sydney, NSW: Law Book Co.

Whipple, R.T.M. (2006). *Property valuation and analysis* (2nd ed.). Pyrmont, NSW: Law Book Co.

Wilson, D.C. (1995). Highest and best use analysis: Appraisal heuristics versus economic theory. *The Appraisal Journal* [11] 63(1), 11–26.

Youngman, J.M., & Malme, J.H. (1994). *An international survey of taxes on land and buildings.* Kluwer Law and Taxation Publishers, for the Lincoln Institute of Land Policy, Organization for Economic Cooperation and Development and International Association of Assessing Officers.

Cases

Department of Lands v Webster (1995) 89 LGERA 341

Gollan v Randwick Municipal Council [1961] AC 82

Maurici v Office of State Revenue [2003] HCA 8

McNally and Anor v Commissioner of State Revenue [2003] NSWSC

Port Macquarie West Bowling Club v The Minister [1972] 2 NSWLR 63

PT Limited & Westfield Management Limited v The Department of Natural Resources and Mines [2007] (Queensland Land Appeal Court, Brisbane, 17 October 2007)

Ryan & Anor v Commissioner of Land Tax [1982] 1 NSWLR 30

Spencer v The Commonwealth of Australia (1907) 5 C.L.R. 418

Sutton v Blue Mountains City Council (1977) 40 LGRA 51

Tooheys Ltd v Valuer-General [1925] AC 439, Department of Lands v Webster (1995) 89 LGERA 341

Triguboff v Valuer General [2009] NSWLEC 9 (13 February 2009)

Index

Note: Page numbers in italic indicate tables and figures.

www.ingramcontent.com/pod-product-compliance
Ingram Content Group UK Ltd.
Pitfield, Milton Keynes, MK11 3LW, UK
UKHW021835240425
457818UK00006B/203

* 9 7 8 1 1 3 8 8 3 1 2 5 4 *